Lecture Notes in Artificial Intelligence 11081

Subseries of Lecture Notes in Computer Science

More information about this series at http://www.springer.com/series/1244

Luca Pancioni · Friedhelm Schwenker
Edmondo Trentin (Eds.)

Artificial Neural Networks in Pattern Recognition

8th IAPR TC3 Workshop, ANNPR 2018
Siena, Italy, September 19–21, 2018
Proceedings

 Springer

Editors
Luca Pancioni
University of Siena
Siena
Italy

Edmondo Trentin
University of Siena
Siena
Italy

Friedhelm Schwenker
Ulm University
Ulm
Germany

ISSN 0302-9743   ISSN 1611-3349   (electronic)
Lecture Notes in Artificial Intelligence
ISBN 978-3-319-99977-7   ISBN 978-3-319-99978-4   (eBook)
https://doi.org/10.1007/978-3-319-99978-4

Library of Congress Control Number: 2018952642

LNCS Sublibrary: SL7 – Artificial Intelligence

This Springer imprint is published by the registered company Springer Nature Switzerland AG
The registered company address is: Gewerbestrasse 11, 6330 Cham, Switzerland

Preface

This volume contains the papers presented at the 8th IAPR TC3 workshop on Artificial Neural Networks for Pattern Recognition (ANNPR 2018), held at the Università di Siena, Siena, Italy, during September 19–21, 2018. ANNPR 2018 follows the success of the ANNPR workshops of 2003 (Florence), 2006 (Ulm), 2008 (Paris), 2010 (Cairo), 2012 (Trento), 2014 (Montreal), and 2016 (Ulm). The series of ANNPR workshops has served as a major forum for international researchers and practitioners from the communities of pattern recognition and machine learning based on artificial neural networks.

From the 35 manuscripts submitted, the Program Committee of the ANNPR 2018 workshop selected 29 papers for the scientific program, organized in regular oral presentations and one poster session. The workshop was enriched by three IAPR invited sessions: What's Wrong with Computer Vision? given by Prof. Marco Gori, Università di Siena, Italy; Deep Learning in the Wild presented by Prof. Thilo Stadelmann, ZHAW Datalab & School of Engineering, Winterthur, Switzerland; and an invited talk given by Prof. Marcello Pellilio, Università Cà Foscari, Venice, Italy

The workshop would not have been possible without the help of many people and organizations. First of all, we are grateful to all the authors who submitted their contributions to the workshop. We thank the members of the Program Committee and the many additional reviewers for performing the difficult task of selecting the best papers from a large number of high-quality submissions. We hope that readers of this volume may enjoy it and get inspired from its contributions. ANNPR 2018 was supported by the International Association for Pattern Recognition (IAPR), by the IAPR Technical Committee on Neural Networks and Computational Intelligence (TC3), and by the DIISM of the University of Siena, Italy. Finally, we wish to express our gratitude to Springer for publishing our workshop proceedings within their LNCS/LNAI series.

July 2018

Luca Pancioni
Friedhelm Schwenker
Edmondo Trentin

Organization

Organization Committee

Luca Pancioni	Università di Siena, Italy
Friedhelm Schwenker	Ulm University, Germany
Edmondo Trentin	Università di Siena, Italy

Program Committee

Hazem Abbas	German University in Cairo, Egypt
Shigeo Abe	Kobe University, Japan
Amir Atiya	Cairo University, Egypt
Erwin M. Bakker	Leiden Institute of Advanced Computer Science, The Netherlands
Jinde Cao	Southeast University Nanjing, China
Ludovic Denoyer	Université Pierre et Marie Curie, France
Mauro Forti	Università di Siena, Italy
Sayan Ghosh	Washington State University, USA
Eric Granger	École de Technologie Supérieure, Canada
Markus Hagenbuchner	University of Wollongong, Australia
Barbara Hammer	Bielefeld University, Germany
Lakhmi Jain	University of Canberra, Australia
Nikola Kasabov	Auckland University of Technology, New Zealand
Hans A. Kestler	Ulm University, Germany
Wenlian Lu	Fudan University Shanghai, China
Marco Maggini	University of Siena, Italy
Nadia Mana	Fondazione Bruno Kessler, Italy
Simone Marinai	University of Florence, Italy
Stefan Scherer	University of Southern California, USA
Eugene Semenkin	Siberian State Aerospace University, Russia
Alessandro Sperduti	University of Padua, Italy
Thilo Stadelmann	ZHAW Datalab & School of Engineering Winterthur, Switzerland
Ah-Chung Tsoi	Macau University of Science, Macau, China
Ian Witten	Otago University, New Zealand
Zhi-Hua Zhou	Nanjing University, China

Sponsoring Institutions

International Association for Pattern Recognition (IAPR)
Technical Committee 3 (TC 3) of the IAPR
DIISM, Siena, Italy
Università di Siena, Siena, Italy
Ulm University, Ulm, Germany

Contents

Applications

Invited Papers

What's Wrong with Computer Vision?

Marco Gori[✉]

Department of Information Engineering and Mathematics,
University of Siena, Siena, Italy
marcoxgori@gmail.com
http://sailab.diism.unisi.it/people/marco-gori/

Abstract. By and large, the remarkable progress in visual object recognition in the last few years is attributed to the availability of huge labelled data paired with strong and suitable computational resources. This has opened the doors to the massive use of deep learning which has led to remarkable improvements on common benchmarks. While subscribing this view, in this paper we claim that the time has come to begin working towards a deeper understanding of visual computational processes, that instead of being regarded as applications of general purpose machine learning algorithms, are likely to require appropriate learning schemes. The major claim is that while facing nowadays object recognition problems we have been working a problem that is significantly more difficult than the one offered by nature. This is due to learning algorithms that are working on images while neglecting the crucial role of frame temporal coherence. We address the limitations and discuss how the evolution of the tradition of image recognition towards visual recognition might give rise to remarkable advances in the field of computer vision.

Keywords: Computer vision · Object recognition · Machine learning
Motion invariance

1 Introduction

While the emphasis on a general theory of vision was already the main objective at the dawn of the discipline [13], it has evolved without a systematic exploration of foundations in machine learning. When the target is moved to unrestricted visual environments and the emphasis is shifted from huge labelled databases to a human-like protocol of interaction, we need to go beyond the current peaceful interlude that we are experimenting in vision and machine learning. A fundamental question a good theory is expected to answer is why children can learn to recognize objects and actions from a few supervised examples, whereas nowadays supervised learning approaches strive to achieve this task. In particular, why are they so thirsty for supervised examples? Interestingly, this fundamental difference seems to be deeply rooted in the different communication protocol at the basis of the acquisition of visual skills in children and machines.

© Springer Nature Switzerland AG 2018
L. Pancioni et al. (Eds.): ANNPR 2018, LNAI 11081, pp. 3–16, 2018.
https://doi.org/10.1007/978-3-319-99978-4_1

So far, the semantic labeling of pixels of a given video stream has been mostly carried out at frame level. This seems to be the natural outcome of well-established pattern recognition methods working on images, which have given rise to nowadays emphasis on collecting big labelled image databases (e.g. [4]) with the purpose of devising and testing challenging machine learning algorithms. While this framework is the one in which most of nowadays state of the art object recognition approaches have been developing, we argue that there are strong arguments to start exploring the more natural visual interaction that animals experiment in their own environment.

This suggests to process video instead of image collection, that naturally leads to a paradigm-shift in the associated processes of learning to see. The idea of shifting to video is very much related to the growing interest of *learning in the wild* that has been explored in the last few years[1]. The learning processes that take place in this kind of environments has a different nature with respect to those that are typically considered in machine learning. Learning convolutional nets on ImageNet typically consists of updating the weights from the processing of temporally unrelated images, whereas a video carries out information where we pass from one frame to the previous one by smooth changes. While ImageNet is a collection of unrelated images, a video supports information only when motion is involved. In presence of fixed images that last for awhile, the corresponding stream of equal frames basically supports only the information of a single image. As a consequence, it is clear that visual environments diffuse information only when motion is involved. There is no transition from one image to the next one—like in ImageNet—but, as time goes by, the information is only carried out by motion, which modifies one frame to the next one according to the optical flow. Once we deeply capture this fundamental features of visual environment, we early realize that we need a different theory of machine learning that naturally processes streams that cannot be regarded just as collection of independent images anymore.

A crucial problem that was recognized by Poggio and Anselmi [15] is the need to incorporate visual invariances into deep nets that go beyond simple translation invariance that is currently characterizing convolutional networks. They propose an elegant mathematical framework on visual invariance and enlightened some intriguing neurobiological connections. Overall, the ambition of extracting distinctive features from vision poses a challenging task. While we are typically concerned with feature extraction that is independent of classic geometric transformation, it looks like we are still missing the fantastic human skill of capturing distinctive features to recognize ironed and rumpled shirts! There is no apparent difficulty to recognize shirts by keeping the recognition coherence in case we roll up the sleeves, or we simply curl them up into a ball for the laundry basket. Of course, there are neither rigid transformations, like translations and rotation, nor scale maps that transforms an ironed shirt into the same shirt thrown into the laundry basket. Is there any natural invariance?

---

[1] See. e.g. https://sites.google.com/site/wildml2017icml/.

In this paper, we claim that motion invariance is in fact the only one that we need. Translation and scale invariance, that have been the subject of many studies, are in fact examples of invariances that can be fully gained whenever we develop the ability to detect features that are invariant under motion. If my inch moves closer and closer to my eyes then any of its representing features that is motion invariant will also be scale invariant. The finger will become bigger and bigger as it approaches my face, but it is still my inch! Clearly, translation, rotation, and complex deformation invariances derive from motion invariance. Humans life always experiments motion, so as the gained visual invariances naturally arise from motion invariance. Animals with foveal eyes also move quickly the focus of attention when looking at fixed objects, which means that they continually experiment motion. Hence, also in case of fixed images, conjugate, vergence, saccadic, smooth pursuit, and vestibulo-ocular movements lead to acquire visual information from relative motion. We claim that the production of such a continuous visual stream naturally drives feature extraction, since the corresponding convolutional filters are expected not to change during motion. The enforcement of this consistency condition creates a mine of visual data during animal life. Interestingly, the same can happen for machines. Of course, we need to compute the optical flow at pixel level so as to enforce the consistency of all the extracted features. Early studies on this problem [8], along with recent related improvements (see e.g. [2]) suggests to determine the velocity field by enforcing brightness invariance. As the optical flow is gained, it is used to enforce motion consistency on the visual features. Interestingly, the theory we propose is quite related to the variational approach that is used to determine the optical flow in [8]. It is worth mentioning that an effective visual system must also develop features that do not follow motion invariance. These kind of features can be conveniently combined with those that are discussed in this paper with the purpose of carrying out high level visual tasks. Early studies driven by these ideas are reported in [6], where the authors propose the extraction of visual features as a constraint satisfaction problem, mostly based on information-based principles and early ideas on motion invariance.

In this paper we mostly deal with an in-depth discussion of the principles that one should follow to construct a sound theory of vision that, later on, can likely be also applied to computer vision. In addition, we discuss some of the reasons of the limitations of current approaches, where perceptual and linguistic tasks interwound with vision are not properly covered. This issue is enlighten by proposing a hierarchy of cognitive tasks connected to vision that contributes to shed light on the intriguing connection between gaining perceptual and linguistic skills. The discussion suggests that most problems of computer vision are likely to be posed according to the historical evolution of the applications more than on a formal analysis of the underlying computational processes. While this choice has been proven to be successful in many real-world cases, stressing this research guideline might lead, on the long run, to wrong directions.

2   Top Ten Questions a Theory on Vision Should Address

The extraction of informative and robust cues from visual scenes has been attracting more and more interest in computer vision. For many years, scientists and engineers have contributed to the construction of solutions to extract visual features, that are mostly based on clever heuristics (see e.g. [12]). However, the remarkable achievements of the last few years have been mostly based on the accumulation of huge visual collections enriched by crowdsourcing. It has created labels to carry out massive supervised learning in deep convolutional networks, that has given rise to very effective internal representations of visual features. The have been successfully used in an impressive number of application (see e.g. [10,17]).

In this paper, we argue that while stressing this issue we have been facing artificial problems that, from a pure computational point of view, are likely to be significantly more complex than natural visual tasks that are daily faced by animals. In humans, the emergence of cognition from visual environments is interwound with language. This often leads to attack the interplay between visual and linguistic skills by simple models that, like for supervised learning, strongly rely on linguistic attachment. However, when observing the spectacular skills of the eagle that catches the pray, one promptly realizes that for an in-depth understanding of vision, that likely yields also an impact in computer implementation, one should begin with a neat separation with language! This paper is mostly motivated by the curiosity of addressing a number of questions that arise when looking at natural visual processes [3]. While they come from natural observation, they are mostly regarded as general issues strongly rooted in information-based principles, that we conjecture are of primary importance also in computer vision.

Q1 *How can animals conquer visual skills without requiring "intensive supervision"?*

Recent remarkable achievements in computer vision are mostly based on tons of supervised examples—of the order of millions! This does not explain how can animals conquer visual skills with scarse "supervision" from the environment. Hence, there is plenty of evidence and motivations for invoking a theory of truly unsupervised learning capable of explaining the process of extraction of features from visual data collections. While the need for theories of unsupervised learning in computer vision has been advocated in a number of papers (see e.g. [7,11,16,19]), so far, the powerful representations that arise from supervised learning, because of many recent successful applications, seem to attract much more interest. While information-based principles could themselves suffice to construct visual features, the absence of any feedback from the environment make those methods quite limited with respect to supervised learning. Interestingly, the claim of this paper is that motion invariance offers a huge amount of free supervisions from the visual environment, thus explaining the reason why humans do not need the massive supervision process that is dominating feature extraction in convolutional neural networks.

Q2 *How can animals gradually conquer visual skills in a visual environments?*
Animals, including primates, not only receive a scarse supervision, but they also conquer visual skills by living in their own visual environment. This is gradually achieved without needing to separate learning from test environments. At any stage of their evolution, it looks like they acquire the skills that are required to face the current tasks. On the opposite, most approaches to computer vision do not really grasp the notion of time. The typical ideas behind on-line learning do not necessarily capture the natural temporal structure of the visual tasks. Time plays a crucial role in any cognitive process. One might believe that this is restricted to human life, but more careful analyses lead us to conclude that the temporal dimension plays a crucial role in the well-positioning of most challenging cognitive tasks, regardless of whether they are faced by humans or machines. Interestingly, while many people struggle for the acquisition of huge labeled databases, the truly incorporation of time leads to a paradigm shift in the interpretation of the learning and test environment. In a sense, such a distinction ceases to apply, and we can regard unrestricted visual collections as the information accumulated during all the agent life, that can likely surpass any attempt to collect image collection. The theory proposed in this paper is framed in the context of agent life characterized by the ordinary notion of time, which emerges in all its facets. We are not concerned with huge visual data repositories, but merely with the agent life in its own visual environments.

Q3 *Can animals see in a world of shuffled frames?*
One might figure out what human life could have been in a world of visual information with shuffled frames. Could children really acquire visual skills in such an artificial world, which is the one we are presenting to machines? Notice that in a world of shuffled frames, a video requires order of magnitude more information for its storing than the corresponding temporally coherent visual stream. This is a serious warning that is typically neglected; any recognition process is remarkably more difficult when shuffling frames, which clearly indicates the importance of keeping the spatiotemporal structure that is offered by nature. This calls for the formulation of a new theory of learning capable of capturing spatiotemporal structures. Basically, we need to abandon the safe model of restricting computer vision to the processing of images. The reason for formulating a theory of learning on video instead of on images is not only rooted in the curiosity of grasping the computational mechanisms that take place in nature. It looks like that, while ignoring the crucial role of temporal coherence, the formulation of most of nowadays current computer vision tasks leads to tackle a problem that is remarkably more difficult than the one nature has prepared for humans! We conjecture that animals could not see in a world of shuffled frames, which indicates that such an artificial formulation might led to a very hard problem. In a sense, the very good results that we already can experiment

8   M. Gori

nowadays are quite surprising, but they are mostly due to the stress of the computational power. The theory proposed in this paper relies of the choice of capturing temporal structures in natural visual environments, which is claimed to simplify dramatically the problem at hand, and to give rise to lighter computation.

Q4 *How can humans attach semantic labels at pixel level?*

Humans provide scene interpretation thanks to linguistic descriptions. This requires a deep integration of visual and linguistic skills, that are required to come up with compact, yet effective visual descriptions. However, amongst these high level visual skills, it is worth mentioning that humans can attach semantic labels to a single pixel in the retina. While this decision process is inherently interwound with a certain degree of ambiguity, it is remarkably effective. The linguistic attributes that are extracted are related to the context of the pixel that is taken into account for label attachment, while the ambiguity is mostly a linguistic more than a visual issue. The theory proposed in this paper addresses directly this visual skill since the labels are extracted for a given pixel at different levels of abstraction. Unlike classic convolutional networks, there is no pooling; the connection between the single pixels and their corresponding features is kept also when the extracted features involve high degree of abstraction, that is due to the processing over large contexts. The focus on single pixels allows us to go beyond object segmentation based sliding windows, which somewhat reverses the pooling process. Instead of dealing with object proposals [21], we focus on the attachment of symbols at single pixels in the retina. The bottom line is that human-like linguistic descriptions of visual scenes is gained on top of pixel-based feature descriptions that, as a byproduct, must allow us to perform semantic labeling. Interestingly, there is more; as it will be shown in the following, there are in fact computational issues that lead us to promote the idea of carrying our the feature extraction process while focussing attention on salient pixels.

Q5 *Why are there two mainstream different systems in the visual cortex (ventral and dorsal mainstream)?*

It has been pointed out that the visual cortex of humans and other primates is composed of two main information pathways that are referred to as the ventral stream and dorsal stream [5]. The traditional distinction distinguishes the ventral "what" and the dorsal "where/how" visual pathways, so as the ventral stream is devoted to perceptual analysis of the visual input, such as object recognition, whereas the dorsal stream is concerned with providing motion ability in the interaction with the environment. The enforcement of motion invariance is clearly conceived for extracting features that are useful for object recognition to assolve the "what" task. Of course, neurons with built-in motion invariance are not adeguate to make spatial estimations. A good model for learning of the convolutional need to access to velocity estimation, which is consistent with neuroanatomical evidence.

Q6 *Why is the ventral mainstream organized according to a hierarchical architecture with receptive fields?*

Beginning from early studies by Hubel and Wiesel [9], neuroscientists have gradually gained evidence of that the visual cortex presents a hierarchical structure and that the neurons process the visual information on the basis of inputs restricted to receptive field. Is there a reason why this solution has been developed? We can promptly realize that, even though the neurons are restricted to compute over receptive fields, deep structures easily conquer the possibility of taking large contexts into account for their decision. Is this biological solution driven by computational laws of vision? In [3], the authors provide evidence of the fact that receptive fields do favor the acquisition of motion invariance which, as already stated, is the fundamental invariance of vision. Since hierarchical architectures is the natural solution for developing more abstract representations by using receptive fields, it turns out that motion invariance is in fact at the basis of the biological structure of the visual cortex. The computation at different layers yields features with progressive degree of abstraction, so as higher computational processes are expected to use all the information extracted in the layers.

Q7 *Why do animals focus attention?*
The retina of animals with well-developed visual system is organized in such a way that there are very high resolution receptors in a restricted area, whereas lower resolution receptors are present in the rest of the retina. Why is this convenient? One can easily argue that any action typically takes place in a relatively small zone in front of the animals, which suggests that the evolution has led to develop high resolution in a limited portion of the retina. On the other hand, this leads to the detriment of the peripheral vision, that is also very important. In addition, this could apply for the dorsal system whose neurons are expected to provide information that is useful to support movement and actions in the visual environment. The ventral mainstream, with neurons involved in the "what" function does not seem to benefit from foveal eyes. From the theory proposed in this paper, the need of foveal retinas is strongly supported for achieving efficient computation for the construction of visual features. However, it will be argued that the most important reason for focussing attention is that of dramatically simplifying the computation and limit the ambiguities that come from the need to sustaining a parallel computation over each frame.

Q8 *Why do foveal animals perform eye movements?*
Human eyes make jerky saccadic movements during ordinary visual acquisition. One reason for these movements is that the fovea provides high-resolution in portions of about 1, 2 degrees. Because of such a small high resolution portions, the overall sensing of a scene does require intensive movements of the fovea. Hence, the foveal movements do represent a good alternative to eyes with uniformly high resolution retina. On the other hand, the preference of the solution of foveal eyes with saccadic movements is arguable, since while a uniformly high resolution retina is more complex to achieve than foveal retina, saccadic movements are less important. The information-based theory presented in this paper makes it possible to conclude that foveal retina with saccadic movements is in fact a solution that is computationally sustainable and very effective.

Q9 Why does it take 8–12 months for newborns to achieve adult visual acuity?
There are surprising results that come from developmental psychology on
what a newborn see. Charles Darwin came up with the following remark:
> It was surprising how slowly he acquired the power of following with
> his eyes an object if swinging at all rapidly; for he could not do this
> well when seven and a half months old.

At the end of the seventies, this early remark was given a technically sound
basis [20]. In the paper, three techniques,—optokinetic nystagmus (OKN),
preferential looking (PL), and the visually evoked potential (VEP)—were
used to assess visual acuity in infants between birth and 6 months of age.
More recently, the survey by Braddick and Atkinson [14] provides an in-
depth discussion on the state of the art in the field. It is clearly stated that
for newborns to gain adult visual acuity, depending on the specific visual
test, several months are required. Is the development of adult visual acuity
a biological issue or does it come from higher level computational laws?

Q10 Causality and Non Rapid Eye Movements (NREM) sleep phases
Computer vision is mostly based on huge training sets of images, whereas
humans use video streams for learning visual skills. Notice that because of
the alternation of the biological rhythm of sleep, humans somewhat process
collections of visual streams pasted with relaxing segments composed of
"null" video signal. This happens mostly during NREM phases of sleep, in
which also eye movements and connection with visual memory are nearly
absent. Interestingly, the Rapid Eye Movements (REM) phase is, on the
opposite, similar to ordinary visual processing, the only difference being
that the construction of visual features during the dream is based on the
visual internal memory representations [18]. As a matter of fact, the process
of learning the filters experiments an alternation of visual information with
the reset of the signal. A good theory of learning visual features should
provide evidence to claim that such a relaxation coming from the reset
of the signal nicely fits the purpose of optimizing an overall optimization
index based on the previously stated principles. In particular, in [3], the
authors point out that periodic resetting of the visual information favors the
optimization under causality requirements. Hence, the role of eye movement
and of sleep seem to be important for the optimal development of visual
features.

3 Hierarchical Description of Visual Tasks

In this section we discuss visual tasks and their intriguing connection with language. This analysis is motivated by the evidence provided in nature of excellent
visual skills that arise regardless of language. At the light of the following analysis, one should consider to start go beyond the tradition of computer vision
of emphasizing classification tasks. Visual perception drives different functional
tasks in animals, so as the human intersection with language must properly be
analyzed.

Let $\mathcal{T} = [t_0, t_1]$ be the *temporal domain* and let $\mathcal{X} \subset \mathbb{R}^2$ be the retina. We consider the video domain $\mathcal{D} := \mathcal{T} \times \mathcal{X}$ so as

$$v : \mathcal{D} \to \mathbb{R}^d : \ (t, x) \to [v_1(t, x), \dots, v_d(t, x)]'$$

is the video signal on $\mathcal{D}$. In the classic case of RGB coding, we have $d = 3$. Throughout the paper, $v(\mathcal{D})$ denotes any video, while we use $\mathcal{V}$ to denote the universal set of videos, where any video belongs to. Likewise, $v(t, \mathcal{X})$ denotes the frame at t and $\mathcal{I}$ denotes the universal set of images with values $v(t, x) \in \mathbb{R}^d$. Clearly, we have $v(\mathcal{D}) \in \mathcal{V}$. Now, humans are capable of providing sophisticated linguistic representations from video $v(\mathcal{D}) \in \mathcal{V}$, which involve both local and global features. Clearly, abstract descriptions of a visual scene do require considerable linguistic skills, which emerge also at local level when specific words can also be attached to any pixel of a given visual frame. Basically, humans are capable of providing a linguistic description of $v(\mathcal{D})$ that goes well beyond object classification. The amount of visual information is typically so huge that for an appropriate cognitive transcription at linguistic level to take place one cannot rely on classification, but must necessarily involve the compositional structure of language. This kind of difficulty clearly emerges when trying to provide a linguistic description to blind people, a task which is quite difficult also for humans.

3.1 Pixel-Wise and Abstract Visual Interpretations

One of the most challenging issues in vision is human ability to jump easily from pixel-wise to recognition processes and more abstract visual interpretations that involve frames as well as portions of a video. When focussing attention on a certain pixel in a picture, humans can easily make a list of "consistent objects" that reflects the visual information around that pixel. Interestingly, that process takes place by automatically adapting a sort of "virtual window" used for the decision. This results in the typical detection of objects with dimension which is growing as that virtual window gets larger and larger. More structured objects detected at a given pixel are clearly described by more categories than simple primitive objects, but, for humans, the resulting *pixel-wise* process is surprisingly well-posed from a pure cognitive point of view. However, such a pixel-wise process seems to emerge upon request; apparently, humans do not carry out such a massive computation over all the retina. In addition, there are abstract visual skills that are unlikely to be attacked by pixel-wise computation. Humans provide visual interpretations that goes beyond the truly visual pattern (see e.g. Kanizsa's illusions). This happens because of the focus of attention, which somehow locates the object to be processed. As the focus is on the pixel $f(t)$, the corresponding object can be given an abstract geometrical interpretation by its shape expressed in term of its contour. While pixel-based processes are based on all the visual information of the retina associated with a given pixel, shape-based recognition emerges when recognizing objects on the basis of their contour, once we focus attention of a point of the object.

Pixel-wise processes can only lead to the emergence of decisions on objects, which is fact a static concept. It cannot allow us to draw conclusions on actions, whose understanding does require to involve portions of video. However, like for objects, the detection of the "contour of actions" yields a very useful abstraction. The notion *object affordance* has a strict connection with that of action. We carry out many object recognition processes on the basis of actions in which they are involved, so as objects are detected because of their role in the scene. In other words, the affordance involves the *functional role* of objects, which is used for the emergence of abstract categories.

3.2   The Interwound Story of Vision and Language

In the previous section, we have discussed pixel-wise versus abstract computational processes aimed at generating labels to be attached to objects and actions. We can think of two different alphabets Σ_p and Σ_s which refer to words related to *pixel-wise and shape-based recognition processes*, respectively. For instance, while terms like `eye`, `mouth`, and `face` are typical elements of Σ_p, their geometrical description is based on terms in Σ_s. So we say that the `face` has an `oval` shape, where `oval` is a typical elements of Σ_s.

Overall, a visual agent performs cognitive tasks by working on $\Sigma_a = \Sigma_p \vee \Sigma_s$. It is important to point out that Σ_a is only the alphabet of *primitive terms*, since when dealing with structured objects and actions, visual agents play with concepts described by additional terms

Basically, the extraction of semantics from video requires linguistic descriptions, even at local level, where one is asked to select words from the alphabet $\omega \in \Sigma_s$. Here we regard any word ω as a symbol with attached semantics, like in the case of any natural language.

The most abstract task that humans are capable to face is that of constructing a function χ_0 as follows

$$\chi_0 : \mathscr{V} \to \mathcal{L}_0 : v(\mathscr{D}) \to \chi_0(v(\mathscr{D})), \tag{1}$$

where $\mathcal{L}_0 \subset \Sigma_s^\star$ is a type zero language in Chomsky's hierarchy. This embraces any linguistic report from visual scenes, like, for instance, movie review. In addition to the ability of extracting information from visual sources, a remarkable specific problem in the construction of χ_0 is that of properly handing the temporal flow of the frames and to provide a semantic representation of the movie actions. Clearly, a movie review does not only require the ability of extracting a visual representation, but also to properly understand the actions so as to produce a corresponding descriptions. While cats and eagles are commonly regarded as animals with very good visual skills, they cannot produce movie reports. Basically, the sentence $\chi_0(v(\mathscr{D})) \in \Sigma_s^\star$ is expected to be taken from a language $\mathcal{L}_0$ of highest level in Chomsky classification, which is denoted by $\mathcal{L}_0$.

Another fundamental visual task is that of *query answer*, that can be regarded as

$$\chi_0 : \mathscr{V} \times \mathcal{L}_0 \to \mathcal{L}_0 : v(\mathscr{D}) \to \chi_0(v(\mathscr{D})), \tag{2}$$

Table 1. Hierarchical structure of semantic labeling.

input $\leadsto$ semantic description	Remarks
$\chi_0(v(\mathscr{D})) \leadsto \mathcal{L}_0$	The language involves ordinary human scene descriptions. Spatial and knowledge levels are both involved.
$\chi_1(v(t, \mathscr{X})) \leadsto \mathcal{L}_0$	The language involves ordinary human picture descriptions. Spatial knowledge is only involved.
$\chi_2(t, x, v(t, \mathscr{X})) \leadsto \mathcal{L}_{-1}$	The language consists of a list of words (language degeneration, no ordering), that is $\mathcal{L}_{-1} \subset \Sigma_s^\star$ only.
$\chi_3(t, x, v(t, \mathscr{X})) \leadsto \mathcal{L}_{-2}$	The language consists of a vector of words (language degeneration, no order). Unlike $\mathcal{L}_{-1}$, the number of symbols is known in advance.

A simplified version and more realistic formulation of semantic labeling, when actions are not the target, is the one in which

$$\chi_1 : \mathscr{I} \to \mathcal{L}_0 : (v(t, \mathscr{X})) \to \chi_1(v(t, \mathscr{X})). \tag{3}$$

This tasks still requires $\mathcal{L}_0$ for linguistic description, but only spatial knowledge $\mathcal{K}_s$ is needed, since, unlike the previous case, there is no temporal processing required (Table 1).

A dramatic drop of complexity arises when asking the agent to provide visual skills on $v(t, \mathscr{X})$ while focussing attention to (t, x). This is described by

$$\chi_2 : \mathscr{D} \times \mathscr{I} \to \Sigma_s : (t, x, v(t, \mathscr{X})) \to \chi_2(t, x, v(t, \mathscr{D})), \tag{4}$$

Basically, while the decision is based on $u(t, x) = (t, x, v(t, \mathscr{X})) \in \mathscr{U}$, which represents quite an unusual granule of information with respect to what is typically processed in machine learning and pattern recognition, this time there is no linguistic description, since we only expected the agent to return a list of symbols of Σ_s. This simplifies dramatically the overall problem, thus contributing to decoupling visual and semantic processes. It is worth mentioning that the dramatic reduction of complexity in the semantic processes is paired with the emergence of focus of attention, namely with decisions based on $u(t, x) \in \mathscr{U}$. In principle, one can expect semantic labeling of (t, x) by means of a single $\omega \in \Sigma_s$, but in some cases dozens of words might be associated with $u(t, x)$. While the linguistic structure degenerates, we are still in presence of a compositional structure, so as the agent might generate remarkable lengthy sentences of pertinent words $\Sigma_s^\star$.

3.3   When Vision Collapses to Classification

An additional simplification on the semantic level arises when considering that the process of generating the words $\omega \in \Sigma^\star$ can be thought of as a compositional process based on a set Σ_d of "dummy symbols", so as

$$\chi_3 : \mathscr{D} \times \mathscr{I} \to \Sigma_d : (t, x, v(t, \mathscr{X})) \to \chi_3(t, x, v(t, \mathscr{D})), \tag{5}$$

Basically, the transition from $\chi_2(\cdot)$ to $\chi_3(\cdot)$ involves a further definitive linguistic simplification, which restricts the symbolic description from $\Sigma_s^\star$ to Σ_d. In so doing, all the complexity is now on the visual side, which requires decisions based on $u(t, x)$, so as we are finally in front of a *classification problem*. This description of visual tasks makes it clear that in order to conquer abstract computer vision skills, any agent does require to address both issues of input representation and linguistic descriptions. We claim that any systematic approach to vision cannot avoid to face the issue of decoupling the classification of visual features, with symbols in Σ_d, and the appropriate linguistic description.

Let us analyze the problems connected with the construction of χ_0 and χ_1, which both operate on global input representation, thus disregarding any focus of attention mechanism. The complexity can be promptly appreciated also in the simplest task χ_1. Clearly, it cannot be regarded as a classification since the agent is expected to provide truly linguistic descriptions. On top of that, when dealing with unrestricted visual environments, the interpretation of $v(t, \mathscr{X})$ is trapped into the chicken-egg dilemma on whether classification of objects or segmentation must take place first. This is due to the absence of any focus of attention, which necessarily leads to holistic mechanisms of information extraction. Unfortunately, while holistic mechanisms are required at a certain level of abstraction, the direct process of $v(t, \mathscr{X})$ do not offer the right source for their activation. Basically, there is no decoupling between the visual source and its linguistic description.

Interestingly, this decoupling takes place when separating $\chi_3(\cdot)$ with respect to the others. The development of abstract levels of description can follow the chaining process

$$\boxed{\mathscr{U} \xrightarrow{\chi_3} \Sigma_d} \xrightarrow{\chi_2} \Sigma_s^\star \xrightarrow{\chi_1} (\Sigma_s^\star, \mathcal{L}_0, \mathcal{K}_s) \xrightarrow{\chi_0} (\Sigma_s^\star, \mathcal{L}_0, \mathcal{K}_s, \mathcal{K}_t), \tag{6}$$

where $\chi_3(\cdot)$ is the only one which deals with the visual signal. All the other functions involve symbolic processes at different levels of abstraction. From one side, $\chi_3(\cdot)$ exploits the focus of attention on $(t, x) \in \mathscr{D}$ to better process the visual information, and, from the other side, it gets rid of any linguistic structure by relying on the classification of dummy symbols.

4   Conclusions

By and large, there is a lot of excitement around computer vision that is definitely motivated by the successful results obtained in the last few years by deep learning. While recognizing the fundamental progress gained under this new wave of connectionist models, this paper claims that the bullish sentiment behind these achievements might not be fully motivated and that the time has come to address a number of fundamental questions that, once properly addressed, could dramatically improve nowadays technology. The discussion is stimulated by the remark that the construction of learning theories of vision properly conceived for intelligent agents working on video instead of large image collections simplifies any visual task. In particular, the paper promotes the principle of developing visual features invariant under motion, which is claimed to be the only significant invariance that is required to gain the "what" function typical of the ventral mainstream.

References

1. Anderson, J.A., Rosenfeld, E. (eds.): Neurocomputing: Foundations of Research. MIT Press, Cambridge (1988)
2. Baker, S., Scharstein, D., Lewis, J.P., Roth, S., Black, M.J., Szeliski, R.: A database and evaluation methodology for optical flow. Int. J. Comput. Vis. **92**(1), 1–31 (2011)
3. Betti, A., Gori, M.: Convolutional networks in visual environments. Arxiv preprint arXiv:1801.07110v1 (2018)
4. Deng, J., Dong, W., Socher, R., Li, L.-J., Li, K., Fei-Fei, L.: ImageNet: a large-scale hierarchical image database. InL CVPR 2009 (2009)
5. Goodale, M.A., Milner, A.D.: Separate visual pathways for perception and action. Trends Neurosci. **15**(1), 20–25 (1992)
6. Gori, M., Lippi, M., Maggini, M., Melacci, S.: Semantic video labeling by developmental visual agents. Comput. Vis. Image Underst. **146**, 9–26 (2016)
7. Goroshin, R., Brun, J., Tompson, J., Eigen, D., LeCun, Y.: Unsupervised learning of spatiotemporally coherent metrics. In: 2015 IEEE International Conference on Computer Vision, ICCV 2015, Santiago, Chile, 7–13 December 2015, pp. 4086–4093 (2015)
8. Horn, B.K.P., Schunck, B.G.: Determining optical flow. Artif. Intell. **17**(1–3), 185–203 (1981)
9. Hubel, D.H., Wiesel, T.N.: Receptive fields, binocular interaction, and functional architecture in the cat's visual cortex. J. Physiol. (Lond.) **160**, 106–154 (1962)
10. Krizhevsky, A., Sutskever, I., Hinton, G.E.: Imagenet classification with deep convolutional neural networks. In: Pereira, F., Burges, C.J.C., Bottou, L., Weinberger, K.Q. (eds.) Advances in Neural Information Processing Systems, vol. 25, pp. 1097–1105. Curran Associates Inc., (2012)
11. Lee, H., Gross, R., Ranganat, R., Ng, A.Y.: Convolutional deep belief networks for scalable unsupervised learning of hierarchical representations. In: Proceedings of the 26th Annual International Conference on Machine Learning, ICML 2009, pp. 609–616. ACM. New York (2009)
12. Lowe, D.G.: Distinctive image features from scale-invariant keypoints. Int. J. Comput. Vis. **60**(2), 91–110 (2004)
13. Marr, D.: Vision. Freeman, San Francisco (1982). Partially reprinted in [1]
14. Braddick, O., Atkinson, J.: Development of human visual function. Vis. Res. **51**, 1588–1609 (2011)
15. Poggio, T.A., Anselmi, F.: Visual Cortex and Deep Networks: Learning Invariant Representations, 1st edn. The MIT Press, Cambridge (2016)
16. Ranzato, M., Huang, F.J., Boureau, Y.-L., LeCun, Y.: Unsupervised learning of invariant feature hierarchies with applications to object recognition. In: 2007 IEEE Computer Society Conference on Computer Vision and Pattern Recognition (CVPR 2007), 18–23 June 2007, Minneapolis, Minnesota, USA (2007)
17. Simonyan, K., Zisserman, A.: Very deep convolutional networks for large-scale image recognition. CoRR, abs/1409.1556 (2014)

18. Andrillon, T.N., Yuval, N., Cirelli, C., Tononi, G., Itzhak, F.: Single-neuron activity and eye movements during human REM sleep and awake vision. Nature (2014)
19. Tavanaei, A., Masquelier, T., Maida, A.S.: Acquisition of visual features through probabilistic spike-timing-dependent plasticity. CoRR, abs/1606.01102 (2016)
20. Dobson, V., Teller, D.Y.: Visual acuity in human infants: a review and comparison of behavioral and electrophysiological studies. Vis. Res. **18**, 1469–1483 (1978)
21. Zitnick, C.L., Dollár, P.: Edge boxes: locating object proposals from edges. In: Fleet, D., Pajdla, T., Schiele, B., Tuytelaars, T. (eds.) ECCV 2014. LNCS, vol. 8693, pp. 391–405. Springer, Cham (2014). https://doi.org/10.1007/978-3-319-10602-1_26

Deep Learning in the Wild

Thilo Stadelmann[1]([✉]), Mohammadreza Amirian[1,2], Ismail Arabaci[3],
Marek Arnold[1,3], Gilbert François Duivesteijn[4], Ismail Elezi[1,5],
Melanie Geiger[1,6], Stefan Lörwald[7], Benjamin Bruno Meier[3],
Katharina Rombach[1], and Lukas Tuggener[1,8]

[1] ZHAW Datalab & School of Engineering, Winterthur, Switzerland
stdm@zhaw.ch
[2] Institute of Neural Information Processing, Ulm University, Ulm, Germany
[3] ARGUS DATA INSIGHTS Schweiz AG, Zürich, Switzerland
[4] Deep Impact AG, Winterthur, Switzerland
[5] DAIS, Ca' Foscari University of Venice, Venezia Mestre, Italy
[6] Institut d'Informatique, Université de Neuchâtel, Neuchâtel, Switzerland
[7] PricewaterhouseCoopers AG, Zürich, Switzerland
[8] IDSIA Dalle Molle Institute for Artificial Intelligence, Manno, Switzerland

Abstract. Deep learning with neural networks is applied by an increasing number of people outside of classic research environments, due to the vast success of the methodology on a wide range of machine perception tasks. While this interest is fueled by beautiful success stories, practical work in deep learning on novel tasks without existing baselines remains challenging. This paper explores the specific challenges arising in the realm of real world tasks, based on case studies from research & development in conjunction with industry, and extracts lessons learned from them. It thus fills a gap between the publication of latest algorithmic and methodical developments, and the usually omitted nitty-gritty of how to make them work. Specifically, we give insight into deep learning projects on face matching, print media monitoring, industrial quality control, music scanning, strategy game playing, and automated machine learning, thereby providing best practices for deep learning in practice.

Keywords: Data availability · Deployment
Loss & reward shaping · Real world tasks

1 Introduction

Measured for example by the interest and participation of industry at the annual NIPS conference[1], it is save to say that deep learning [49] has successfully transitioned from pure research to application [32]. Major research challenges still exist, e.g. in the areas of model interpretability [39] and robustness [1], or general understanding [53] and stability [25,67] of the learning process, to name

---

[1] See https://medium.com/syncedreview/a-statistical-tour-of-nips-2017-438201fb6c8a.

© Springer Nature Switzerland AG 2018
L. Pancioni et al. (Eds.): ANNPR 2018, LNAI 11081, pp. 17–38, 2018.
https://doi.org/10.1007/978-3-319-99978-4_2

a few. Yet, and in addition, another challenge is quickly becoming relevant: in the light of more than 180 deep learning publications per day in the last year[2], the growing number of deep learning engineers as well as prospective researchers in the field need to get educated on best practices and what works and what doesn't *"in the wild"*. This information is usually underrepresented in publications of a field that is very competitive and thus striving above all for novelty and benchmark-beating results [38]. Adding to this fact, with a notable exception [20], the field lacks authoritative and detailed textbooks by leading representatives. Learners are thus left with preprints [37,57], cookbooks [44], code[3] and older gems [28,29,58] to find much needed practical advice.

In this paper, we contribute to closing this gap between cutting edge research and application in the wild by presenting case-based best practices. Based on a number of successful industry-academic research & development collaborations, we report what specifically enabled success in each case alongside open challenges. The presented findings (a) come from real-world and business case-backed use cases beyond purely academic competitions; (b) go deliberately beyond what is usually reported in our research papers in terms of tips & tricks, thus complementing them by the stories behind the scenes; (c) include also what didn't work despite contrary intuition; and (d) have been selected to be transferable as lessons learned to other use cases and application domains. The intended effect is twofold: more successful applications, and increased applied research in the areas of the remaining challenges.

We organize the main part of this paper by case studies to tell the story behind each undertaking. Per case, we briefly introduce the application as well as the specific (research) challenge behind it; sketch the solution (referring details to elsewhere, as the final model architecture etc. is not the focus of this work); highlight what measures beyond textbook knowledge and published results where necessary to arrive at the solution; and show, wherever possible, examples of the arising difficulties to exemplify the challenges. Section 2 introduces a *face matching* application and the amount of surrounding models needed to make it practically applicable. Likewise, Sect. 3 describes the additional amount of work to deploy a state-of-the-art machine learning system into the wider IT system landscape of an *automated print media monitoring* application. Section 4 discusses interpretability and class imbalance issues when applying deep learning for *images-based industrial quality control*. In Sect. 5, measures to cope with the instability of the training process of a complex model architecture for large-scale *optical music recognition* are presented, and the class imbalance problem has a second appearance. Section 6 reports on practical ways for deep reinforcement learning in *complex strategy game play* with huge action and state spaces in non-stationary environments. Finally, Sect. 7 presents first results on comparing practical *automated machine learning* systems with the scientific state of the art, hinting at the use of simple baseline experiments. Section 8 summarizes the lessons learned and gives an outlook on future work on deep learning in practice.

---

[2] Google scholar counts $> 68,000$ articles for the year 2017 as of June 11, 2018.

[3] See e.g. https://modelzoo.co/.

2   Face Matching

Designing, training and testing deep learning models for application in face recognition comes with all the well known challenges like choosing the architecture, setting hyperparameters, creating a representative training/dev/test dataset, preventing bias or overfitting of the trained model, and more. Anyway, very good results have been reported in the literature [9,42,50]. Although the challenges in lab conditions are not to be taken lightly, a new set of difficulties emerges when deploying these models in a real product. Specifically, during development, it is known what to expect as input in the controlled environment. When the models are integrated in a product that is used "in the wild", however, all kinds of input can reach the system, making it hard to maintain a consistent and reliable prediction. In this section, we report on approaches to deal with related challenges in developing an actual face-ID verification product.

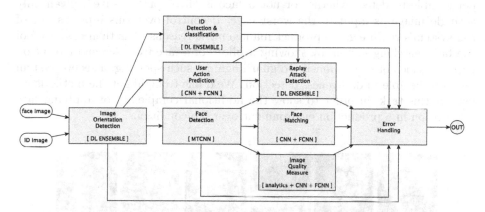

Fig. 1. Schematic representation of a face matching application with ID detection, anti-spoofing and image quality assessment. For any pair of input images (selfie and ID document), the output is the match probability and type of ID document, if no anomaly or attack has been detected. Note that all boxes contain at least one or several deep learning (DL) models with many different (convolutional) architectures.

Although the core functionality of such a product is to quantify the match between a person's face and the photo on the given ID, more functionality is needed to make the system perform its task well, most of it hidden from the user. Thus, in addition to the actual face matching module, the final system contains at least the following machine learnable modules (see Fig. 1):

Image orientation detection When a user takes a photo of the ID on a flat surface using a mobile phone, in many cases the image orientation is random. A deep learning method is applied to predict the orientation angle, used to rotate the image in the correct orientation.

Image quality assessment consists of an ensemble of analytical functions and
deep learning models to test if the photo quality is sufficient for a reliable
match. It also guides the user to improve the picture taking process in case
of bad quality.

User action prediction uses deep learning to predict the action performed by
the user to guide the system's workflow, e.g. making a selfie, presenting an
ID or if the user is doing something wrong during the sequence.

Anti-Spoofing is an essential module that uses various methods to detect if
a person is showing his "real" face or tries to fool the system with a photo,
video or mask. It consists of an ensemble of deep learning models.

For a commercial face-ID product, the anti-spoofing module is both most crucial for success, and technically most challenging; thus, the following discussion
will focus on anti-spoofing in practice. Face matching and recognition systems
are vulnerable to spoofing attacks made by non-real faces, because they are not
per se able to detect whether or not a face is "live" or "not-live", given only
a single image as input in the worst case. If control over this input is out of
the system's reach e.g. for product management reasons, it is then easy to fool
the face matching system by showing a photo of a face from screen or print on
paper, a video or even a mask. To guard against such spoofing, a secure system
needs to be able to do liveness detection. We'd like to highlight the methods we
use for this task, in order to show the additional complexity of applying face
recognition in a production environment over lab conditions.

Fig. 2. Samples from the CASIA dataset [66], where photo 1, 2, and 3 on the left hand
side show a real face, photo 4 shows a replay attack from a digital screen, and photos
5 and 6 show replay attacks from print.

One of the key features of spoofed images is that they usually can be detected
because of degraded image quality: when taking a photo of a photo, the quality deteriorates. However, with high quality cameras in modern mobile phones,
looking at image quality only is not sufficient in the real world. How then can a
spoof detector be designed that approves a real face from a low quality grainy
underexposed photo taken by an old 640 × 480 web cam, and rejects a replay
attack using a photo from a retina display in front of a 4K video camera (compare
Fig. 2)?

Most of the many spoofing detection methods proposed in the literature
use hand crafted features, followed by shallow learning techniques, e.g. SVM
[18,30,34]. These techniques mainly focus on texture differences between real
and spoofed images, differences in color space [7], Fourier spectra [30], or optical

flow maps [6]. In more recent work, deep learning methods have been introduced [3,31,63,64]. Most methods have in common that they attempt to be a one-size-fits-all solution, classifying all incoming cases with one method. This might be facilitated by the available datasets: to develop and evaluate anti-spoofing tools, amongst others CASIA [66], MSU-USSA [43], and the Replay Attack Database [12] exist. Although these datasets are challenging, they turn out to be too easy compared to the input in a production environment.

The main differences between real cases and training examples from these benchmark databases are that the latter ones have been created with a low variety of hardware devices and only use few different locations and light conditions. Moreover, the quality of images throughout the training sets is quite consistent, which does not reflect real input. In contrast, the images that the system receives "in the wild" have the most wide range of possible used hardware and environmental conditions, making the anticipation of new cases difficult. Designing a single system that can classify all such cases with high accuracy seems therefore unrealistic.

We thus create an ensemble of experts, forming a final verdict from 3 independent predictions: the first method consists of 2 patch-based CNNs, one for low resolution images, the other one for high resolution images. They operate on fixed-size tiles from the unscaled input image using a sliding window. This technique proves to be effective for low and high quality input. The second method uses over 20 image quality measures as features combined with a classifier. This method is still very effective when the input quality is low. The third method uses a RNN with LSTM cells to conduct a joint prediction over multiple frames (if available). It is effective in discriminating micro movements of a real face against (simple) translations and rotations of a fake face, e.g. from a photo on paper or screen. All methods return a real vs. fake probability. The outputs of all 3 methods are fed as input features to the final decision tree classifier. This ensemble of deep learning models is experimentally determined to be much more accurate than using any known method individually.

Note that as attackers are inventive and come up with new ways to fool the system quickly, it is important to update the models with new data quickly and regularly.

3 Print Media Monitoring

Content-based print media monitoring serves the task of delivering cropped digital articles from printed newspapers to customers based on their pre-formulated information need (e.g., articles about their own coverage in the media). For this form of article-based information retrieval, it is necessary to segment tens of thousands of newspaper pages into articles daily. We successfully developed neural network-based models to learn how to segment pages into their constituting articles and described their details elsewhere [35,57] (see example results in Fig. 3a–b). In this section, we present challenges faced and learnings gained from integrating a respective model into a production environment with strict performance and reliability requirements.

(a)   (b)   (c)

Fig. 3. Good (a) and bad (b) segmentations (blue lines denote crop marks) for realistic pages, depending on the freedom in the layout. Image (c) shows a non-article page that is excluded from automatic segmentation. (Color figure online)

Exclusion of Non-article Pages. A common problem in print segmentation are special pages that contain content that doesn't represent articles in the common sense, for example classified ads, reader's letters, TV program, share prices, or sports results (see Fig. 3c). Segmentation rules for such pages can be complicated, subjective, and provide little value for general use cases. We thus utilize a random forest-based classifier on handcrafted features to detect such content and avoid feeding respective pages to the general segmentation system to save compute time.

Model Management. One advantage of an existing manual segmentation pipeline is the abundance of high quality, labeled training data being produced daily. To utilize this constant flow of data, we have started implementing an online learning system [52] where results of the automatic segmentation can be corrected within the regular workflow of the segmentation process and fed back to the system as training data.

After training, an important business decision is the final configuration of a model, e.g. determining a good threshold for cuts to weigh between precision and recall, or the decision on how many different models should be used for the production system. We determined experimentally that it is more effective to train different models for different publishers: the same publisher often uses a similar layout even for different newspapers and magazines, while differences between publishers are considerable. To simplify the management of these different models, they are decoupled from the code. This is helpful for rapid development and experimentation.

Technological Integration. For smooth development and operation of the neural network application we have chosen to use a containerized microservices architecture [14] utilizing Docker [62] and RabbitMQ [26]. This decoupled architecture

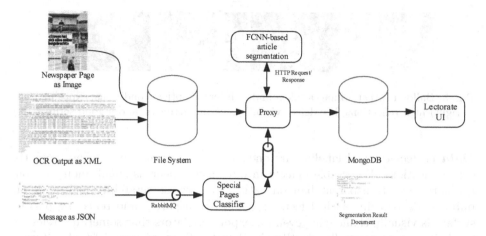

Fig. 4. Architecture of the overall pipeline: the actual model is encapsulated in the "FCNN-based article segmentation" block. Several other systems are required to warrant full functionality: (a) the *Proxy* is responsible to control data input and output from the segmentation model; (b) *RabbitMQ* controls the workflow as a message broker; (c) *MongoDB* stores all segmentation results and metrics; (d) the *Lectorate* UI visualizes results for human assessment and is used to create training data.

(see Fig. 4) brings several benefits especially for machine learning applications: (a) a *separation of concerns* between research, ops and engineering tasks; (b) *decoupling of models/data from code*, allowing for rapid experimentation and high flexibility when deploying the individual components of the system. This is further improved by a modern devops pipeline consisting of continuous integration (CI), continuous deployment (CD), and automated testing; (c) *infrastructure flexibility*, as the entire pipeline can be deployed to an on-premise data center or in the cloud with little effort. Furthermore, the use of Nvidia-docker [62] allows to utilize GPU-computing easily on any infrastructure; (d) precise *controlling and monitoring* of every component in the system is made easy by data streams that enable the injection and extraction of data such as streaming event arguments, log files, and metrics at any stage of the pipeline; and (e) easy *scaling* of the various components to fit different use cases (e.g. training, testing, experimenting, production). Every scenario requires a certain configuration of the system for optimal performance and resource utilization.

4 Visual Quality Control

Manual inspection of medical products for in-body use like balloon catheters is time-consuming, tiring and thus error-prone. A semi-automatic solution with high precision is thus sought. In this section, we present a case study of deep learning for visual quality control of industrial products. While this seems to be a standard use case for a CNN-based approach, the task differs in several interesting respects from standard image classification settings:

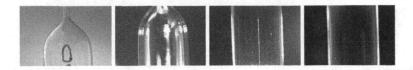

Fig. 5. Balloon catheter images taken under different optical conditions, exposing (left to right) high reflections, low defect visibility, strong artifacts, and a good setup.

Data collection and labeling are one the most critical issues in most practical applications. Detectable defects in our case appear as small anomalies on the surface of transparent balloon catheters, such as scratches, inclusions or bubbles. Recognizing such defects on a thin, transparent and reflecting plastic surface is visually challenging even for expert operators that sometimes refer to a microscope to manually identify the defects. Thus, approx. 50% of a 2-year project duration was used on finding and verifying the optimal optical settings for image acquisition. Figure 5 depicts the results of different optical configurations for such photo shootings. Finally, operators have to be trained to produce consistent labels usable for a machine learning system. In our experience, the labeling quality rises if all involved parties have a basic understanding of the methods. This helps considerably to avoid errors like e.g. only to label a defect on the first image of a series of shots while rotating a balloon: while this is perfectly reasonable from a human perspective (once spotted, the human easily tracks the defect while the balloon moves), it is a no-go for the episodic application of a CNN.

Network and training design for practical applications experiences challenges such as class imbalance, small data regimes, and use case-specific learning targets apart from standard classification settings, making non-standard loss functions necessary (see also Sect. 5). For instance, in the current application, we are looking for relatively small defects on technical images. Therefore, architectures proposed for large-scale natural image classification such as AlexNet [27], GoogLeNet [59], ResNet [24] and modern variants are not necessarily successful, and respective architectures have to be adapted to learn the relevant task. Potential solutions for the class imbalance problem are for example:

- Down-sampling the majority class
- Up-sampling the minority class via image augmentation [13]
- Using pre-trained networks and applying transfer learning [41]
- Increasing the weight of the minority class in the optimization loss [8]
- Generating synthetic data for the minority class using SMOTE [11] or GANs [21]

Selecting a suitable data augmentation approach according for the task is a necessity for its success. For instance, in the present case, axial scratches are more important than radial ones, as they can lead to a tearing of the balloon and its subsequent potentially lethal remaining in a patient's body. Thus, using

	Image	Feature response	Image	Feature response

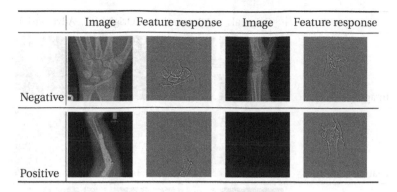

Fig. 6. Visualizing VGG19 feature responses: the first row contains two negative examples (healthy patient) and the second row positives (containing anomalies). All depicted samples are correctly classified.

90° rotation for data augmentation could be fatal. Information like this is only gained in close collaboration with domain experts.

Interpretability of models received considerable attention recently, spurring hopes both of users for transparent decisions, and of experts for "debugging" the learning process. The latter might lead for instance to improved learning from few labeled examples through semantic understanding of the middle layers and intermediate representations in a network. Figure 6 illustrates some human-interpretable representations of the inner workings of a CNN on the recently published MUsculoskeletal RAdiographs (MURA) dataset [45] that we use here as a proxy for the balloon dataset. We used guided-backpropagation [56] and a standard VGG19 network [55] to visualize the feature responses, i.e. the part of the X-ray image on which the network focuses for its decision on "defect" (e.g., broken bone, foreign object) or "ok" (natural and healthy body part). It can be seen that the network mostly decides based on joints and detected defects, strengthening trust in its usefulness. We described elsewhere [2] that this visualization can be extended to an automatic defense against adversarial attacks [21] on deployed neural networks by thresholding the local spatial entropy [10] of the feature response. As Fig. 7 depicts, the focus of a model under attack widens considerably, suggesting that it "doesn't know where to look" anymore.

5 Music Scanning

Optical music recognition (OMR) [46] is the process of translating an image of a page of sheet music into a machine-readable structured format like MusicXML. Existing products exhibit a symbol recognition error rate that is an order of magnitude too high for automatic transcription under professional standards, but don't leverage deep learning computer vision capabilities yet. In this section, we therefore report on the implementation of a deep learning approach to detect

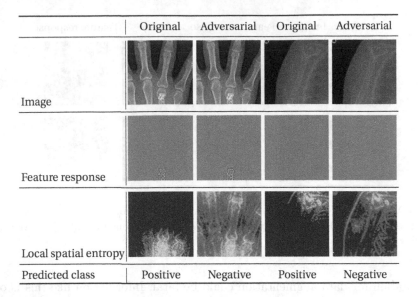

	Original	Adversarial	Original	Adversarial
Image				
Feature response				
Local spatial entropy				
Predicted class	Positive	Negative	Positive	Negative

Fig. 7. Input, feature response and local spatial entropy for clean and adversarial images, respectively. We used VGG19 to estimate predictions and the fast gradient sign attack (FGSM) method [21] to compute the adversarial perturbation.

and classify all musical symbols on a full page of written music in one go, and integrate our model into the open source system Audiveris[4] for the semantic reconstruction of the music. This enables products like digital music stands based on active sheets, as most of todays music is stored in image-based PDF files or on paper.

We highlight four typical issues when applying deep learning techniques to practical OMR: (a) the absence of a comprehensive dataset; (b) the extreme class imbalance present in written music with respect to symbols; (c) the issues of state-of-the-art object detectors with music notation (many tiny and compound symbols on large images); and (d) the transfer from synthetic data to real world examples.

Synthesizing Training Data. The notorious data hunger of deep learning has lead to a strong dependence of results on large, well annotated datasets, such as ImageNet [48] or PASCAL VOC [16]. For music object recognition, no such dataset has been readily available. Since labeling data by hand is no feasible option, we put a one-year effort in synthesizing realistic (i.e., semantically and syntactically correct music notation) data and the corresponding labeling from renderings of publicly available MusicXML files and recently open sourced the resulting *DeepScores* dataset [60].

Dealing with Imbalanced Data. While typical academic training datasets are nicely balanced [16,48], this is rarely the case in datasets sourced from real

---

[4] See http://audiveris.org.

Fig. 8. Symbol classes in *DeepScores* with their relative frequencies (red) in the dataset. (Color figure online)

world tasks. Music notation (and therefore *DeepScores*) shows an extreme class imbalance (see Fig. 8). For example, the most common class (note head black) contains more than 55% of the symbols in the entire dataset, and the top 10 classes contain more than 85% of the symbols. At the other extreme, there is a class which is present only once in the entire dataset, making its detection by pattern recognition methods nearly impossible (a "black swan" is no pattern). However, symbols that are rare are often of high importance in the specific pieces of music where they appear, so simply ignoring the rare symbols in the training data is not an option. A common way to address such imbalance is the use of a weighted loss function, as described in Sect. 4.

This is not enough in our case: first, the imbalance is so extreme that naively reweighing loss components leads to numerical instability; second, the signal of these rare symbols is so sparse that it will get lost in the noise of the stochastic gradient descent method [61], as many symbols will only be present in a tiny fraction of the mini batches. Our current answer to this problem is *data synthesis* [37], using a three-fold approach to synthesize image patches with rare symbols (cp. Fig. 8): (a) we locate rare symbols which are present at least 300 times in the dataset, and crop the parts containing those symbols including their local context (other symbols, staff lines etc.); (b) for rarer symbols, we locate a semantically similar but more common symbol in the dataset (based on some

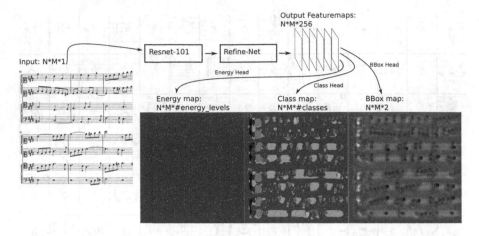

Fig. 9. Schematic of the Deep Watershed Detector model with three distinct output heads. N and M are the height and width of the input image, #classes denotes the number of symbols and #energy_levels is a hyperparameter of the system.

expert-devised notion of symbol similarity), replace this common symbol with the rare symbol and add the resulting page to the dataset. This way, synthesized sheets still have semantic sense, and the network can learn from syntactically correct context symbols. We then crop patches around the rare symbols similar to the previous approach; (c) for rare symbols without similar common symbols, we automatically "compose" music containing those symbols.

Then, during training, we augment each input page in a mini batch with 12 randomly selected synthesized crops of rare symbols (of size 130×80 pixels) by putting them in the margins at the top of the page. This way, that the neural network (on expectation) does not need to wait for more than 10 iterations to see every class which is present in the dataset. Preliminary results show improvement, though more investigation is needed: overfitting on extreme rare symbols is still likely, and questions remain regarding how to integrate the concept of patches (in the margins) with the idea of a full page classifier that considers all context.

Enabling and Stabilizing Training. We initially used state-of-the-art object detection models like Faster R-CNN [47] to attempt detection and classification of musical symbols on *DeepScores*. These algorithms are designed to work well on the prevalent datasets that are characterized by containing low-resolution images with a few big objects. In contrast, *DeepScores* consists of high resolution musical sheets containing hundreds of very small objects, amounting to a very different problem [60]. This disconnect lead to very poor out-of-the-box performance of said systems.

Region proposal-based systems scale badly with the number of objects present on a given image, by design. Hence, we designed the *Deep Watershed Detector* as an entirely new object detection system based on the deep water-

Fig. 10. Top: part of a synthesized image from *DeepScores*; middle: the same part, printed on old paper and photographed using a cell phone; bottom: the same image, automatically retrofitted (based on the dark green lines) to the original image coordinates for ground truth matching (ground truth overlayed in neon green boxes). (Color figure online)

shed transform [4] and described it in detail elsewhere [61]. It detects raw musical symbols (e.g., not a compound note, but note head, stem and flag individually) in their context with a full sheet music page as input. As depicted in Fig. 9, the underlying neural network architecture has three output heads on the last layer, each pertaining to a separate (pixel wise) task: (a) predicting the underlying symbol's class; (b) predicting the energy level (i.e., the degree of belonging of a given pixel location to an object center, also called "objectness"); and (c) predicting the bounding box of the object.

Initially, the training was unstable, and we observed that the network did not learn well if it was directly trained on the combined weighted loss. Therefore, we now train the network on each of the three tasks separately. We further observed that while the network gets trained on the bounding box prediction and classification, the energy level predictions get worse. To avoid this, the network is fine-tuned only for the energy level loss after being trained on all three tasks. Finally, the network is retrained on the combined task (the sum of all three losses, normalized by their respective running means) for a few thousand iterations, giving excellent results on common symbols.

Generalizing to Real-World Data. The basic assumption in machine learning for training and test data to stem from the same distribution is often violated in field applications. In the present case, domain adaptation is crucial: our training set consists of synthetic sheets created by LilyPond scripts [60], while the final product will work on scans or photographs of printed sheet music. These test pictures can have a wide variety of impairments, such as bad printer quality, torn or stained paper etc. While some work has been published on the topic of *domain transfer* [19], the results are non-satisfactory. The core idea to address this problem here is transfer learning [65]: the neural network shall learn the core task of the full complexity of music notation from the synthetic dataset (symbols in context due to full page input), and use a much smaller dataset to adapt to the real world distributions of lighting, printing and defect.

We construct this post-training dataset by carefully choosing several hundred representative musical sheets, printing them with different types of printers on different types of paper, and finally scanning or photographing them. We then use the `BFMatcher` function from OpenCV to align these images with the original musical sheets to use all the ground truth annotation of the original musical sheet for the real-world images (see Fig. 10). This way, we get annotated real-looking images "for free" that have much closer statistics to real-world images than images from *DeepScores*. With careful tuning of the hyperparameters (especially the regularization coefficient), we get promising - but not perfect - results during the inference stage.

6   Game Playing

In this case study, deep reinforcement learning (DRL) is applied to an agent in a multi-player business simulation video game with steadily increasing complexity, comparable to StarCraft or SimCity. The agent is expected to compete with human players in this environment, i.e. to continuously adapt its strategy to challenge evolving opponents. Thus, the agent is required to mimic somewhat general intelligent behavior by transferring knowledge to an increasingly complex environment and adapting its behavior and strategies in a non-stationary, multi-agent environment with large action and state spaces. DRL is a general paradigm, theoretically able to learn any complex task in (almost) any environment. In this section, we share our experiences with applying DRL to the above described competitive environment. Specifically, the performance of a value-based algorithm using Deep Q-Networks (DQN) [36] is compared to a policy gradient method called PPO [51].

Dealing with Competitive Environments. In recent years, astounding results have been achieved by applying DRL in gaming environments. Examples are Atari games [36] and AlphaGo [54], where agents learn human or superhuman performance purely from scratch. In both examples, the environments are either stationary or, if an evolving opponent is present, it did not act simultaneously in the environment; instead, actions were taken in turns. In our environment, multiple evolving players act simultaneously, making changes to the environment that can not be explained solely based on changes in the agent's own policy. Thus, the environment is perceived as non-stationary from the agent's perspective, resulting in stability issues in RL [33]. Another source of complexity in our setting is a huge action and state space (see below). In our experiments, we observed that DQN got problems learning successful control policies as soon as the environment became more complex in this respect, even without non-stationarity

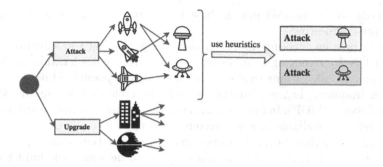

Fig. 11. Heuristic encoding of actions to prevent combinatorial explosion.

induced by opponents. On the other hand, PPO's performance is generally less sensitive to increasing state and action spaces. The impact of non-stationarity to these algorithms is subject of ongoing work.

Reward Shaping. An obvious rewarding choice is the current score of the game (or its gain). Yet, in the given environment, scoring and thus any reward based on it is sparse, since it is dependent on a long sequence of correct actions on the operational, tactical and strategic level. As any rollout of the agent without scoring is not contributing to any gain in knowledge, the learning curve is flat initially. To avoid this initial phase of no information gain, intermediate rewards are given to individual actions, leading to faster learning progress in both DQN and PPO.

Additionally, it is not sufficient for the agent to find a control policy eventually, but it is crucial to find a good policy *quickly*, as training times are anyhow very long. Usually, comparable agents for learning complex behaviors in competitive environments are trained using self-play [5], i.e., the agents are always trained with "equally good" competitors to be able to succeed eventually. In our setting, self play is not a straightforward first option, for several reasons: first, to jump-start learning, it is easier in our setting to play without an opponent first and only learn the art of competition later when a stable ability to act is reached; second, different from other settings, our agents should be entertaining to human opponents, not necessarily winning. It is thus not desirable to learn completely new strategies that are successful yet frustrating to human opponents. Therefore, we will investigate self-play only after stable initializations from (scripted) human opponents on different levels.

Complex State and Action Spaces. Taking the screen frame (i.e., pixels) as input to the control policy is not applicable in our case. First, the policy's input needs to be independent of rendering and thus of hardware, game settings, game version etc. Furthermore, a current frame does not satisfy the Markov property, since attributes like "I own item x" are not necessarily visible in it. Instead, some attributes need to be concluded from past experiences. Thus, the state space needs to be encoded into sufficient features, a task we approach with manual pre-engineering.

Next, a post-engineering approach helps in decreasing the learning time in case of DQN by removing unnecessary actions from consideration as follows: in principal, RL algorithms explore any theoretically possible state-action pair in the environment, i.e., any mathematically possible decision in the Markov Decision Process (MDP). In our environment, the available actions are dependent on the currently available in-game resources of the player, i.e., on the current state. Thus, exploring currently impossible regions in the action space is not efficient and is thus prevented by a post-engineered decision logic built to block these actions from being selected. This reduces the size of the action space per time stamp considerably. These rules where crucial in producing first satisfying learning results in our environment using DQN in a stationary setting of the game. However, when training the agent with PPO, hand-engineered rules where not necessary for proper learning.

The major problem however is the huge action and state space, as it leads to ever longer training times and thus long development cycles. It results from the fact that one single action in our environment might consist of a sequence of sub-decisions. Think e.g. of an action called "attack" in the game of StarCraft, answering the question of WHAT to do (see Fig. 11). It is incompletely defined as long as it does not state WHICH opponent is to be attack using WHICH unit. In other words, each action itself requires a number of different decisions, chosen from different subcategories. To avoid the combinatorial explosion of all possible completely defined actions, we perform another post-processing on the resource management: WHICH unit to choose on WHICH type of enemy, for example, is hard-coded into heuristic rules.

This case study is work in progress, but what becomes evident already is that the combination of the complexity of the task (i.e., acting simultaneously on the operational, tactical and strategic level with exponentially increasing time horizons, as well as a huge state and action space) and the non-stationary environment prevent successful end-to-end learning as in "Pong from pixels"[5]. Rather, it takes manual pre- and post-engineering to arrive at a first agent that learns, and it does so better with policy-based rather than DQN-based algorithms. A next step will explore an explicitly hierarchical learner to cope with the combinatorial explosion of the action space on the three time scales (operational/tactical/strategic) without using hard-coded rules, but instead factorizing the action space into subcategories.

7   Automated Machine Learning

One of the challenging tasks in applying machine learning successfully is to select a suitable algorithm and set of hyperparameters for a given dataset. Recent research in automated machine learning [17,40] and respective academic challenges [22] accurately aimed at finding a solution to this problem for sets of

---

[5] Compare http://karpathy.github.io/2016/05/31/rl/.

practically relevant use cases. The respective Combined Algorithm Selection and Hyperparameter (CASH) optimization problem is defined as finding the best algorithm A^* and set of hyperparameters λ_* with respect to an arbitrary cross-validation loss $\mathscr{L}$ as follows:

$$A^*, \lambda_* = \operatorname*{argmin}_{A \in \mathscr{A}, \lambda \in \Lambda_A} \frac{1}{K} \sum_{i=1}^{K} \mathscr{L}(A_\lambda, D^{(i)}_{train}, D^{(i)}_{valid})$$

where $\mathscr{A}$ is a set of algorithms, Λ_A the set of hyperparameters per algorithm A (together they form the hypothesis space), K is the number of cross validation folds and D are datasets. In this section, we compare two methods from the scientific state-of-the-art (one uses Bayesian optimization, the other genetic programming) with a commercial automated machine learning prototype based on random search.

Scientific State-of-the-Art. Auto-sklearn [17] is the most successful automated machine learning framework in past competitions [23]. The algorithm starts with extracting meta-features from the given dataset and finds models which perform well on similar datasets (according to the meta-features) in a fixed pool of stored successful machine learning endeavors. Auto-sklearn then performs meta-learning by initializing a set of model candidates with the model and hyperparameter choices of k nearest neighbors in dataset space; subsequently, it optimizes their hyperparameters and feature preprocessing pipeline using Bayesian optimization. Finally, an ensemble of the optimized models is build using a greedy search. On the other side, Tree-based Pipeline Optimization Tool (TPOT) [40] is toolbox based on genetic programming. The algorithm starts with random initial configurations including feature preprocessing, feature selection and a supervised classifier. At every step, the top 20% best models are retained and randomly modified to generate offspring. The offspring competes with the parent, and winning models proceed to the next iteration of the algorithm.

Commercial Prototype. The Data Science Machine (DSM) is currently used inhouse for data science projects by a business partner. It uses random sampling of the solution space for optimization. Machine learning algorithms in this system are leveraged from Microsoft Azure, scikit-learn and can be user-enhanced. DSM can be deployed in the cloud, on-premise, as well as standalone. The pipeline of DSM includes data preparation, feature reduction, automatic model optimization, evaluation and final ensemble creation. The question is: can it prevail against much more sophisticated systems even at this early stage of development?

Evaluation is performed using the protocol of the AutoML challenge [22] for comparability, confined to a subset of ten datasets that is processable for the current DSM prototype (i.e., non-sparse, non-big). It spans the tasks of regression, binary and multi-class classification. For applicability, we constrain the time budget of the searches by the required time for DSM to train 100 models

Table 1. Comparison of different automated machine learning algorithms.

Dataset	Task	Metric	Auto-Sklearn		TPOT		DSM	
			Validation	Test	Validation	Test	Validation	Test
Cadata	Regression	Coefficient of determination	0.7913	0.7801	**0.8245**	**0.8017**	0.7078	0.7119
Christine	Binary classification	Balanced accuracy score	0.7380	0.7405	**0.7435**	**0.7454**	0.7362	0.7146
Digits	Multiclass classification	Balanced accuracy score	**0.9560**	**0.9556**	0.9500	0.9458	0.8900	0.8751
Fabert	Multiclass classification	Accuracy score	**0.7245**	**0.7193**	0.7172	0.7006	0.7112	0.6942
Helena	Multiclass classification	Balanced accuracy score	**0.3404**	**0.3434**	0.2654	0.2667	0.2085	0.2103
Jasmine	Binary classification	Balanced accuracy score	0.7987	0.8348	**0.8188**	0.8281	0.8020	**0.8371**
Madeline	Binary classification	Balanced accuracy score	**0.8917**	**0.8769**	0.8885	0.8620	0.7707	0.7686
Philippine	Binary classification	Balanced accuracy score	0.7787	0.7486	**0.7839**	**0.7646**	0.7581	0.7406
Sylvine	Binary classification	Balanced accuracy score	0.9414	0.9454	**0.9512**	**0.9493**	0.9414	0.9233
Volkert	Multiclass classification	Accuracy score	**0.7174**	**0.7101**	0.6429	0.6327	0.5220	0.5153
Average performance			**0.7678**	**0.7654**	0.7586	0.7497	0.7048	0.6991

using random algorithm selection. A performance comparison is given in Table 1, suggesting that Bayesian optimization and genetic programming are superior to random search. However, random parameter search lead to reasonably good models and useful results as well (also in commercial practice). This suggests room for improvement in actual meta-*learning*.

8   Conclusions

Does deep learning work in the wild, in business and industry? In the light of the presented case studies, a better questions is: *what does it take to make it work?* Apparently, the challenges are different compared to academic competitions: instead of a given task and known (but still arbitrarily challenging) environment, given by data and evaluation metric, real-world applications are characterized by (a) data quality and quantity issues; and (b) unprecedented (thus: unclear) learning targets. This reflects the different nature of the problems: competitions provide a controlled but unexplored environment to facilitate the discovery of new methods; real-world tasks on the other hand build on the knowledge of a zoo of methods (network architectures, training methods) to solve a specific, yet still unspecified (in formal terms) task, thereby enhancing the method zoo in return in case of success. The following lessons learned can be drawn from our six case studies (section numbers given in parentheses refer to respective details):

Data acquisition usually needs much more time than expected Sect. 4, yet is the basis for all subsequent success Sect. 5. Class imbalance and covariate shift are usual Sects. 2, 4, 5.

Understanding of what has been learned and how decisions emerge help both the user and the developer of neural networks to build trust and improve quality Sects. 4, 5. Operators and business owners need a basic understanding of used methods to produce usable ground truth and provide relevant subject matter expertise Sect. 4.

Deployment should include online learning Sect. 3 and might involve the buildup of up to dozens of other machine learning models Sects. 2, 3 to flank the original core part.

Loss/reward shaping is usually necessary to enable learning of very complex target functions in the first place Sects. 5, 6. This includes encoding expert knowledge manually into the model architecture or training setup Sects. 4, 6, and handling special cases separately Sect. 3 using some automatic pre-classification.

Simple baselines do a good job in determining the feasibility as well as the potential of the task at hand when final datasets or novel methods are not yet seen Sects. 4, 7. Increasing the complexity of methods and (toy-)tasks in small increments helps monitoring progress, which is important to effectively debug failure cases Sect. 6.

Specialized models for identifiable sub-problems increase the accuracy in production systems over all-in-one solutions Sects. 2, 3, and ensembles of experts help where no single method reaches adequate performance Sect. 2.

Best practices are straightforward to extract on the general level ("plan enough resources for data acquisition"), yet quickly get very specific when broken down to technicalities ("prefer policy-based RL given that ..."). An overarching scheme seems to be that the challenges in real-world tasks need similar amounts of creativity and knowledge to get solved as fundamental research tasks, suggesting they need similar development methodologies on top of proper engineering and business planning.

We identified specific areas for future applied research: (a) *anti-spoofing* for face verification; (b) the *class imbalance* problem in OMR; and (c) the slow learning and poor performance of *RL agents* in non-stationary environments with large action and state spaces. The latter is partially addressed by new challenges like Dota 2[6], Pommerman or VizDoom[7], but for example doesn't address hierarchical actions. Generally, future work should include (d) making deep learning more *sample efficient* to cope with smaller training sets (e.g. by one-shot learning, data or label generation [15], or architecture learning); (e) finding *suitable architectures* and *loss designs* to cope with the complexity of real-world tasks; and (f) improving the *stability* of training and *robustness* of predictions along with (d) the *interpretability* of neural nets.

Acknowledgements. We are grateful for the invitation by the ANNPR chairs and the support of our business partners in Innosuisse grants 17719.1 "PANOPTES", 17963.1

---

[6] See e.g. https://blog.openai.com/dota-2/.
[7] See https://www.pommerman.com/competitions and http://vizdoom.cs.put.edu.pl.

"DeepScore", 25256.1 "Libra", 25335.1 "FarmAI", 25948.1 "Ada" and 26025.1 "Qual-itAI".

References

1. Akhtar, N., Mian, A.: Threat of adversarial attacks on deep learning in computer vision: a survey. arXiv preprint arXiv:1801.00553 (2018)
2. Amirian, M., Schwenker, F., Stadelmann, T.: Trace and detect adversarial attacks on CNNs using feature response maps. In: ANNPR (2018)
3. Atoum, Y., Liu, Y., Jourabloo, A., Liu, X.: Face anti-spoofing using patch and depth-based CNNs. In: IEEE International Joint Conference on Biometrics (IJCB) (2017)
4. Bai, M., Urtasun, R.: Deep watershed transform for instance segmentation. In: CVPR (2017)
5. Bansal, T., Pachocki, J., Sidor, S., Sutskever, I., Mordatch, I.: Emergent complexity via multi-agent competition. arXiv preprint arXiv:1710.03748 (2017)
6. Bao, W., Li, H., Li, N., Jiang, W.: A liveness detection method for face recognition based on optical flow field. In: International Conference on Image Analysis and Signal Processing (2009)
7. Boulkenafet, Z., Komulainen, J., Hadid, A.: Face anti-spoofing based on color texture analysis. In: International Conference on Image Processing (ICIP) (2015)
8. Buda, M., Maki, A., Mazurowski, M.A.: A systematic study of the class imbalance problem in convolutional neural networks. arXiv preprint arXiv:1710.05381 (2017)
9. Cao, Q., Shen, L., Xie, W., Parkhi, O.M., Zisserman, A.: VGGFace2: a dataset for recognising faces across pose and age. arXiv preprint arXiv:1710.08092 (2017)
10. Chanwimaluang, T., Fan, G.: An efficient blood vessel detection algorithm for retinal images using local entropy thresholding. In: International Symposium on Circuits and Systems (ISCAS), vol. 5 (2003)
11. Chawla, N.V., Bowyer, K.W., Hall, L.O., Kegelmeyer, W.P.: SMOTE: synthetic minority over-sampling technique. J. Artif. Intell. Res. **16**, 321–357 (2002)
12. Chingovska, I., Anjos, A., Marcel, S.: On the effectiveness of local binary patterns in face anti-spoofing. In: BIOSIG (2012)
13. Ciresan, D.C., Meier, U., Schmidhuber, J.: Multi-column deep neural networks for image classification. In: CVPR (2012)
14. Dragoni, N., Lanese, I., Larsen, S.T., Mazzara, M., Mustafin, R., Safina, L.: Microservices: how to make your application scale. In: Petrenko, A.K., Voronkov, A. (eds.) PSI 2017. LNCS, vol. 10742, pp. 95–104. Springer, Cham (2018). https://doi.org/10.1007/978-3-319-74313-4_8
15. Elezi, I., Torcinovich, A., Vascon, S., Pelillo, M.: Transductive label augmentation for improved deep network learning. In: ICPR (2018)
16. Everingham, M., Gool, L.J.V., Williams, C.K.I., Winn, J.M., Zisserman, A.: The PASCAL visual object classes (VOC) challenge. Int. J. Comput. Vis. **88**(2), 303–338 (2010)
17. Feurer, M., Klein, A., Eggensperger, K., Springenberg, J., Blum, M., Hutter, F.: Efficient and robust automated machine learning. In: NIPS (2015)
18. Galbally, J., Marcel, S., Fiérrez, J.: Image quality assessment for fake biometric detection: application to iris, fingerprint, and face recognition. IEEE Trans. Image Process. **23**(2), 710–724 (2014)
19. Gebru, T., Hoffman, J., Fei-Fei, L.: Fine-grained recognition in the wild: a multi-task domain adaptation approach. In: ICCV (2017)

20. Goodfellow, I., Bengio, Y., Courville, A.: Deep Learning. MIT Press, Cambridge (2016)
21. Goodfellow, I.J., Shlens, J., Szegedy, C.: Explaining and harnessing adversarial examples. In: ICLR (2015)
22. Guyon, I., et al.: Design of the 2015 ChaLearn AutoML challenge. In: IJCNN (2015)
23. Guyon, I., et al.: A brief review of the ChaLearn AutoML challenge. In: AutoML workshop@ICML (2016)
24. He, K., Zhang, X., Ren, S., Sun, J.: Deep residual learning for image recognition. In: CVPR (2016)
25. Irpan, A.: Deep reinforcement learning doesn't work yet, 14 February 2018. https://www.alexirpan.com/2018/02/14/rl-hard.html
26. John, V., Liu, X.: A survey of distributed message broker queues. arXiv preprint arXiv:1704.00411 (2017)
27. Krizhevsky, A., Sutskever, I., Hinton, G.E.: ImageNet classification with deep convolutional neural networks. In: NIPS (2012)
28. Larochelle, H., Bengio, Y., Louradour, J., Lamblin, P.: Exploring strategies for training deep neural networks. JMLR 10(1), 1–40 (2009)
29. LeCun, Y., Bottou, L., Orr, G.B., Müller, K.-R.: Efficient backprop. In: Orr, G.B., Müller, K.-R. (eds.) Neural Networks: Tricks of the Trade. LNCS, vol. 1524, pp. 9–50. Springer, Heidelberg (1998). https://doi.org/10.1007/3-540-49430-8_2
30. Li, J., Wang, Y., Tan, T., Jain, A.K.: Live face detection based on the analysis of Fourier spectra. In: Biometric Technology for Human Identification (2004)
31. Li, L., Feng, X., Boulkenafet, Z., Xia, Z., Li, M., Hadid, A.: An original face anti-spoofing approach using partial convolutional neural network. In: International Conference on Image Processing Theory, Tools and Applications (IPTA) (2016)
32. Liu, W., Wang, Z., Liu, X., Zeng, N., Liu, Y., Alsaadi, F.E.: A survey of deep neural network architectures and their applications. Neurocomputing 234, 11–26 (2017)
33. Lowe, R., Wu, Y., Tamar, A., Harb, J., Abbeel, O.P., Mordatch, I.: Multi-agent actor-critic for mixed cooperative-competitive environments. In: NIPS (2017)
34. Määttä, J., Hadid, A., Pietikäinen, M.: Face spoofing detection from single images using micro-texture analysis. In: International Joint Conference on Biometrics (IJCB) (2011)
35. Meier, B., Stadelmann, T., Stampfli, J., Arnold, M., Cieliebak, M.: Fully convolutional neural networks for newspaper article segmentation. In: ICDAR (2017)
36. Mnih, V., et al.: Playing Atari with deep reinforcement learning. arXiv preprint arXiv:1312.5602 (2013)
37. Ng, A.: Machine learning yearning - technical strategy for AI engineers in the era of deep learning (2018, to appear)
38. Olah, C., Carter, S.: Research debt. Distill (2017)
39. Olah, C., et al.: The building blocks of interpretability. Distill (2018)
40. Olson, R.S., Urbanowicz, R.J., Andrews, P.C., Lavender, N.A., Kidd, L.C., Moore, J.H.: Automating biomedical data science through tree-based pipeline optimization. In: Squillero, G., Burelli, P. (eds.) EvoApplications 2016. LNCS, vol. 9597, pp. 123–137. Springer, Cham (2016). https://doi.org/10.1007/978-3-319-31204-0_9
41. Pan, S.J., Yang, Q.: A survey on transfer learning. IEEE Trans. Knowl. Data Eng. 22(10), 1345–1359 (2010)
42. Parkhi, O.M., Vedaldi, A., Zisserman, A.: Deep face recognition. In: BMVC (2015)
43. Patel, K., Han, H., Jain, A.K.: Secure face unlock: spoof detection on smartphones. IEEE Trans. Inf. Forensics Secur. 11(10), 2268–2283 (2016)

44. Perez, C.E.: The deep learning AI playbook - strategy for disruptive artificial intelligence (2017)
45. Rajpurkar, P., et al.: MURA dataset: towards radiologist-level abnormality detection in musculoskeletal radiographs. arXiv preprint arXiv:1712.06957 (2017)
46. Rebelo, A., Fujinaga, I., Paszkiewicz, F., Marçal, A.R.S., Guedes, C., Cardoso, J.S.: Optical music recognition: state-of-the-art and open issues. Int. J. Multimed. Inf. Retr. **1**(3), 173–190 (2012)
47. Ren, S., He, K., Girshick, R.B., Sun, J.: Faster R-CNN: towards real-time object detection with region proposal networks. In: NIPS (2015)
48. Russakovsky, O., et al.: ImageNet large scale visual recognition challenge. Int. J. Comput. Vis. **115**(3), 211–252 (2015)
49. Schmidhuber, J.: Deep learning in neural networks: an overview. Neural Netw. **61**, 85–117 (2015)
50. Schroff, F., Kalenichenko, D., Philbin, J.: FaceNet: a unified embedding for face recognition and clustering. In: CVPR (2015)
51. Schulman, J., Wolski, F., Dhariwal, P., Radford, A., Klimov, O.: Proximal policy optimization algorithms. arXiv preprint arXiv:1707.06347 (2017)
52. Shalev-Shwartz, S.: Online learning and online convex optimization. Found. Trends Mach. Learn. **4**(2), 107–194 (2012)
53. Shwartz-Ziv, R., Tishby, N.: Opening the black box of deep neural networks via information. arXiv preprint arXiv:1703.00810 (2017)
54. Silver, D., et al.: Mastering the game of go with deep neural networks and tree search. Nature **529**(7587), 484–489 (2016)
55. Simonyan, K., Zisserman, A.: Very deep convolutional networks for large-scale image recognition. In: ICLR (2015)
56. Springenberg, J.T., Dosovitskiy, A., Brox, T., Riedmiller, M.: Striving for simplicity: the all convolutional net. arXiv preprint arXiv:1412.6806 (2014)
57. Stadelmann, T., Tolkachev, V., Sick, B., Stampfli, J., Dürr, O.: Beyond ImageNet - deep learning in industrial practice. In: Braschler, M., Stadelmann, T., Stockinger, K. (eds.) Applied Data Science - Lessons Learned for the Data-Driven Business. Springer (2018, to appear)
58. Sutskever, I., Martens, J., Dahl, G., Hinton, G.: On the importance of initialization and momentum in deep learning. In: ICML (2013)
59. Szegedy, C., et al.: Going deeper with convolutions. In: CVPR (2015)
60. Tuggener, L., Elezi, I., Schmidhuber, J., Pelillo, M., Stadelmann, T.: DeepScores - a dataset for segmentation, detection and classification of tiny objects. In: ICPR (2018)
61. Tuggener, L., Elezi, I., Schmidhuber, J., Stadelmann, T.: Deep watershed detector for music object recognition. In: ISMIR (2018)
62. Xu, P., Shi, S., Chu, X.: Performance evaluation of deep learning tools in Docker containers. arXiv preprint arXiv:1711.03386 (2017)
63. Xu, Z., Li, S., Deng, W.: Learning temporal features using LSTM-CNN architecture for face anti-spoofing. In: ACPR (2015)
64. Yang, J., Lei, Z., Li, S.Z.: Learn convolutional neural network for face anti-spoofing. arXiv preprint arXiv:1408.5601 (2014)
65. Yosinski, J., Clune, J., Bengio, Y., Lipson, H.: How transferable are features in deep neural networks? In: NIPS (2014)
66. Zhang, Z., Yan, J., Liu, S., Lei, Z., Yi, D., Li, S.Z.: A face antispoofing database with diverse attacks. In: International Conference on Biometrics (ICB) (2012)
67. Zheng, S., Song, Y., Leung, T., Goodfellow, I.: Improving the robustness of deep neural networks via stability training. In: CVPR (2016)

Learning Algorithms and Architectures

Effect of Equality Constraints to Unconstrained Large Margin Distribution Machines

Shigeo Abe[✉]

Kobe University, Rokkodai, Nada, Kobe, Japan
abe@kobe-u.ac.jp
http://www2.kobe-u.ac.jp/~abe

Abstract. Unconstrained large margin distribution machines (ULDMs) maximize the margin mean and minimize the margin variance without constraints. In this paper, we first reformulate ULDMs as a special case of least squares (LS) LDMs, which are a least squares version of LDMs. By setting a hyperparameter to control the trade-off between the generalization ability and the training error to zero, LS LDMs reduce to ULDMs. In the computer experiments, we include the zero value of the hyperparameter as a candidate value for model selection. According to the experiments using two-class problems, in most cases LS LDMs reduce to ULDMs and their generalization abilities are comparable. Therefore, ULDMs are sufficient to realize high generalization abilities without equality constraints.

1 Introduction

In a classification problem, margins between data and the separating hyperplane play an important role. Here, margin is defined as the distance between a data point and the separating hyperplane and it is nonnegative when correctly classified, and negative, when misclassified. In the support vector machine (SVM) [1,2], the minimum margin is maximized.

Because the SVM does not assume a specific data distribution, the obtained separating hyperplane is optimal under the assumption that the data obey an unknown but fixed distribution. Therefore, if prior knowledge is available, it can improve the generalization ability.

The central idea of SVMs, maximizing the minimum margin, has been applied to improving generalization performance of other classifiers. However, for AdaBoost, instead of the minimum margin, directly controlling the margin distribution has been known to improve the generalization ability [3,4].

Among several classifiers to control the margin distribution [5–12], in [6], the margin mean for the training data is maximized without constraints. This approach is extended in [11]: the bias and slope of the separating hyperplanes are optimized and then equality constraints are introduced. This introduction results in the least squares SVM. According to the computer experiments, without equality constraints, the generalization ability is inferior to that of the SVM.

© Springer Nature Switzerland AG 2018
L. Pancioni et al. (Eds.): ANNPR 2018, LNAI 11081, pp. 41–53, 2018.
https://doi.org/10.1007/978-3-319-99978-4_3

In [9,10], in addition to maximizing the margin mean, the margin variance is minimized and the classifier is called large margin distribution machine (LDM). The advantage of the LDM is that the generalization ability is better than or comparable to that of the SVM, but one of the disadvantages is that two hyperparameters are added to the SVM. This will lengthen model selection. To solve this problem, in [12], an unconstrained LDM (ULDM) is developed, where the number of hyperparameters is the same as that of the SVM.

In this paper, we reformulate the ULDM as a special case of the least squares LDM (LS LDM). As in [12], we formulate the LS LDM as maximizing the margin mean and minimizing the margin variance, in addition to minimizing the square norm of the coefficient vector of the hyperplane and the square sum of slack variables. As in the LS SVM, we impose the equality constraints for training data. Because the hyperparameters are necessary for the square sum of slack variables and the margin variance, one hyperparameter is added to the LS SVM. Eliminating the square sum of slack variables in the objective function and the equality constraints, we obtain the ULDM.

By computer experiments we perform model selection of the LS LDM including the parameter value of zero for the slack variables, which results in the ULDM. Checking the number that the parameter value of zero is taken, we judge whether the equality constraints are necessary for improving the generalization ability.

In Sect. 2, we summarize the LS SVM. And in Sect. 3, we explain the LDM and then discuss its variants: the LS LDM and ULDM. In Sect. 4, we evaluate the effect of equality constraints to the ULDM using two-class problems.

2 Least Squares Support Vector Machines

Let the decision function in the feature space be

$$f(\mathbf{x}) = \mathbf{w}^\top \boldsymbol{\phi}(\mathbf{x}) + b, \tag{1}$$

where $\boldsymbol{\phi}(\mathbf{x})$ maps the m-dimensional input vector $\mathbf{x}$ into the l-dimensional feature space, $\mathbf{w}$ is the l-dimensional coefficient vector, $\top$ denotes the transpose of a vector, and b is the bias term.

Let the M training input-output pairs be $\{\mathbf{x}_i, y_i\}$ $(i = 1, \ldots, M)$, where $\mathbf{x}_i$ are training inputs and y_i are the associated labels and $y_i = 1$ for Class 1 and -1 for Class 2.

The margin of $\mathbf{x}_i$, δ_i, is defined as the distance from the separating hyperplane $f(\mathbf{x}) = 0$, and is given by

$$\delta_i = y_i \, f(\mathbf{x}_i) / \|\mathbf{w}\|. \tag{2}$$

If $\delta \, \|\mathbf{w}\| = 1$, where δ is the minimum margin among δ_i $(i = 1, \ldots, M)$, maximizing δ is equivalent to minimizing $\|\mathbf{w}\|$. To make δ_i larger than or equal to 1, $\mathbf{x}_i$ need to satisfy $y_i \, f(\mathbf{x}_i) \geq 1$. Then allowing misclassification, the LS

SVM is formulated in the primal form as follows:

$$\text{minimize} \quad Q(\mathbf{w}, b, \boldsymbol{\xi}) = \frac{1}{2}\mathbf{w}^\top \mathbf{w} + \frac{C}{2}\sum_{i=1}^{M}\xi_i^2 \tag{3}$$

$$\text{subject to} \quad y_i\, f(\mathbf{x}_i) = 1 - \xi_i \quad \text{for} \quad i = 1, \ldots, M, \tag{4}$$

where $Q(\mathbf{w}, b, \boldsymbol{\xi})$ is the objective function, C is the margin parameter that controls the trade-off between the training error and the generalization ability, ξ_i are the slack variables for $\mathbf{x}_i$, and $\boldsymbol{\xi} = (\xi_1, \ldots, \xi_M)^\top$. If we change ξ_i^2 to ξ_i, and $C/2$ to C in (3), and the equality constraints in (4) to inequality constraints, we obtain the L1 SVM.

Solving the equation in (4) for ξ_i and substituting it to the objective function in (3), we obtain the unconstrained optimization problem:

$$\text{minimize} \quad Q(\mathbf{w}, b) = \frac{1}{2}\mathbf{w}^\top \mathbf{w} + \frac{C}{2}\sum_{i=1}^{M}(1 - y_i\, f(\mathbf{x}_i))^2. \tag{5}$$

The solution of the LS SVM can be obtained by solving a set of linear equations and generalization performance is known to be comparable to the L1 SVM [2], but unlike the L1 SVM, the solution is not sparse.

In the following we use the LS SVM to derive an LS LDM, which is a variant of the LDM, and also use to compare performance of the ULDM.

3 Large Margin Distribution Machines and Their Variants

In this section, first we briefly summarize the LDM. Then, we define the LS LDM and ULDM in a way slightly different from [12].

3.1 Large Margin Distribution Machines

The LDM [9] maximizes the margin mean and minimizes the margin variance.

The margin mean $\bar{\delta}$ and margin variance $\hat{\delta}$ are given, respectively, by

$$\bar{\delta} = \frac{1}{M}\sum_{i=1}^{M}\delta_i, \tag{6}$$

$$\hat{\delta} = \frac{1}{M}\sum_{i=1}^{M}\left(\delta_i - \bar{\delta}\right)^2 = \frac{1}{M}\sum_{i=1}^{M}\delta_i^2 - \bar{\delta}^2. \tag{7}$$

Here, instead of (2), we consider the margin as

$$\delta_i = y_i\, f(\mathbf{x}_i). \tag{8}$$

Similar to the L1 SVM, the LDM is formulated as follows:

$$\text{minimize} \quad Q(\mathbf{w}, b, \boldsymbol{\xi}) = \frac{1}{2}\mathbf{w}^\top\mathbf{w} - \lambda_1\,\bar{\delta} + \frac{1}{2}\lambda_2\,\hat{\delta} + C\sum_{i=1}^{M}\xi_i \tag{9}$$

$$\text{subject to} \quad y_i\,f(\mathbf{x}_i) \geq 1 - \xi_i \qquad \text{for } i = 1,\dots,M, \tag{10}$$

where λ_1 and λ_2 are parameters to control maximization of the margin mean and minimization of the margin variance, respectively. In the objective function, the second and the third terms are added to the L1 SVM.

Because the LDM uses all the training data to calculate the margin mean and the margin variance, the solution is dense. Furthermore, because four parameter values (including one kernel parameter value), instead of two, need to be determined by model selection, model selection requires more time than the L1 SVM does.

3.2 Least Squares Large Margin Distribution Machines

The LS LDM that maximizes the margin mean and minimizes the margin variance is given by replacing the slack sum in (9) with the square sum and the inequality constraints in (10) with the equality constraints as follows:

$$\text{minimize} \quad Q(\mathbf{w}, b, \boldsymbol{\xi}) = \frac{1}{2}\mathbf{w}^\top\mathbf{w} - \lambda_1\,\bar{\delta} + \frac{1}{2}\lambda_2\,\hat{\delta} + \frac{C}{2}\sum_{i=1}^{M}\xi_i^2 \tag{11}$$

$$\text{subject to} \quad y_i\,f(\mathbf{x}_i) = 1 - \xi_i \qquad \text{for } i = 1,\dots,M. \tag{12}$$

Solving the equation in (12) for ξ_i and substituting it to the objective function in (11) yield

$$\begin{aligned}\text{minimize} \quad Q(\mathbf{w}, b) &= \frac{1}{2}\mathbf{w}^\top\mathbf{w} - \lambda_1\,\bar{\delta} + \frac{1}{2}\lambda_2\,\hat{\delta} + \frac{C}{2}\sum_{i=1}^{M}(1 - y_i\,f(\mathbf{x}_i))^2 \\ &= \frac{1}{2}\mathbf{w}^\top\mathbf{w} - \lambda_1\,\bar{\delta} + \frac{1}{2}\lambda_2\,\hat{\delta} + \frac{C}{2}\sum_{i=1}^{M}(\delta_i - 1)^2. \end{aligned} \tag{13}$$

In the above objective function, the last term, which is the variance of margins around the minimum margin works similarly to the third term, which is the variance of margin around the margin mean, $\hat{\delta}$.

Now substituting (6), (7), and (8) into the objective function of (13) and deleting the constant term, we obtain

$$\begin{aligned}Q(\mathbf{w}, b) = \frac{1}{2}\mathbf{w}^\top\mathbf{w} + \frac{\lambda_2}{2M}\left(1 + \frac{MC}{\lambda_2}\right)\sum_{i=1}^{M}f^2(\mathbf{x}_i) - \frac{\lambda_2}{2}\left(\frac{1}{M}\sum_{i=1}^{M}y_i\,f(\mathbf{x}_i)\right)^2 \\ -\left(\frac{\lambda_1}{M} + C\right)\sum_{i=1}^{M}y_i\,f(\mathbf{x}_i). \end{aligned} \tag{14}$$

The first three terms in the above objective function are quadratic and the last term is linear with respect to $\mathbf{w}$ and b. Therefore, the coefficient of the linear term is a scaling factor of the decision function obtained by minimizing (14) with respect to $\mathbf{w}$ and b. Dividing (14) by λ_2 and eliminating the coefficient of the last term, we obtain

$$Q(\mathbf{w}, b) = \frac{1}{2\,C_{\mathrm{m}}}\mathbf{w}^{\top}\mathbf{w} + \frac{1 + C_{\mathrm{e}}}{2\,M}\sum_{i=1}^{M} f^2(\mathbf{x}_i)$$

$$-\frac{1}{2}\left(\frac{1}{M}\sum_{i=1}^{M} y_i\, f(\mathbf{x}_i)\right)^2 - \sum_{i=1}^{M} y_i\, f(\mathbf{x}_i). \tag{15}$$

Here, $C_{\mathrm{m}} = \lambda_2$ and $C_{\mathrm{e}} = M\,C/\lambda_2$.

According to the above formulation of the LS LDM, the parameter λ_1 in (13) does not work for controlling the margin mean. Therefore, the three hyperparameters in (11) and (12) are reduced to two.

3.3   Unconstrained Large Margin Distribution Machines

Deleting the square sum of the slack variables in (11) and equality constraints in (12), we consider the unconstrained LDM (ULDM) as follows:

$$\text{minimize}\quad Q(\mathbf{w}, b) = \frac{1}{2\,C_{\mathrm{m}}}\mathbf{w}^{\top}\mathbf{w} - M\,\bar{\delta} + \frac{1}{2}\hat{\delta}$$

$$= \frac{1}{2\,C_{\mathrm{m}}}\mathbf{w}^{\top}\mathbf{w} + \frac{1}{2\,M}\sum_{i=1}^{M} f^2(\mathbf{x}_i) - \frac{1}{2}\left(\frac{1}{M}\sum_{i=1}^{M} y_i\, f(\mathbf{x}_i)\right)^2$$

$$-\sum_{i=1}^{M} y_i\, f(\mathbf{x}_i). \tag{16}$$

Here, we multiply $\bar{\delta}$ with M so that the coefficient of the linear term is 1.

Comparing (15) and (16), the ULDM is obtained by setting $C_{\mathrm{e}} = 0\,(C = 0)$.

Because the LS LDM includes the ULDM, we derive the optimality conditions for (15) in the empirical feature space [2]. Let $\{\mathbf{z}_1, \ldots, \mathbf{z}_N\}$ be a subset of $\{\mathbf{x}_1, \ldots, \mathbf{x}_M\}$, where $N \leq M$ and let $\{\boldsymbol{\phi}(\mathbf{z}_1), \ldots, \boldsymbol{\phi}(\mathbf{z}_N)\}$ span the empirical feature space. Then the mapping function that maps the input space into the empirical feature space is expressed by

$$\mathbf{h}(\mathbf{x}) = (K(\mathbf{x}, \mathbf{z}_1), \ldots, K(\mathbf{x}, \mathbf{z}_N))^{\top}, \tag{17}$$

where $K(\mathbf{x}, \mathbf{z}_j) = \boldsymbol{\phi}^{\top}(\mathbf{x})\,\boldsymbol{\phi}(\mathbf{z}_j)$. Then the decision function (1) is expressed by

$$f(\mathbf{x}) = \mathbf{w}^{\top}\mathbf{h}(\mathbf{x}) + b. \tag{18}$$

For a linear kernel with $m < N$, to improve sparsity, we use the Euclidean coordinates: $\mathbf{z}_1 = \{1, 0, \ldots, 0\}, \cdots, \mathbf{z}_m = \{0, \cdots, 0, 1\}$, and use the identity mapping: $\mathbf{h}(\mathbf{x}) = \mathbf{x}$.

We derive the optimality condition of the LS LDM given by (15), using (18):

$$\frac{\partial Q(\mathbf{w}, b)}{\partial \mathbf{w}} = \left(\frac{1}{C_{\mathrm{m}}} I_N + (1 + C_{\mathrm{e}}) \overline{K^2} - \overline{K^y}^\top \overline{K^y} \right) \mathbf{w}$$
$$+ \left((1 + C_{\mathrm{e}}) \bar{K}^\top - \bar{y} \overline{K^y}^\top \right) b - \overline{K^y}^\top = 0, \qquad (19)$$

$$\frac{\partial Q(\mathbf{w}, b)}{\partial b} = \left((1 + C_{\mathrm{e}}) \bar{K} - \bar{y} \overline{K^y} \right) \mathbf{w} + \left(1 + C_{\mathrm{e}} - \bar{y}^2 \right) b - \bar{y} = 0, \qquad (20)$$

where I_N is the $N \times N$ unit matrix,

$$\overline{K^2} = \frac{1}{M} \sum_{i=1}^{M} K_i^\top K_i, \qquad \bar{K} = \frac{1}{M} \sum_{i=1}^{M} K_i, \qquad \overline{K^y} = \frac{1}{M} \sum_{i=1}^{M} y_i K_i, \qquad \bar{y} = \frac{1}{M} \sum_{i=1}^{M} y_i,$$

$$K_i = (K_{i1}, \ldots, K_{iN}) = \mathbf{h}^\top(\mathbf{x}_i),$$
$$K_{ij} = K(\mathbf{x}_i, \mathbf{z}_j) \quad \text{for } i = 1, \ldots, M, \ j = 1, \ldots, N. \qquad (21)$$

In a matrix form, (19) and (20) are given by

$$\begin{pmatrix} \frac{1}{C_{\mathrm{m}}} I_N + (1 + C_{\mathrm{e}}) \overline{K^2} - \overline{K^y}^\top \overline{K^y} & (1 + C_{\mathrm{e}}) \bar{K}^\top - \bar{y} \overline{K^y}^\top \\ (1 + C_{\mathrm{e}}) \bar{K} - \bar{y} \overline{K^y} & 1 + C_{\mathrm{e}} - \bar{y}^2 \end{pmatrix} \begin{pmatrix} \mathbf{w} \\ b \end{pmatrix}$$
$$= \begin{pmatrix} \overline{K^y}^\top \\ \bar{y} \end{pmatrix} \qquad (22)$$

If $C = 0$, (22) reduces to the ULDM. The difference between (22) with $C = 0$ and the ULDM in [12] is that $1/C_{\mathrm{m}}$ is used in (22) instead of C_{m}.

Because the coefficient matrix of (22) is positive definite, we can solve (22) for $\mathbf{w}$ and b by the coordinate descent method [13] as well as by matrix inversion.

In model selection, we need to determine the values of C_{m}, C in C_{e}, and γ in the kernel. To speed up model selection, as well as grid search of three values, we consider line search: after determining the values of C_{m} and γ with $C = 0$ by grid search, we determine the C value fixing the values of C_{m} and γ with the determined values.

4   Performance Evaluation

We compare performance of the ULDM with that of the LS LDM to clarify whether the equality constraints in the LS LDM are necessary. We also compare the ULDM with the LS SVM and the L1 SVM. Because of the space limitation, we only use two-class problems.

4.1   Conditions for Experiment

Because the coefficient matrix of (22) is positive definite, (22) can be solved by the coordinate descent method [9]. But to avoid the imprecise accuracy caused by the improper convergence, we train the ULDM and LS LDM by matrix inversion. We also train the LS SVM given by (3) and (4) by matrix inversion. For the L1 SVM, we use SMO-NM [14], which fuses SMO (Sequential minimal optimization) and NM (Newton's method).

We use the radial basis function (RBF) kernels: $K(\mathbf{x}, \mathbf{x}') = \exp(-\gamma \|\mathbf{x} - \mathbf{x}'\|^2/m)$, where m is the number of inputs for normalization and γ is used to control a spread of a radius. We carry out model selection by fivefold cross-validation. To speed up cross-validation for the LS LDM, which has three hyper-parameters including γ for the RBF kernel, we use line search in addition to grid search of the optimal values of C, C_m and γ. In line search, after determining the values of C_m and γ by grid search, we determine the optimal value of C by cross-validation. Therefore C_m and γ for the ULDM give the same values for the LS LDM by line search.

We select the γ value from {0.01, 0.1, 0.5, 1, 5, 10, 15, 20, 50, 100, 200}, for the C value from {0.1, 1, 10, 50, 100, 500, 1000, 2000}, and for the C_m value from {0.1, 1, 10, 100, 1000, 10^4, 10^6, 10^8}. In the LS LDM, we also include 0 as a candidate of the C value. Then if 0 is selected, the LS LDM reduces to the ULDM.

We measure the average CPU time per data set including model selection by fivefold cross-validation, training a classifier, and classifying the test data by the trained classifier. We used a personal computer with 3.4 GHz CPU and 16 GB memory.

4.2 Results for Two-Class Problems

Table 1 lists the numbers of inputs, training data, test data, and data set pairs of two-class problems [15]. Each data set pair consists of the training data set and the test data set. Using the training data set, we determine parameter values by cross-validation, train classifiers with the determined parameter values and evaluate the performance using the test data set. Then we calculate the average accuracy and the standard deviation for all the test data sets.

Table 2 lists the parameter values determined by cross-validation. In the first row, (l) and (g) show that the three hyperparameters of the LS LDM are determined by linear search and grid search, respectively. Because each classification problem consists of 100 or 20 training and test data pairs, we show the most frequently selected parameter values. For the LS LDM, most selected value for C is 0. Thus, in the table, we show the number that $C \neq 0$ is selected in the parentheses.

As we discussed before, the C_m and γ values for the ULDM and the LS LDM (l) are the same. Therefore, if the number that $C \neq 0$ is selected is 0, the LS LDM (l) reduces to ULDM for all the training data sets. This happens for seven problems. Except for the german problem, the C value of zero is selected frequently. For the LS LDM (g) also, the C value of zero is frequently selected. Therefore, LS LDM (g) reduces to ULDM frequently. These results indicate that the equality constraints are not important in the LS LDM.

Table 1. Benchmark data for two-class problems

Data	Inputs	Train	Test	Sets
Banana	2	400	4,900	100
Breast cancer	9	200	77	100
Diabetes	8	468	300	100
Flare-solar	9	666	400	100
German	20	700	300	100
Heart	13	170	100	100
Image	18	1,300	1,010	20
Ringnorm	20	400	7,000	100
Splice	60	1,000	2,175	20
Thyroid	5	140	75	100
Titanic	3	150	2,051	100
Twonorm	20	400	7,000	100
Waveform	21	400	4,600	100

The γ values for the three classifiers are very similar and so are the C values for the LS and L1 SVMs.

In the following we show the distributions of C, C_e, and γ values for the german data, in which $C = 0$ is least frequently selected for the LS LDM.

Table 3 shows the C value distributions for the german data. The distributions for the LS LDM by line search and by grid search are very similar. The values of C smaller than or equal to 1 are selected 93 times and 90 times for the LS LDM (l) and LS LDM (g), respectively. Therefore, C does not affect much to the generalization ability. The distributions for the LS SVM and L1 SVM are similar and although the value of 1 is frequently selected, the larger values are also selected. This means that the value of C affect directly on the generalization ability.

Table 4 shows the distributions of C_m values for the ULDM and LS LDM (g). The both distributions are similar. The distribution for the LS LDM (l) is the same as that for the ULDM.

Table 5 lists the γ value distributions for the german data. The γ values larger than 20 are not selected for the four classifiers. The distributions of the ULDM (LS LDM (l)) and LS LDM (g) are similar although smaller values are selected for the ULDM (LS LDM (l)). The distributions of the LS SVM and L1 SVM are similar and tend to gather towards smaller values than those of the ULDM (LS LDM (l)) and LS LDM (g).

Table 6 shows the average accuracies and their standard deviations of the five classifiers with RBF kernels. Among the five classifiers the best average accuracy is shown in bold and the worst average accuracy is underlined. The "Average" row shows the average accuracy of the 13 average accuracies and

Table 2. Most-frequently-selected parameter values for the two-class problems. The numeral in the parentheses shows the number that $C \neq 0$ is selected.

Data	ULDM C_{m}, γ	LS LDM (l) $C_{\mathrm{m}}, \gamma \, (C)$	LS LDM (g) $C_{\mathrm{m}}, \gamma \, (C)$	LS SVM C, γ	L1 SVM C, γ
Banana	10^4, 50	10^4, 50 (1)	10^4, 100 (1)	10, 50	1, 20
B. cancer	10, 0.01	10, 0.01 (17)	10, 10 (30)	1, 5	1, 0.5
Diabetes	100, 5	100, 5 (10)	100, 5 (22)	1, 0.5	500, 0.1
Flare-solar	10, 0.01	10, 0.01 (0)	10, 1 (0)	10, 0.01	50, 0.01
German	100, 10	100, 10 (31)	100, 10 (38)	1, 0.1	1, 0.1
Heart	100, 0.01	100, 0.01 (0)	10^4, 0.5 (1)	10, 0.01	100, 0.01
Image	10^8, 15	10^8, 15 (1)	10^8, 20 (1)	50, 50	50, 100
Ringnorm	10, 50	10, 50 (0)	10, 100 (0)	0.1, 50	1, 50
Splice	10^4, 10	10^4, 10 (0)	10^6, 10 (0)	10, 10	10, 10
Thyroid	10, 100	10, 100 (0)	10, 200 (6)	1, 100	50, 5
Titanic	10^4, 0.01	10^4, 0.01 (0)	10, 1 (3)	10, 0.01	50, 0.01
Twonorm	1000, 0.01	1000, 0.01 (0)	100, 5 (1)	50, 0.01	1, 0.01
Waveform	100, 50	100, 50 (10)	100, 50 (21)	1, 20	1, 15

Table 3. Distribution of C values for the german data

C	LS LDM (l)	LS LDM (g)	LS SVM	L1 SVM
0.0	69	62	—	—
0.1	11	11	0	0
1	13	17	42	32
10	3	4	11	9
50	2	2	14	20
100	1	2	7	8
500	0	1	9	8
1000	0	0	5	7
2000	1	1	12	16

the two numerals in the parentheses show the numbers of the best and worst accuracies in the order. We performed Welch's t test with the confidence intervals of 95%. The "W/T/L" row shows the results; W, T, and L denote the numbers that the ULDM shows statistically better than, the same as, and worse than the LS LDM (l), LS LDM (g), LS SVM, and L1 SVM, respectively. Symbols "+" and "−" in the L1 SVM column show that the ULDM is statistically better and worse than the L1 SVM, respectively.

Ignore the difference of 0.01 for the average accuracies and the standard deviations. Then the results of the ULDM and those of the LS LDM (l) are

Table 4. Distribution of C_m values for the german data

C_e	ULDM	LS LDM (g)
0.1	7	5
1	44	50
10	18	21
100	12	13
10^3	6	5
10^4	2	4
10^6	5	2
10^8	6	0

Table 5. Distribution of γ values for the german data

γ value	ULDM	LS LDM (g)	LS SVM	L1 SVM
0.01	11	0	10	12
0.1	2	6	23	24
0.5	9	8	16	16
1	8	9	11	13
5	23	26	22	15
10	27	30	12	8
15	9	10	5	8
20	11	11	1	4
50	0	0	0	0
100	0	0	0	0
200	0	0	0	0

different only for the german problem. Whereas for the ULDM and LS LDM (g), only the ringnorm problem gives the same results.

From the table, from the standpoint of the average accuracy, the ULDM and LS LDM (l) performed best and the LS SVM, the worst. But from the standpoint of statistical analysis the ULDM is statistically comparable with the remaining four classifiers.

Therefore, because the LS LDM frequently reduces to the ULDM and the ULDM is comparable with the LS LDM, the LS LDM can be replaced with the ULDM.

Table 7 shows the average CPU time per data set for calculating the accuracies. The last row shows the numbers that each classifier shows best/worst execution time. In average, the LS SVM is the fastest and the LS LDM (g) the slowest because of the slow model selection by grid search of three hyperparameters. Because the ULDM and LS SVM are trained by solving the sets

Table 6. Accuracy comparison of the two-class problems for RBF kernels

Data	ULDM	LS LDM (l)	LS LDM (g)	LS SVM	L1 SVM
Banana	89.13±0.69	89.13±0.69	89.16±0.59	**89.17±0.66**	**89.17±0.72**
B. cancer	**73.73±4.34**	**73.73±4.35**	**73.73±4.48**	73.13±4.68	73.03±4.51
Diabetes	**76.52±1.95**	**76.52±1.95**	76.32±2.00	76.19±2.00	76.29±1.70
Flare-solar	66.33±2.02	66.33±2.02	66.18±1.94	66.25±1.98	⁻66.99±2.12
German	76.14±2.30	76.10±2.30	**76.25±2.17**	76.10±2.10	75.95±2.24
Heart	82.61±3.61	82.61±3.61	82.33±3.77	82.49±3.60	**82.82±3.37**
Image	97.16±0.68	97.17±0.68	97.23±0.53	**97.52±0.54**	97.16±0.41
Ringnorm	98.16±0.35	98.16±0.35	98.17±0.34	**98.19±0.33**	98.14±0.35
Splice	89.13±0.60	89.13±0.60	**89.17±0.55**	88.98±0.70	88.89±0.91
Thyroid	95.28±2.28	95.28±2.28	95.25±2.42	95.08±2.55	**95.35±2.44**
Titanic	77.45±0.89	77.45±0.89	**77.48±0.87**	77.39±0.83	77.39±0.74
Twonorm	**97.43±0.25**	**97.43±0.25**	97.37±0.28	**97.43±0.27**	97.38±0.26
Waveform	90.19±0.52	90.19±0.53	**90.22±0.51**	90.05±0.59	⁺89.76±0.66
W/T/L	—	0/13/0	0/13/0	0/13/0	1/11/1
Average	**85.33** (3/2)	**85.33** (3/1)	85.30 (5/3)	85.23 (4/3)	85.26 (4/7)

Table 7. Execution time comparison of the two-class problems (in seconds)

Data	ULDM	LS LDM(l)	LS LDM(g)	LS SVM	L1 SVM
Banana	28.13	30.67	249.08	12.03	**4.92**
B. cancer	2.91	3.17	25.83	**1.69**	7.08
Diabetes	44.13	48.63	428.30	**20.3**	22.96
Flare-solar	223.96	249.05	2067.59	**67.28**	218.67
German	383.45	431.55	3387.80	**98.72**	776.53
Heart	1.66	1.87	15.04	**1.12**	1.75
Image	4813.18	5419.68	46138.67	1826.86	**56.7**
Ringnorm	26.68	29.42	237.83	13.15	**12.57**
Splice	1919.64	1986.73	15747.32	740.76	**30.71**
Thyroid	0.96	1.06	8.68	0.69	**0.33**
Titanic	1.20	1.33	10.93	**0.75**	21.25
Twonorm	27.81	30.83	271.14	13.33	**10.46**
Waveform	26.64	29.96	246.24	**13.64**	35.61
B/W	0/0	0/0	0/12	7/0	6/1

of linear equations with the equal number of variables, slower training by the ULDM is due to more complex calculation in setting the coefficients of the linear equations. Because the matrix size is the number of training data plus one and because the numbers of training data are smaller than 1000 except for the image and splice data sets, the execution time is relatively short.

The L1 SVM is trained by iterative method. Therefore the training speed depends on the parameter values and for the titanic data, training of the L1 SVM is the slowest. For the ULDM, LS LDM, and LS SVM, the execution time depends on the number of training data not on the parameter values.

5   Conclusions

The unconstrained large margin distribution machine (ULDM) maximizes the margin mean and minimizes the margin variance without constraints.

In this paper, we investigated the effect of the constraints to the ULDM. To do this, we derived the ULDM as a special case of the least squares (LS) LDM, which is the least squares version of the LDM. If the hyperparameter associated with the constraints is set to be zero, the LS LDM reduces to the ULDM. In computer experiments, we carried out model selection of the LS LDM including the zero value of the hyperparameter as a candidate value. For the two-class problems with 100 or 20 data set pairs, in most cases, the LS LDM reduced to the ULDM and if not, there was no statistical difference of generalization abilities. According to the results, the effect of the equality constraints to the generalization ability of the LS LDM is considered to be small and the ULDM can be used instead of the LS LDM.

References

1. Vapnik, V.N.: Statistical Learning Theory. Wiley, New York (1998)
2. Abe, S.: Support Vector Machines for Pattern Classification, 2nd edn. Springer, London (2010). https://doi.org/10.1007/978-1-84996-098-4
3. Reyzin, L., Schapire, R.E.: How boosting the margin can also boost classifier complexity. In: Proceedings of the 23rd International Conference on Machine learning, pp. 753–760. ACM (2006)
4. Gao, W., Zhou, Z.-H.: On the doubt about margin explanation of boosting. Artif. Intell. **203**, 1–18 (2013)
5. Garg, A., Roth, D.: Margin distribution and learning. In: Proceedings of the Twentieth International Conference on Machine Learning (ICML 2003), Washington, DC, USA, pp. 210–217 (2003)
6. Pelckmans, K., Suykens, J., Moor, B.D.: A risk minimization principle for a class of Parzen estimators. In: Platt, J.C., Koller, D., Singer, Y., Roweis, S.T. (eds.) Advances in Neural Information Processing Systems 20, pp. 1137–1144. Curran Associates Inc. (2008)
7. Aiolli, F., Da San Martino, G., Sperduti, A.: A kernel method for the optimization of the margin distribution. In: Kůrková, V., Neruda, R., Koutník, J. (eds.) ICANN 2008. LNCS, vol. 5163, pp. 305–314. Springer, Heidelberg (2008). https://doi.org/10.1007/978-3-540-87536-9_32
8. Zhang, L., Zhou, W.-D.: Density-induced margin support vector machines. Pattern Recognit. **44**(7), 1448–1460 (2011)

9. Zhang, T., Zhou, Z.-H.: Large margin distribution machine. In: Twentieth ACM SIGKDD Conference on Knowledge Discovery and Data Mining, pp. 313–322 (2014)
10. Zhou, Z.-H.: Large margin distribution learning. In: El Gayar, N., Schwenker, F., Suen, C. (eds.) ANNPR 2014. LNCS (LNAI), vol. 8774, pp. 1–11. Springer, Cham (2014). https://doi.org/10.1007/978-3-319-11656-3_1
11. Abe, S.: Improving generalization abilities of maximal average margin classifiers. In: Schwenker, F., Abbas, H.M., El Gayar, N., Trentin, E. (eds.) ANNPR 2016. LNCS (LNAI), vol. 9896, pp. 29–41. Springer, Cham (2016). https://doi.org/10.1007/978-3-319-46182-3_3
12. Abe, S.: Unconstrained large margin distribution machines. Pattern Recognit. Lett. **98**, 96–102 (2017)
13. Hsieh, C.-J., Chang, K.-W., Lin, C.-J., Keerthi, S.S., Sundararajan, S.: A dual coordinate descent method for large-scale linear SVM. In: Proceedings of the 25th International Conference on Machine Learning, ICML 2008, New York, pp. 408–415. ACM (2008)
14. Abe, S.: Fusing sequential minimal optimization and Newton's method for support vector training. Int. J. Mach. Learn. Cybern. **7**(3), 345–364 (2016)
15. Rätsch, G., Onoda, T., Müller, K.-R.: Soft margins for AdaBoost. Mach. Learn. **42**(3), 287–320 (2001)

DLL: A Fast Deep Neural Network Library

Baptiste Wicht[1,2(✉)], Andreas Fischer[1,2], and Jean Hennebert[1,2]

[1] HES-SO, University of Applied Science of Western, Switzerland, Delémont, Switzerland
[2] University of Fribourg, Fribourg, Switzerland
baptiste.wicht@gmail.com

Abstract. Deep Learning Library (DLL) is a library for machine learning with deep neural networks that focuses on speed. It supports feed-forward neural networks such as fully-connected Artificial Neural Networks (ANNs) and Convolutional Neural Networks (CNNs). Our main motivation for this work was to propose and evaluate novel software engineering strategies with potential to accelerate runtime for training and inference'. Such strategies are mostly independent of the underlying deep learning algorithms. On three different datasets and for four different neural network models, we compared DLL to five popular deep learning libraries. Experimentally, it is shown that the proposed library is systematically and significantly faster on CPU and GPU. In terms of classification performance, similar accuracies as the other libraries are reported.

1 Introduction

In recent years, neural networks have regained a large deal of attention with deep learning approaches. Such approaches rely on the use of bigger and deeper networks, typically by using larger input dimensions to incorporate more context and by increasing the number of layers to extract information at different levels of granularity. The success of deep learning can be attributed mainly to three factors. First, there is the advent of *big data*, meaning the availability of larger quantities of training data. Second, new training strategies have been developed, such as unsupervised pre-training that allows deep networks to initialize well and also to learn efficient feature extractors on large sets of unlabelled data. Finally, better and faster hardware has helped dealing with the training of such networks. Deep systems are currently improving the state-of-the-art in many domains. Successful deep learning applications include near-human performance at recognizing objects in images [27], generating detailed image descriptions [13], adding colors to grayscale images [3] or generating highly-realistic images [7]. Moreover, the availability of free and easy-to-use libraries, as well as the availability of detailed implementation examples on public datasets, have contributed to the widespread use of deep learning technologies.

From a practical point of view, an ideal deep learning library would be easy to use, would offer fast training with good precision and would be versatile

L. Pancioni et al. (Eds.): ANNPR 2018, LNAI 11081, pp. 54–65, 2018.
https://doi.org/10.1007/978-3-319-99978-4_4

with many configuration options. Reaching all these qualities is difficult as some are contradictory. For this reason, we may observe large differences among the available libraries.

In this work, we report on the development of a deep learning library where we have clearly opted to focus on efficient computation, targeting specific network models and algorithm configurations. While we are aware of these limitations, we believe that the different optimizations we have implemented may be of interest to the scientific community. Our library, Deep Learning Library (DLL), is freely available, with source code[1]. This library can be used to train standard Artificial Neural Networks (ANNs) and Convolutional Neural Networks (CNNs) [18], as well as Restricted Boltzmann Machine (RBM) [26] and Convolutional RBM (CRBM) [20].

While speedups are also observed on the GPU, the proposed library has been especially optimized for speed on Central Processing Unit (CPU). Although GPUs are beginning to be the de-facto standard for training deep networks, they are not always available and some deployments are still targeting existing CPU implementations. Moreover, inference is generally performed on CPU once the network has been trained. Therefore, we believe that it remains important to be able to both train neural networks in reasonable time and achieve fast inference on CPUs. In this work, we also report successful optimizations on GPU, but we have to note that advanced parallelization capabilities of GPU where already well used [28], especially for convolutional networks [16].

Further to our speedup contributions, a special contribution of this paper is a comprehensive evaluation against several important state of the art libraries. The evaluation is carried on four models and three data sets. Comparisons are performed in terms of computation time on both CPU and GPU. This shows that state of the art libraries have still some large margin of optimization.

The rest of this paper is organized as follows. The DLL library is described in details in Sect. 2. The evaluation is presented in Sect. 3. Section 4 is presenting the results of the experiments on MNIST, Sect. 5 on CIFAR-10 and Sect. 6 on ImageNet. Finally, conclusions are drawn in Sect. 7.

2   DLL: Deep Learning Library

Deep Learning Library (DLL) is a Machine Learning library originally focused on RBM and CRBM support. It was developed and used in the context of several research work [29–32]. It also has support for various neural network layers and backpropagation techniques. It is written in C++ and its main interface is C++ (example in Sect. 2.2). The library can also be used by describing the task in a simple descriptor language, to make it easier for researchers.

The library supports conventional neural network. As such, ANNs and CNNs can be trained. Max Pooling and Average Pooling layers are also supported for CNNs. These networks can be trained with mini-batch gradient descent. The

---

[1] URL https://github.com/wichtounet/dll.

basic learning options such as momentum and weight decay are supported. The library also support advanced techniques such as Dropout [10] and Batch Normalization [11]. Finally, optimizers with adaptive learning rates such as Adagrad [6], Adadelta [33] and Adam [14] are also integrated. The library also supports Auto-Encoders [2] and Convolutional Auto-Encoders [21].

Also, the library has complete support for the RBM model [26]. The model can be trained using Contrastive Divergence (CD) [9]. The implementation was designed following the model from [8]. It also supports Deep Belief Network (DBN), pretrained layer by layer and then fine-tuned using gradient descent. The RBM supports a wide range of visible and hidden unit types, such as binary, Gaussian and Rectified Linear Unit (ReLU) [23]. Support for CRBM is also integrated, following the two models from [20].

The DLL library is available online[2], free of charge, under the terms of the MIT open source license. Details of the project as well as some tutorials are available on the home page.

2.1 Performance

The focus of the library is runtime performance, for training and for inference.

The implementation uses several techniques to optimize as much as possible the runtime performance for training and inference. First, all the computations are performed using single-precision floating point numbers. This leads to a better data locality and an increased potential for vectorization. On GPU, it would even be possible to use half-precision, but modern processors do not have native capabilities for such computations. Another simple optimization is that all the computations are performed on a batch rather than on one sample at the time. This has the advantage of leveraging the necessary operations to higher level computations. Since this is also generally advantageous for the quality of the training, this is currently the most common way to train a neural network.

The forward activation of a dense layer for a mini-batch can be computed with a single matrix-matrix multiplication [31]. This is also possible for the backward pass, by transposing the weight matrix. Finally, the gradients for the dense layer can also be computed using one matrix-matrix multiplication. Thus, such a network mainly needs a good implementation of this operation to be fast.

The Basic Linear Algebra Subprograms (BLAS) interface contains a set of small and highly-optimized kernels for matrix and vector computation [17]. When using an efficient BLAS library, the matrix-matrix multiplication operation can be very efficient. Moreover, using a parallel BLAS library also leads to significantly increased performance for large layers. Moreover, although BLAS libraries are highly optimized for very large matrices, they are not as fast as possible for small matrices. Therefore, we automatically detect such cases and use custom vectorized kernels for small matrix multiplications.

Optimization is more complicated for CNNs. Indeed, the dense layers only account for a small portion of the training time. Convolutional layers use two

---

[2] URL https://github.com/wichtounet/dll.

forms of convolution. A valid convolution for the forward pass, which shrinks the representation and a full convolution for the backward pass to expand it. Every image batch is convolved with K kernels. It is possible to rearrange an image into columns so that a matrix-matrix multiplication can be used to compute the K valid convolutions of the image at once [24,31]. This proved to be very efficient for large images or large kernels. When images or kernels are small, it is not efficient since the rearranging of the input matrix is a slow operation. Therefore, in these cases, we observed that it is more interesting to perform a real convolution using an highly-optimized implementation. First, several floating point operations are computed during the same CPU cycle, using SSE and AVX, a technique known as Single Instruction Multiple Data (SIMD). Then, to ensure the maximum throughput, the matrices are padded so that the last dimension is a multiple of the vector size. Specialized kernels for the most used kernel sizes, such as 3×3 and 5×5, are also used. Finally, most of the convolutions can be performed in parallel since there are no dependencies between them. This proved significantly faster than the reduction to a matrix-matrix multiplication in several configurations.

There are several possible implementations for the full convolution. First, it can be expressed in terms of another operation, the Fast Fourier Transform (FFT) [22]. For this, the input image and the kernel are padded to the size of the output. Then, their transforms are computed, in parallel. The Hadamard product of the input image with the transform of the kernel is computed. The inverse transform of this product is the full convolution. Computing several convolutions of the same image with different kernels is more efficient since the image transform is only computed once. In our experiments, we observed that such implementation is very efficient for large inputs and large kernels, but it is not as interesting for small configurations. With very small kernels, it is more efficient to pad the input and the kernels and perform a valid convolution. Indeed, a full convolution is equivalent to a valid convolution with some amount of padding. When the necessary padding is small enough, it becomes significantly faster than performing the FFTs. The last option is to use an optimized implementation of the full convolution. However, due to the large number of border cases, this would only be faster than the implementation as a valid convolution for large dimensions, in which case the reduction to FFT would be faster.

Since there is no one-size-fits-all implementation for all configurations, heuristics are used to select the most suited implementations. These heuristics are based on the size of the convolution kernels and the size of the batch.

Although most of the time is contained inside the previously mentioned operations, it is still important to optimize the other operations such as activation functions and gradient computations. In our implementation, these operations are vectorized and parallelized to maximize the processor utilization.

Fortunately, when optimizing for GPU, most of the routines are already implemented in highly specialized libraries. DLL uses NVIDIA libraries in order to optimize most kernels. NVIDIA CUBLAS is used for the matrix-matrix multiplications and a few other linear algebra operations and NVIDIA CUDNN [4] is used for the machine learning operations such as convolutions, activation functions and gradients computation. For other operations, CUDA kernels have been written to ensure that most of the time is spent on the GPU. When optimizing for GPU, it is most important to avoid copies between the CPU and GPU. Moreover, most of the kernels are launched asynchronously, without device synchronization. This significantly reduces the overhead of CUDA kernel calls.

2.2 Example

Figure 1 shows the code necessary to train a three-layer fully-connected network on the MNIST data set with the DLL library. The code starts by loading the MNIST data set in memory. Then, the network is declared layer by layer. After that, the network training parameters are set and the training is started. Finally, the accuracy on the test set is computed.

```
using namespace dll;

auto dataset = make_mnist_dataset(batch_size<100>{}, scale_pre<255>{});

using network_type = network_desc<
    network_layers<
        dense_layer<28 * 28, 500, sigmoid>,
        dense_layer<500,     250, sigmoid>,
        dense_layer<250,     10,  softmax>
    >
    , updater<updater_type::MOMENTUM>
    , batch_size<100>
>::network_t;

auto net = std::make_unique<network_type>();

net->learning_rate = 0.1;
net->momentum = 0.9;

net->display();
net->fine_tune(dataset.train(), 50);
net->evaluate(dataset.test());
```

Fig. 1. Example to train and evaluate a dense network on the MNIST data set.

3   Experimental Evaluation

We compared our library against popular libraries on four experiments. The time to train each model is compared for each library, on CPU and on GPU. Each experiment was run five times. And for each library, the best time is kept as the final measure. There is no significant different between the different runs. Their accuracy was also computed. It was shown that all the tested libraries were all exhibiting comparable accuracy when trained with the same parameters. For lack of space, these results are not shown here.

The following reference libraries have been selected:

1. Caffe [12]: A high-level Machine Learning library, focusing on speed and expression, developed in C++ and used through a text descriptor language. Caffe 1.0 was installed from the sources with GPU and MKL support.

2. TensorFlow [1]: A general low-level library, allowing expressing a data flow graph to perform numerical computation. The core of the system is written in C++, but the features are used in Python. Tensorflow 1.3.1 was installed from the sources with CUDA, CUDNN and MKL support.
3. Keras[3]: A high-level Machine Learning library, providing a frontend for Tensorflow and Theano, written in Python. It provides a large number of high-level models, easing the development of Machine Learning models. The version 2.0.8 was installed using the official package with Tensorflow 1.3.1.
4. Torch [5]: Torch is another low-level Machine Learning library, one of the earliest, started in 2002. It is used through a Lua front-end. Although it is a low-level library, it also contains high-level modules for Machine Learning. It was installed from the sources, from Git commit 3e9e141 with CUDA and MKL support.
5. DeepLearning4J[4]: DeepLearning4J is a deep learning library for Java, written in Java, C and C++. It has a very large set of features and focuses on distributed computing. The version 0.9.1 was used, from Maven.

The libraries have been selected based on their popularity and also to have a broad range of programming languages. DLL is used directly from the sources, with the latest version available at this time (Git commit 2f3c62c).

We are underlying here that the goal of these experiments is not to reach state of the art performance on the tested data sets. The models are kept simple to allow comparison with a wider range of libraries. Moreover, the networks are not always trained for as many epochs as they would be, if achieving high accuracy was the goal. Finally and very importantly, we are not aware of the full details of all the libraries. We did our best to have similar network architecture and training parameters, but it could be that some implementation details lead to slightly different training, explaining time differences.

All the results presented in this chapter have been computed on a Gentoo Linux machine, on an Intel® Core™ i7-2600, running at 3.4 GHz (CPU frequency scaling has been disabled for the purpose of these tests). Both SSE and AVX vectorization extensions were enabled on the machine. BLAS operations are executed with the Intel® Math Kernel Library (MKL), in parallel mode. The GPU used is a NVIDIA Geforce® GTX 960 card. CUDA 8.0.4.4 and CUDNN 5.0.5 are used. The source code used for these experiments is available online[5].

All the experiments are trained using mini-batch gradient descent. The last layer of each network is always a softmax layer. The loss is a softmax cross entropy loss.

4 MNIST

The first experiment is performed on the MNIST data set [19]. It is a digit recognition task. The data set is made of 60'000 28×28 grayscale images for

---

[3] https://github.com/fchollet/keras.
[4] http://deeplearning4j.org.
[5] https://github.com/wichtounet/frameworks.

training and 10'000 images for testing. It is a very well-known data set and has been repeatedly used with most of the existing Machine Learning algorithms. Although it is considered an easy task, it remains an excellent problem for comparing libraries since most of them use it as example and have code available.

4.1 Fully-Connected Neural Network

The first tested network is a fully-connected three-layer ANN with 500 units in the first layer, 250 in the second layer and 10 final output units for classification. The first two layers are using the sigmoid function. The network is trained with mini-batches of 100 images, for 50 epochs, with a learning rate of 0.1 and a momentum of 0.9. The training accuracy is computed after each epoch and the test accuracy is computed after the end of the complete training. As an example, the code using the DLL library is presented in Fig. 1.

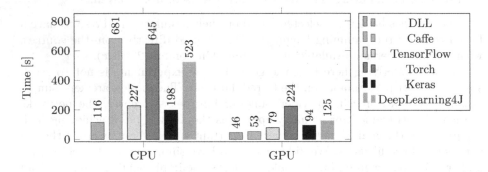

Fig. 2. Training time performance of the libraries for an ANN, on MNIST

Figure 2 presents the performance of each of the libraries. In CPU mode, DLL outperforms all the other libraries, being around 40% faster than TensorFlow and Keras, 4.5 times faster than DeepLearning4J and 5.5 times faster than Torch and Caffe. On GPU, DLL is the fastest library, closely followed by Caffe. DLL is about 40% faster than TensorFlow and twice faster than Keras. DeepLearning4J and Torch are respectively 2.5 and 5 times slower than DLL.

4.2 Convolutional Neural Network

The second network, for the same task, is a small CNN with six layers. The first layer is a convolutional layer using 8 5×5 kernels and followed by a max pooling layer with a 2×2 kernel. The third and fourth layers are using the same configuration. The last layers are fully-connected, the first with 150 units and the last with 10 units for classification. The two convolutional layers and the first fully-connected layer use a sigmoid activation function. The full network is trained in the same manner as the first network.

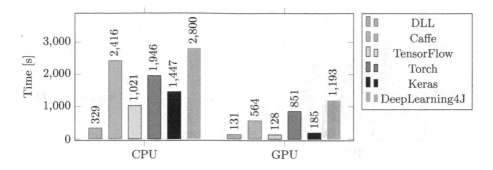

Fig. 3. Training time performance of the libraries for a CNN, on MNIST

Figure 3 presents the results obtained on this experiment. Again, DLL is the fastest library on CPU, by a significant margin, three times faster than TensorFlow and almost four times faster than Keras. DLL is more than 8 times faster than the slowest library, DeepLearning4J. This shows the effects of the in-depth CPU optimization of the convolutions. On GPU, TensorFlow and DLL are the fastest libraries, about 30% faster than Keras and significantly faster than Caffe (4 times), Torch (6.5 times) and DeepLearning4J (9 times).

5   CIFAR-10

The second data set that is tested is CIFAR-10 [15], a data set for object recognition, consisting of 50'000 images for training and 10'000 for testing, in 10 different classes. The data set is composed of colour images of 32 × 32 pixels.

A larger CNN is used for this task. The first layer is convolutional with 12 5 × 5 kernels, followed by a 2 × 2 max pooling layer. They are followed by another convolutional layer with 24 3 × 3 kernels and a 2 × 2 max pooling layer. A dense layer with 64 hidden units is then used, followed by a softmax layer with 10 output units. All the layers but the last one are using ReLUs. The network is trained similarly to the previous networks, with a learning rate of 0.001.

In Fig. 4, the training times for this task are presented. The speedups are less significant than for the previous CNN. Nevertheless, DLL still manages to be the fastest library on CPU. It is about twice faster than TensorFlow, Keras, DeepLearning4J and Torch and about three times faster than Caffe. On GPU, DLL is also the fastest library on this experiment, about 30% faster than TensorFlow and 40% faster than Keras. It is three times faster than Caffe and about 4.5 times faster than Torch and ten times faster than DeepLearning4J. This network is significantly larger than in the MNIST experiment. This seems to indicate that most libraries are more optimized for larger networks. This shows that GPU performance is better when a lot of data is available.

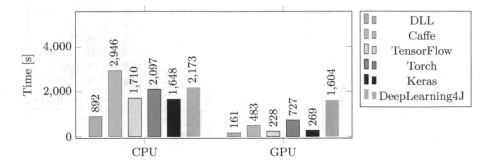

Fig. 4. Training time performance of the libraries on the CIFAR-10 task

6   ImageNet

The last experiment is performed on ImageNet, a large data set for image classification. We consider the sub part of the ImageNet Large Scale Visual Recognition Challenge (ILSVRC) 2012 [25], there are 50'000 validation images, 100'000 test images, around 1.2 million training images and 1000 categories. All the images have been resized to 256×256 images.

The entire data set cannot be kept in memory. Therefore, the images are loaded from the disk for each epoch. For this experiment, only Caffe provides an official, up-to-date, code for this data set. The DeepLearning4J reader was based on existing official reader for structures similar to ImageNet. For Keras, TensorFlow and Torch, a simple data reader has been written with the image loading tools available in each library.

The network is significantly larger than the previous networks. It is made of five convolutional layers, with 16 3×3 kernels for the first two layers and 32 3×3 kernels for the next three layers. Each of these layers is followed by a ReLU activation function and a 2×2 max pooling layer. All the convolutional layers are using zero-padding so that their output is the same size as their input The last two layers are a dense layer with 2048 hidden units, with a ReLU function and a dense layer with 1000 outputs. The training is different than for the other data sets. The full network is only trained for five epochs with each library. The networks are trained using a batch size of 128. However, Torch and DeepLearning4J models were trained with a batch size of 64, respectively 16, samples. Indeed, both of these libraries needed more than 12GB of RAM to train with a batch size of 128 images. This may lead to some small degradation of the performance for those two libraries.

For the sake of comparison, the average time to train one batch of samples is used as results. For Torch and DeepLearning4J, the results are the times for several batches, to make up for 128 samples. These results are presented in Fig. 5. DLL shows to be again the fastest library on CPU for training this large model, 35% faster than Keras, about 45% faster than TensorFlow and twice faster than Caffe. Torch is already more than 3 times slower than DLL and DeepLearning4J

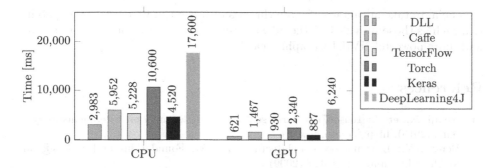

Fig. 5. Training time performance of the libraries, on ImageNet. The time is the average time necessary for the training of one batch of 128 elements.

around 6 times slower. On GPU, DLL is, also, the fastest library. Comparisons with Keras and TensorFlow show that most of the difference comes from the poor performance of reading the ImageNet data from the Python code. Once this is taken into account, the three libraries have comparable performance. DLL is more than twice faster than Caffe and almost four times faster than Torch and almost 10 times faster than DeepLearning4J.

7   Conclusion and Future Work

For all the experiments and the different neural networks models that were tested, the DLL library has shown to be the fastest gradient descent based library for training the model when using CPU and GPU. For each test, the accuracies of the models trained with DLL are similar to the models trained by the other five Machine Learning libraries.

The speedups provided by the library on CPU mode are especially important for convolutional layers for which advanced optimization was performed. The library was especially optimized for small convolutions, but is still able to bring significant speedups for large images such as the images from the ImageNet data set. Moreover, while some libraries are mostly optimized for the convolutional and fully-connected parts of the computation, every part of the training in the DLL library was tuned. However, since DLL is written in C++, programs using it need to be compiled. This may make it more complicated for researchers to use. Finally, while the language itself is very common about performance software developers, it is not very common for machine learning researchers. Therefore, there is more of a barrier for use compared to libraries using more common languages for machine learning.

A few DLL routines are not optimized enough for GPU, such as Dropout and Batch Normalization. Future work could also include better support for Recurrent Neural Networks (RNNs), which would be a great advantage for the library. Finally, the library has currently been optimized only on few machines and especially consumer grade processors and graphics cards. It would be greatly beneficial to take advantage of more threads or advanced vectorization capabilities such as those provided by the latest Intel® Xeon processors or more recent and more powerful NVIDIA graphics cards.

References

1. Abadi, M., et al.: TensorFlow: large-scale machine learning on heterogeneous systems (2015). http://tensorflow.org/
2. Bengio, Y.: Learning deep architectures for AI. Foundations and trends® in Machine Learning, pp. 1–127 (2009)
3. Cheng, Z., Yang, Q., Sheng, B.: Deep colorization. In: Proceedings of the IEEE International Conference on Computer Vision, pp. 415–423 (2015)
4. Chetlur, S., et al.: cuDNN: efficient primitives for deep learning. arXiv preprint arXiv:1410.0759 (2014)
5. Collobert, R., Kavukcuoglu, K., Farabet, C.: Torch7: a matlab-like environment for machine learning. In: BigLearn, NIPS Workshop. No. EPFL-CONF-192376 (2011)
6. Duchi, J., Hazan, E., Singer, Y.: Adaptive subgradient methods for online learning and stochastic optimization. J. Mach. Learn. Res. **12**, 2121–2159 (2011)
7. Goodfellow, I., et al.: Generative adversarial nets. In: Advances in Neural Information Processing Systems, pp. 2672–2680 (2014)
8. Hinton, G.E.: A practical guide to training restricted Boltzmann machines. In: Montavon, G., Orr, G.B., Müller, K.-R. (eds.) Neural Networks: Tricks of the Trade. LNCS, vol. 7700, pp. 599–619. Springer, Heidelberg (2012). https://doi.org/10.1007/978-3-642-35289-8_32
9. Hinton, G.E., Osindero, S., Teh, Y.W.: A fast learning algorithm for deep belief nets. Neural Comput. **18**, 1527–1554 (2006)
10. Hinton, G.E., Srivastava, N., Krizhevsky, A., Sutskever, I., Salakhutdinov, R.R.: Improving neural networks by preventing co-adaptation of feature detectors. arXiv preprint arXiv:1207.0580 (2012)
11. Ioffe, S., Szegedy, C.: Batch normalization: accelerating deep network training by reducing internal covariate shift. arXiv preprint arXiv:1502.03167 (2015)
12. Jia, Y., et al.: Caffe: convolutional architecture for fast feature embedding. arXiv preprint arXiv:1408.5093 (2014)
13. Karpathy, A., Fei-Fei, L.: Deep visual-semantic alignments for generating image descriptions. In: Proceedings of the IEEE Conference on Computer Vision and Pattern Recognition, pp. 3128–3137 (2015)
14. Kingma, D., Ba, J.: Adam: a method for stochastic optimization. arXiv preprint arXiv:1412.6980 (2014)
15. Krizhevsky, A., Hinton, G.E.: Learning multiple layers of features from tiny images. Technical report (2009)
16. Krizhevsky, A., Sutskever, I., Hinton, G.E.: Imagenet classification with deep convolutional neural networks. In: Advances in neural information processing systems, pp. 1097–1105 (2012)
17. Lawson, C.L., Hanson, R.J., Kincaid, D.R., Krogh, F.T.: Basic linear algebra subprograms for FORTRAN usage. ACM Trans. Math. Softw. (TOMS) **5**, 308–323 (1979)
18. LeCun, Y., Bottou, L., Bengio, Y., Haffner, P.: Gradient-based learning applied to document recognition. In: Proceedings of the IEEE, pp. 2278–2324 (1998)

19. LeCun, Y., Cortes, C., Burges, C.J.C.: The mnist database of handwritten digits (1998). http://yann.lecun.com/exdb/mnist/. Accessed 04 Feb 2018
20. Lee, H., Grosse, R., Ranganath, R., Ng, A.Y.: Convolutional deep belief networks for scalable unsupervised learning of hierarchical representations. In: Proceedings of the 26th Annual International Conference on Machine Learning, pp. 609–616. ACM (2009)
21. Masci, J., Meier, U., Cireşan, D., Schmidhuber, J.: Stacked convolutional auto-encoders for hierarchical feature extraction. In: Honkela, T., Duch, W., Girolami, M., Kaski, S. (eds.) ICANN 2011. LNCS, vol. 6791, pp. 52–59. Springer, Heidelberg (2011). https://doi.org/10.1007/978-3-642-21735-7_7
22. Mathieu, M., Henaff, M., LeCun, Y.: Fast training of convolutional networks through FFTs. arXiv preprint arXiv:1312.5851 (2013)
23. Nair, V., Hinton, G.E.: Rectified linear units improve restricted Boltzmann machines. In: Proceedings of the International Conference on Machine Learning, pp. 807–814 (2010)
24. Ren, J.S., Xu, L.: On vectorization of deep convolutional neural networks for vision tasks. arXiv preprint arXiv:1501.07338 (2015)
25. Russakovsky, O., et al.: ImageNet large scale visual recognition challenge. Int. J. Comput. Vis. (IJCV) **115**, 211–252 (2015)
26. Smolensky, P.: Information processing in dynamical systems: foundations of harmony theory. Technical report, Colorado University (1986)
27. Szegedy, C., et al.: Going deeper with convolutions. In: Proceedings of the IEEE Conference on Computer Vision and Pattern Recognition, pp. 1–9 (2015)
28. Upadhyaya, S.R.: Parallel approaches to machine learning: a comprehensive survey. J. Parallel Distrib. Comput. **73**, 284–292 (2013)
29. Wicht, B.: Deep learning features for image processing. Ph.D. thesis, University of Fribourg (2018)
30. Wicht, B., Fischer, A., Hennebert, J.: Deep learning features for handwritten key-word spotting. In: 2016 23rd International Conference on Pattern Recognition (ICPR), pp. 3434–3439. IEEE (2016)
31. Wicht, B., Fischer, A., Hennebert, J.: On CPU performance optimization of restricted Boltzmann machine and convolutional RBM. In: Schwenker, F., Abbas, H.M., El Gayar, N., Trentin, E. (eds.) ANNPR 2016. LNCS (LNAI), vol. 9896, pp. 163–174. Springer, Cham (2016). https://doi.org/10.1007/978-3-319-46182-3_14
32. Wicht, B., Hennebert, J.: Mixed handwritten and printed digit recognition in sudoku with convolutional deep belief network. In: 2015 13th International Conference on Document Analysis and Recognition (ICDAR), pp. 861–865. IEEE (2015)
33. Zeiler, M.D.: Adadelta: an adaptive learning rate method. arXiv preprint arXiv:1212.5701 (2012)

Selecting Features from Foreign Classes

Ludwig Lausser[1], Robin Szekely[1], Viktor Kessler[1], Friedhelm Schwenker[2], and Hans A. Kestler[1(✉)]

[1] Institute of Medical Systems Biology, Ulm University, 89069 Ulm, Germany
`hans.kestler@uni-ulm.de`
[2] Institute of Neural Information Processing, Ulm University, 89069 Ulm, Germany

Abstract. Supervised learning algorithms restrict the training of classification models to the classes of interest. Other related classes are typically neglected in this process and are not involved in the final decision rule. Nevertheless, the analysis of these foreign samples and their labels might provide additional information on the classes of interest. By revealing common patterns in foreign classification tasks it might lead to the identification of structures suitable for the original classes. This principle is used in the field of transfer learning. In this work, we investigate the use of foreign classes for the feature selection process of binary classifiers. While the final classification model is trained according to the traditional supervised learning scheme, its feature signature is designed for separating a pair of foreign classes. We systematically analyse these classifiers in 10×10 cross-validation experiments on microarray datasets with multiple diagnostic classes. For each evaluated classification model, we observed foreign feature combinations that outperformed at least 90% of those feature sets designed for the original diagnostic classes on at least 88.9% of all datasets.

1 Introduction

The design of classification models for molecular diagnostics is mainly influenced by the interest in identifying molecular characteristics or even molecular causes of disease. Starting from high-dimensional profiles, feature selection is a main ingredient in this process [9,23]. Selecting (primary) measurements instead of generating (secondary) feature representations, these methods allow a direct interpretation of a classification model regarding the individual molecules [24]. Although feature selection cannot guarantee an improved accuracy, it directs a classification model to a small set of candidate markers that might be used as potential drug targets [14].

Feature selection is not only of interest for traditional dichotomous classification. It can also be used for characterising the landscape of larger collections of diagnostic classes [19]. It might be used to reveal similarities and differences

---

L. Lausser and R. Szekely—Contributed equally.

© Springer Nature Switzerland AG 2018
L. Pancioni et al. (Eds.): ANNPR 2018, LNAI 11081, pp. 66–77, 2018.
https://doi.org/10.1007/978-3-319-99978-4_5

between individual classes and combinations thereof. In previous work, we analysed feature selection in the context of multi-class fusion architectures for binary base classifiers [18]. One interesting observation of this study was that certain multi-class feature selection strategies mimic the feature signature of one of their binary base classifiers. High overlaps or even complete overlaps were observed. Designed for a particular two-class classification task, signatures excluded most of the features selected for other class combinations. Nevertheless, the corresponding multi-class architectures achieved accuracies comparable to those of other more heterogeneous feature selection strategies, leading to the hypothesis that the feature selections of these base classifiers are informative for the remaining class combinations but were not selected due to other more prominent features.

In this work, we further investigate this phenomenon. Focusing on the underlying binary base classifiers, we now systematically analyse the direct influence of foreign feature selections on the classification accuracy. As we utilize foreign samples for the training of the classification models, our setup might be categorized as transfer learning [21] or learning with semantic domain knowledge [16,27]. Related concepts can also be found in learning schemes for partially labeled datasets [5,15,17].

2   Methods

The manuscript will be based on the following notation. An object $\mathbf{x}$ will be represented as a n-dimensional vector of measurements $(x^{(1)}, \ldots, x^{(n)})^T \in \mathcal{X} \subseteq \mathbb{R}^n$. It is assumed that each object can be categorised into one of $|\mathcal{Y}|$ classes $y \in \mathcal{Y}$. A decision function or classifier is a function

$$c : \mathcal{X} \longrightarrow \mathcal{Y}. \tag{1}$$

A classifier will be called binary if $|\mathcal{Y}| = 2$ otherwise it will be called multi-class classifier.

In classical supervised learning, an untrained classifier $c \in \mathcal{C}$ is adapted to its classification task via a training set of labeled samples of the output classes $\mathcal{T} = \{(\mathbf{x}_i, y_i)\}_{i=1}^{|\mathcal{T}|}$, $y_i \in \mathcal{Y}$. The symbol $\mathcal{C}$ denotes a concept or function class and describes the structural properties of the chosen type of classifier. The notation $\mathcal{T}_y$ will be used to denote a training set of samples of class y. Other samples, such as unlabeled instances (semi-supervised learning) or labeled samples of foreign classes $\mathcal{T}_{y'}$, $y' \notin \mathcal{Y}$ are typically ignored.

Especially in high-dimensional settings, the training of a classifier can incorporate an internal feature selection process which discards a set of input features from the final decision process. Formally this process can be characterised as a function

$$f : \mathcal{C} \times \mathcal{T} \to \mathcal{I} = \{\mathbf{i} \in \mathbb{N}^{\hat{n}} \mid \hat{n} \leq n, \, i_k < i_{k+1}, 1 \leq i_k \leq n\}, \tag{2}$$

which maps to the space of sorted and repetition-free index vectors $\mathcal{I}$. An element of $\mathcal{I}$, a feature signature $(i^{(1)}, \ldots, i^{(\hat{n})})^T = \mathbf{i} \in \mathcal{I}$, provides the indices of the

$\hat{n} \leq n$ remaining features. The final classification of a sample $\mathbf{x}$ will only take into account a reduced feature representation $\mathbf{x}^{(i)} = (x^{i^{(1)}}, \ldots, x^{(i^{(\hat{n})})})^T$.

In this work, we will focus on the Threshold Number of Misclassifications (TNoM) feature selection criterion [1]. The TNoM is a univariate filter criterion that ranks the individual features according to the error rate of an optimal single threshold classifier. A number of selected features $\hat{n}$ is chosen a priori. The TNoM is an independent preprocessing step before starting the training algorithm of the original classifier.

2.1 Learning from Context Classes

Based on our observations for feature selecting multi-class classifiers [18], we extend the standard supervised learning scenario in the following way. Let again $\mathcal{Y}$ denote the set of output classes of a trained classifier. We assume that the classification problem is embedded in a larger *context* in which additional classes $\mathcal{Y} \subset \mathcal{Y}'$ exist. Similar to the standard multi-class scenario, we assume all classes to be pairwise disjoint $\forall y_i, y_j \in \mathcal{Y}' : y_i \neq y_j$. The training set of the over-all learning procedure is allowed to be comprised of samples of all available classes $\mathcal{T} = \{(\mathbf{x}_i, y_i)\}_{i=1}^{|\mathcal{T}|}, y_i \in \mathcal{Y}'$. Note that in contrast to learning approaches for partially labeled data the learning algorithm receives the class labels of all instances here. In general, these additional samples allow to screen for discriminative patterns of the context classes, which might also be suitable for the current classification task. The patterns detected in this process might be hidden in the original classes due to other more prominent patterns or due to a too low amount of available samples. During the prediction phase of the trained classifier only the original classes in $\mathcal{Y}$ are considered.

In this work, we apply the idea of context classes for the selection process of features selecting two-class classifiers. The original classes of the two-class problem y_a and y_b will be indicated as subscript of the trained classification model

$$c_{[y_a, y_b]} : \mathbb{R}^{\hat{n}} \longrightarrow \{y_a, y_b\}. \tag{3}$$

The classifier learns on $\mathcal{T}_{y_a} \cup \mathcal{T}_{y_b}$ and will be used to predict classes y_a and y_b. It operates on a reduced feature signature of $\hat{n}$ features provided by the initial feature selection process

$$f_{[y_c, y_d]} : \mathcal{T}_{y_c} \cup \mathcal{T}_{y_d} \longrightarrow \mathbb{N}^{\hat{n}}. \tag{4}$$

The feature selection criterion is based on the evaluation of two arbitrarily chosen but fixed classes $y_c, y_d \in \mathcal{Y}'$. The training set of the corresponding binary classifier is finally given by

$$\mathcal{T}_{FFS} = \left\{ (\mathbf{x}^{(i)}, y^{(i)}) \mid (\mathbf{x}, y) \in \mathcal{T}_{y_a} \cup \mathcal{T}_{y_b}, \mathbf{i} = f_{[y_c, y_d]} \right\}. \tag{5}$$

We call the resulting feature selection *original* (OFS) if $\{y_a, y_b\} = \{y_c, y_d\}$ and *foreign* (FFS) otherwise.

2.2   Foreign Class Combinations

Assuming a symmetric feature selection strategy ($f_{[y_c, y_d]} = f_{[y_d, y_c]}$), this scheme can be applied to construct $\frac{|\mathcal{Y}|(|\mathcal{Y}|-1)}{2}$ classifiers for a fixed pair of output classes y_a and y_b of a dataset with $|\mathcal{Y}|$ classes. They can be organised according to the number of classes participating in both the classification and the feature selection process (Table 1). A complete overlap leads to the original class combination. The corresponding OFS is clearly related to the original classification task. If both processes share one common class, the FFS can be seen as an alternative characterisation of this class. Nevertheless, its relevance for the original classification task can not be guaranteed. The same is true for completely disjunct class labels.

Table 1. Number of foreign feature combinations for a fixed pair of classes $y_a, y_b \in \mathcal{Y}$. In this context, the symbols y_c, y_d denote foreign classes $y_c, y_d \in \mathcal{Y} \setminus \{y_a, y_b\}$.

	y_a	y_c
y_b	(1) original class combination	($|\mathcal{Y}| - 2$) characterizations of y_b
y_d	($|\mathcal{Y}| - 2$) characterisations of y_a	$\left(\frac{(|\mathcal{Y}|-2)(|\mathcal{Y}|-3)}{2} \right)$ foreign class combinations

The tradeoff of foreign class characterisations and completely foreign class combination depends on the total number of classes $|\mathcal{Y}|$. For datasets comprising a small number of classes ($2 < |\mathcal{Y}| \leq 6$), classes y_a and y_b are included in the majority of the foreign class combinations. An equilibrium is reached at $|\mathcal{Y}| = 7$. For larger number of classes ($|\mathcal{Y}| > 7$), the number of foreign class combinations is dominated by totally foreign class combinations.

3   Experiments

We evaluated FFS in classification experiments with linear Support Vector Machines [28] (SVM, *cost* = 1), Random Forests [4] (RF, *trees* = 500) and k-Nearest Neighbor classifiers [7] (k-NN, $k = 3$). All experiments are designed as 10×10 cross-validation (10×10 CV) experiments [11]. That is the original multi-class dataset $\mathcal{S}$ is splitted into ten folds of approximately equal size. Nine of these folds are combined to a training set for the two-class classifiers. The tenth fold is used to evaluate the performance the trained classifier. The training set is reduced to the samples of those classes which are needed for the feature selection process and the training of the subsequent classification model. The test set is reduced to the samples of the output classes of the final two-class classification problem. Training and evaluation of a feature selecting classifier is repeated for each possible split of the folds. The performance of the classification model is characterised by its empirical accuracy over all test folds. The cross-validation procedure is performed for ten permutations of the initial dataset. All experiments are performed in the TunePareto-Framework [20].

For each multi-class dataset with $|\mathcal{Y}|$ classes, $\frac{|\mathcal{Y}|(|\mathcal{Y}|-1)}{2}$ two-class classification tasks are analysed. The corresponding classifiers are again coupled to $\frac{|\mathcal{Y}|(|\mathcal{Y}|-1)}{2}$ different feature selections. Sticking to our previous multi-class analysis, experiments are conducted for $\hat{n} \in \{25, 50\}$. A pair of classification algorithm and number of features will be called a test setting in the following. For a fixed pair of output classes y_a and y_b, the classifiers that utilise a foreign feature selection (FFS) are compared to following reference classifiers:

1. Baseline (BL): For baseline comparisons, we utilise the accuracies of constant predictors that predict the class label of the majority class

$$ \text{BL} = \frac{\max(|\mathcal{S}_{y_a}|, |\mathcal{S}_{y_b}|)}{|\mathcal{S}_{y_a}| + |\mathcal{S}_{y_b}|}, \tag{6} $$

where $\mathcal{S}_y$ denotes the set of all available samples of class y.

2. Random Feature Selection (RFS): Classifiers that are based on sets of $\hat{n}$ randomly chosen features. Similar to the trained feature signatures, the random signatures are changed in each training phase of the 10×10 CV.
3. Original Feature Selection (OFS): Classifiers that utilise feature sets trained for the original class combination $f_{[y_a, y_b]}$.
4. No Feature Selection (NoFS): Classifiers that operate on the original n-dimensional feature set.

3.1 Datasets

Our analysis is based on 9 multi-class datasets which comprise multiple entities ($m \geq 59$, $|\mathcal{Y}| \geq 4$) of a common biomedical context. An overview on the analysed datasets is given in Table 2. All datasets are gene expression profiles from

Table 2. Datasets used. The number of classes $|\mathcal{Y}|$, features n, samples m and samples per class m_i are reported.

Id: description	$|\mathcal{Y}|$	n	m	$m_1, \ldots, m_{|\mathcal{Y}|}$
d_1: Leukemia [10]	18	54613	12096	$40, 36, 58, 48, 28, 351, 38, 37, 40, 237,$ $122, 448, 76, 13, 206, 74, 70, 174$
d_2: Cancer cell lines [22]	9	54613	174	$18, 15, 21, 26, 18, 21, 23, 26, 6$
d_3: Cancer cell lines [3]	9	54613	1777	$366, 85, 343, 264, 130, 225, 280, 41, 43$
d_4: Colorectal cancer [25]	7	22215	381	$49, 47, 20, 54, 13, 186, 12$
d_5: Renal cell cancer [12]	6	22215	92	$23, 32, 11, 8, 6, 12$
d_6: Liposarcoma [8]	5	22215	140	$40, 11, 17, 20, 52$
d_7: Alcohol [13]	5	8740	59	$11, 12, 12, 12, 12$
d_8: Brain tissue [2]	4	54613	173	$43, 48, 43, 49$
d_9: Colon tumors [26]	4	54613	145	$30, 34, 36, 45$

microarray experiments ($n \geq 8740$). That is each individual feature represents the concentration level of a particular mRNA molecule of the analyzed biological sample. The feature representation can therefore be seen as a set of homogeneous primary measurements and does not contain any secondary derived features.

4   Results

The results of the 10×10 CV experiments are shown in Fig. 1. Each histogram summarizes the experiments of a particular classification algorithm. In each test setting, at least 81.57% (3-NN) of all FFS classifiers outperformed the BL classifiers on 90% of all foreign class combinations. In general, the comparisons against RFS classifiers are harder than the theoretical baseline. Most FFS classifiers outperform RFS classifiers only in a subset of classification experiments. For $\hat{n} = 50$ features, first FFS classifiers appear that show an inferior performance on all foreign class combinations. Nevertheless, in all test settings at least 39.25% (RF)

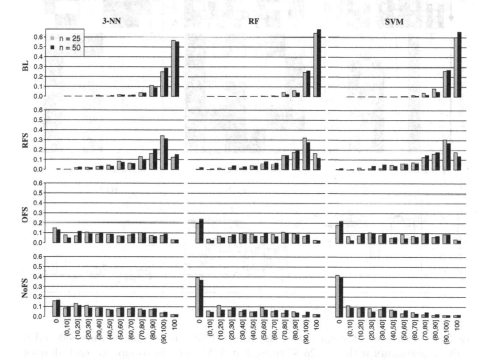

Fig. 1. Evaluation of foreign feature selection (FFS). The overall figure organizes rowwise the comparisons of FFSs to reference feature selection strategies. From the top to the bottom the comparisons against the baseline classifier (BL), the random feature selection (RFS), the original feature selection (OFS) or the classifier without feature selection (NoFS) are shown. The columns give the results for the analysed base classifiers 3-NN, RF and SVM. The y axis of each histogram provides the percentage of all FFSs that outperformed the reference feature selection.

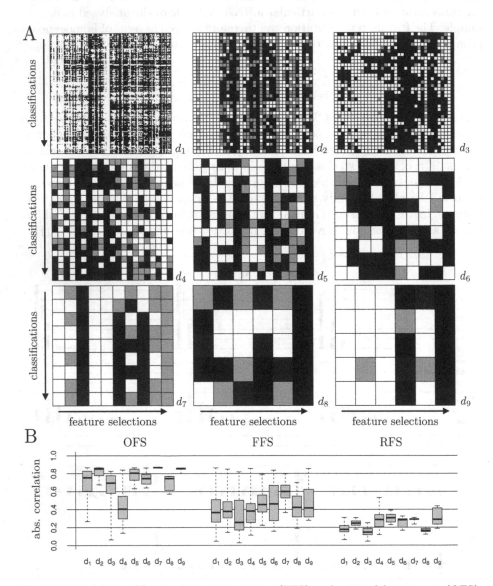

Fig. 2. Comparison of foreign feature signatures (FFS) and original feature sets (OFS) on the experiments with $\hat{n} = 25$ features. Panel A shows heatmaps for each dataset. Each column corresponds to the feature set of a particular class combination. It is compared to the OFS classifiers of all other class combinations. The color of a each patch indicates the number of classification algorithms for which FFS has outperformed the OFS (white: none, black: all). Panel B gives the mean absolute Spearman correlations between the selected features and the labels of the classification task.

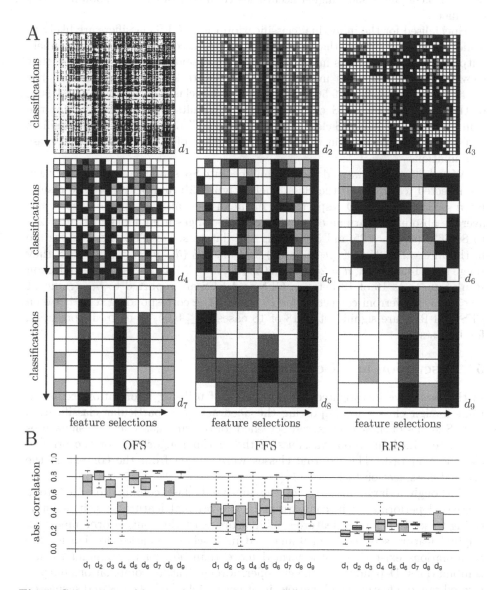

Fig. 3. Comparison of foreign feature signatures (FFS) and original feature sets (OFS) of the experiments with $\hat{n} = 50$ features. Panel A shows heatmaps for each dataset. Each column corresponds to the feature set of a particular class combination. It is compared to the OFS classifiers of all other class combinations. The color of a each patch indicates the number of classification algorithms for which FFS has outperformed the OFS (white: none, black: all). Panel B gives the mean absolute Spearman correlations between the selected features and the labels of the classification task.

of all FFS classifiers show higher accuracies than 90% of the corresponding RFS classifiers.

Specialised for the individual classification tasks, OFS classifiers are likely to achieve higher accuracies than RFS classifiers. As a consequence OFS classifiers outperform FFS classifiers more frequently. That said FFS classifiers exist that show higher accuracies in multiple classification tasks (Figs. 2A and 3A). For all test settings at least 9.56% (RF) of all FFS classifiers outperformed 90% of the corresponding OFS classifiers. Similar results can be observed for the comparisons against NoFS classifiers. At least 3.41% (SVM) of all FFS classifiers win the comparisons against the NoFS classifiers of 90% of the foreign class combinations.

We additionally analysed the median absolute Spearman correlation between the selected features and the class labels. Figures 2B and 3B report the results for $\hat{n} = 25$ and $\hat{n} = 50$ respectively. In mean over all classification tasks, the average absolute Spearman correlation of OFS ranges from 40.53% (d_4, $\hat{n} = 25$) to 86.66% (d_7, $\hat{n} = 25$ and 50) among the datasets. For FFS these values are in the interval of 25.61% (d_3, $\hat{n} = 25$) and 59.93% (d_7, $\hat{n} = 25$). For RFS they range from 14.63% (d_3, $\hat{n} = 25$) to 30.60% (d_5, $\hat{n} = 25$). In dataset-wise Wilcoxon Rank-Sum tests 16 of 18 between OFS and FFS are reported to be significant ($p \leq 0.05$, Bonferroni correction for 18 tests). The corresponding tests between FFS and RFS are significant in 18 of 18 cases ($p \leq 0.05$, Bonferroni correction for 18 tests).

5   Discussion and Conclusion

In this work, we investigated the possibility of utilising information extracted from foreign classes for improving the accuracy of a particular classification task. Samples of these classes were screened for discriminative feature signatures and determined the input variables of the learning algorithm for the original classification task. The external classes were chosen from the context of the original classification task.

Although a large majority of the analysed class combinations led to a decreased performance in comparison to the original feature sets, it is interesting to see that foreign feature combinations exist that outperform almost all original feature sets. For 8 of 9 analysed multi-class scenarios, foreign feature combinations exist that outperformed more than 90% of the original feature combinations. Depending on the classifier, foreign feature combinations outperformed random feature combinations in at least 73.43% of all cases and surpassed the minimal baseline accuracy in at least 95.55% of all cases.

In our experiments, we analysed datasets of technically homogeneous feature representations. All profiles solely comprise measurements of individual gene expression levels which were recorded according to the same technical principals. In particular, these profiles do not comprise secondary derived features that provide the same information as the primary measurements. The benefit of a feature set can therefore not be attributed to its technical superiority over different feature types.

The phenomenon might partially be explained by an overfitting or over-searching of the original feature sets. Showing a high mean correlation to their own class labels, the original feature selections can suffer from a declined variability and do not provide enough diverse information [6]. Nevertheless, this is no simple explanation for high accuracies achieved by the foreign feature sets. One could be that most of the data sets are from oncology and good feature sets capture basic processes in cancer development. A minimal prerequisite of a successful transfer is that a foreign feature signature allows an accurate discrimination of the original classification task. The classifiers trained on this signature are not required to be related to each other. This especially implies that they are allowed to operate on different subspaces or subsets of features. The task of identifying a feature set, where each member is informative for all classification tasks, therefore, might be facilitated by the task of collecting individual features that are informative for particular class combinations.

That said, the selected foreign features must also be informative for their own class combination. A successfully transferred foreign feature set must, therefore, be informative for at least two class combinations. A close relationship between both tasks probably increases the chance for this event. Our experiments are based on publicly available multi-class datasets that comprise distinct diagnostic classes of a common biomedical context. All classes are pairwise mutually exclusive. Individual samples can therefore not be informative for two or more classes. The context information we utilised must be seen as external semantic domain knowledge. It is not guaranteed that this is reflected in feature space. Other experimental setups like multi-label experiments, in which samples can receive multiple labels in parallel, might allow alternative context definitions.

Acknowledgements. The research leading to these results has received funding from the European Community's Seventh Framework Programme (FP7/2007-2013) under grant agreement n°602783, the German Research Foundation (DFG, SFB 1074 project Z1), and the Federal Ministry of Education and Research (BMBF, Gerontosys II, Forschungskern SyStaR, ID 0315894A and e:Med, SYMBOL-HF, ID 01ZX1407A, CONFIRM, ID 01ZX1708C) all to HAK.

References

1. Ben-Dor, A., Bruhn, L., Friedman, N., Nachman, I., Schummer, M., Yakhini, Z.: Tissue classification with gene expression profiles. J. Comput. Biol. **7**(3–4), 559–583 (2000)
2. Berchtold, N.C., et al.: Gene expression changes in the course of normal brain aging are sexually dimorphic. Proc. Natl. Acad. Sci. USA **105**(40), 15605–15610 (2008)
3. Bittner, M.: Expression project for oncology (expO). National Center for Biotechnology Information (2005)
4. Breiman, L.: Random forests. Mach. Learn. **45**(1), 5–32 (2001)
5. Chapelle, O., Schölkopf, B., Zien, A.: Semi-Supervised Learning, 1st edn. The MIT Press, Cambridge (2010)
6. Cover, T.M.: The best two independent measurements are not the two best. IEEE Trans. Syst. Man Cybern. **4**(1), 116–117 (1974)

7. Fix, E., Hodges, J.L.: Discriminatory analysis: nonparametric discrimination: consistency properties. Technical report Project 21-49-004, Report Number 4, USAF School of Aviation Medicine, Randolf Field, Tx (1951)
8. Gobble, R.M., et al.: Expression profiling of liposarcoma yields a multigene predictor of patient outcome and identifies genes that contribute to liposarcomagenesis. Cancer Res. **71**(7), 2697–2705 (2011)
9. Guyon, I., Elisseeff, A.: An introduction to variable and feature selection. J. Mach. Learn. Res. **3**, 1157–1182 (2003)
10. Haferlach, T., et al.: Clinical utility of microarray-based gene expression profiling in the diagnosis and subclassification of leukemia: report from the international microarray innovations in leukemia study group. J. Clin. Oncol. **28**(15), 2529–2537 (2010)
11. Japkowicz, N., Shah, M.: Evaluating Learning Algorithms: A Classification Perspective. Cambridge University Press, New York (2011)
12. Jones, J., et al.: Gene signatures of progression and metastasis in renal cell cancer. Clin. Cancer Res. **11**(16), 5730–5739 (2005)
13. Kimpel, M.W., et al.: Functional gene expression differences between inbred alcohol-preferring and non-preferring rats in five brain regions. Alcohol **41**(2), 95–132 (2007)
14. Lausser, L., Müssel, C., Kestler, H.A.: Measuring and visualizing the stability of biomarker selection techniques. Comput. Stat. **28**(1), 51–65 (2013)
15. Lausser, L., Schmid, F., Kestler, H.A.: On the utility of partially labeled data for classification of microarray data. In: Schwenker, F., Trentin, E. (eds.) PSL 2011. LNCS (LNAI), vol. 7081, pp. 96–109. Springer, Heidelberg (2012). https://doi.org/10.1007/978-3-642-28258-4_11
16. Lausser, L., Schmid, F., Platzer, M., Sillanpää, M.J., Kestler, H.A.: Semantic multi-classifier systems for the analysis of gene expression profiles. Arch. Data Sci. Ser. A (Online First) **1**(1), 157–176 (2016)
17. Lausser, L., Schmid, F., Schmid, M., Kestler, H.A.: Unlabeling data can improve classification accuracy. Pattern Recognit. Lett. **37**, 15–23 (2014)
18. Lausser, L., Szekely, R., Schirra, L.R., Kestler, H.A.: The influence of multi-class feature selection on the prediction of diagnostic phenotypes. Neural Proc. Lett. (2017)
19. Lorena, A., de Carvalho, A., Gama, J.: A review on the combination of binary classifiers in multiclass problems. Artif. Intell. Rev. **30**, 19–37 (2008)
20. Müssel, C., Lausser, L., Maucher, M., Kestler, H.A.: Multi-objective parameter selection for classifiers. J. Stat. Softw. **46**(5), 1–27 (2012)
21. Pan, S.J., Yang, Q.: A survey on transfer learning. IEEE Trans. Knowl. Data Eng. **22**(10), 1345–1359 (2010)
22. Pfister, T.D., et al.: Topoisomerase I levels in the NCI-60 cancer cell line panel determined by validated ELISA and microarray analysis and correlation with indenoisoquinoline sensitivity. Mol. Cancer Ther. **8**(7), 1878–1884 (2009)
23. Saeys, Y., Iñza, I., Larrañaga, P.: A review of feature selection techniques in bioinformatics. Bioinformatics **23**(19), 2507–2517 (2007)
24. Schirra, L.-R., Lausser, L., Kestler, H.A.: Selection stability as a means of biomarker discovery in classification. In: Wilhelm, A.F.X., Kestler, H.A. (eds.) Analysis of Large and Complex Data. SCDAKO, pp. 79–89. Springer, Cham (2016). https://doi.org/10.1007/978-3-319-25226-1_7
25. Sheffer, M., et al.: Association of survival and disease progression with chromosomal instability: a genomic exploration of colorectal cancer. Proc. Natl. Acad. Sci. **106**(17), 7131–7136 (2009)

26. Skrzypczak, M., et al.: Modeling oncogenic signaling in colon tumors by multi-directional analyses of microarray data directed for maximization of analytical reliability. PloS One **5**(10), e13091 (2010)
27. Taudien, S., et al.: Genetic factors of the disease course after sepsis: rare deleterious variants are predictive. EBioMedicine **12**, 227–238 (2016)
28. Vapnik, V.: Statistical Learning Theory. Wiley, New York (1998)

A Refinement Algorithm for Deep Learning via Error-Driven Propagation of Target Outputs

Vincenzo Laveglia[1,2] and Edmondo Trentin[2(✉)]

[1] DINFO, Università di Firenze, Via di S. Marta, 3, 50139 Florence, Italy
vincenzo.laveglia@unifi.it
[2] DIISM, Università di Siena, Via Roma, 56, 53100 Siena, Italy
trentin@dii.unisi.it

Abstract. Target propagation in deep neural networks aims at improving the learning process by determining target outputs for the hidden layers of the network. To date, this has been accomplished via gradient-descent or relying on autoassociative networks applied top-to-bottom in order to synthesize targets at any given layer from the targets available at the adjacent upper layer. This paper proposes a different, error-driven approach, where a regular feed-forward neural net is trained to estimate the relation between the targets at layer ℓ and those at layer $\ell - 1$ given the error observed at layer ℓ. The resulting algorithm is then combined with a pre-training phase based on backpropagation, realizing a proficuous "refinement" strategy. Results on the MNIST database validate the feasibility of the approach.

Keywords: Target propagation · Deep learning
Deep neural network · Refinement learning

1 Introduction

The impressive results attained nowadays in a number of AI applications of neural networks stem mostly from using deep architectures with proper deep learning techniques [10]. Looking under the hood, deep learning still heavily relies (explicitly or implicitly) on the traditional backpropagation (BP) algorithm [18]. While BP works outstandingly on networks having a limited number of hidden layers, several weaknesses of the algorithm emerge when dealing with significantly deep architectures. In particular, due to the non-linearity of the activation functions associated to the units in the hidden layers, the backpropagated gradients tend to vanish in the lower layers of the network, hence hindering the corresponding learning process [8]. Besides its numerical problems, BP is also known to lack any plausible biological interpretation [16].

To overcome these difficulties, researchers proposed improved learning strategies, such as pre-training of the lower layers via auto-encoders [1], the use of rectifier activation functions [9], and the dropout technique [21] to avoid neurons

© Springer Nature Switzerland AG 2018
L. Pancioni et al. (Eds.): ANNPR 2018, LNAI 11081, pp. 78–89, 2018.
https://doi.org/10.1007/978-3-319-99978-4_6

co-adaptation. Amongst these and other potential solutions to the aforementioned difficulties, target propagation has been arousing interest in the last few years [2,5,16], albeit it still remains an under-investigated research area. Originally proposed in [3,4] within the broader framework of learning the form of the activation functions, the idea underlying target propagation goes as follows. While in BP the delta values δ_i to be backpropagated are related to the partial derivatives of the global loss function w.r.t. the layer-specific parameters of the network, in target propagation the real target outputs (naturally defined at the output layer in regular supervised learning) are propagated downward through the network, from the topmost to the bottommost layers. In so doing, each layer gets explicit target output vectors that, in turn, define layer-specific loss functions that can be minimized locally (on a layer by layer basis) without any need to involve explicitly the partial derivatives of the overall loss function defined at the whole network level. Therefore, the learning process gets rid altogether of the troublesome numerical problems determined by repeatedly backpropagating partial derivatives from top to bottom.

To this end, [16] proposed an approach called difference target propagation (DTP) that relies on autoencoders. DTP is aimed at realizing a straight mapping $\hat{\mathbf{y}}_{\ell-1} = \phi(\hat{\mathbf{y}}_\ell)$ from the targets $\hat{\mathbf{y}}_\ell$ at layer ℓ to the expected[1] targets $\hat{\mathbf{y}}_{\ell-1}$ at layer $\ell - 1$. As shown by [16], the technique is effective (it improves over regular gradient-descent in the experiments carried out on the MNIST dataset), although the accuracy yielded by DTP does not compare favorably with the state-of-the-art methods (mostly based on convolutional networks). Moreover, DTP offers the advantages of being readily applied to stochastic and discrete neural nets. The approach is loosely related to the algorithm proposed by [12], where a layer-specific neural network is used to estimate the gradients of the global loss function w.r.t. the weights of the corresponding layer (instead of the target outputs).

Differently from DTP, the core of the present approach is that the backward mapping from layer ℓ to $\ell - 1$ shall be learnt by a regular feed-forward neural network as an explicit function $\varphi(.)$ of the actual error $\mathbf{e}_\ell$ observed at layer ℓ (namely, the signed difference between the target and actual outputs at ℓ), that is $\hat{\mathbf{y}}_{\ell-1} = \varphi(\hat{\mathbf{y}}_\ell, \mathbf{e}_\ell)$. In so doing, after training has been completed, the image of $\varphi(\hat{\mathbf{y}}_\ell, \mathbf{0})$ is an estimated "optimal" value of $\hat{\mathbf{y}}_{\ell-1}$ that is expected to result in a null error $\mathbf{e}_\ell = \mathbf{0}$ when propagated forward (i.e., from $\ell - 1$ to ℓ) through the original network. It is seen that learning $\varphi(.)$ requires that at least a significant fraction of the training samples result in small errors (such that $\mathbf{e}_\ell \simeq \mathbf{0}$). This is the reason why the proposed technique can hardly be expected to be a suitable replacement for the established learning algorithms altogether, but it rather results in an effective refinement method for improving significantly the models realized by pre-trained deep neural networks. The proposed approach is different

---

[1] The term "expected" is herein used according to its statistical notion, since such a $\phi(.)$ is not strictly a function, but it may be reduced to a proper function if we interpret the images in the codomain of $\phi(.)$ as the expected values of $\hat{\mathbf{y}}_{\ell-1}$ given $\hat{\mathbf{y}}_\ell$.

from that introduced in [3,4], as well, since the latter relies on gradient-descent (or, the pseudo-inverse method) and, above all, it does not involve $\mathbf{e}_\ell$.

The error-driven target propagation algorithm is introduced formally in Sect. 2. Section 2.1 presents the details for realizing target propagation via an inversion network used to learn $\varphi(.)$. Section 2.2 hands out the formal procedure for refining pre-trained deep networks relying on the proposed target propagation scheme. Experimental results obtained on the MNIST dataset are presented in Sect. 3, showing that the refinement strategy allows for accuracies that are at least in line with the established results yielded by regular (i.e., non-convolutional) deep networks, relying on much less complex models (i.e., using much fewer free parameters). Finally, preliminary conclusions are drawn in Sect. 4.

2   Error-Driven Target Propagation: Formalization of the Algorithms

Let us consider a deep neural network *dnet* having l layers. When *dnet* is fed with an input vector $\mathbf{x}$, the i-th layer of *dnet* (for $i = 1, \ldots, l$, while $i = 0$ represents the input layer which is not counted) is characterized by a state $\mathbf{h}_i \in \mathbb{R}^{d_i}$, where d_i is the number of units in layer i, $\mathbf{h}_i = \sigma(W_i\mathbf{h}_{i-1} + \mathbf{b}_i)$, and $\mathbf{h}_0 = \mathbf{x}$ as usual. The quantity W_i represents the weights matrix associated to layer i, $W_i \in \mathbb{R}^{d_i \times d_{i-1}}$, $\mathbf{b}_i \in \mathbb{R}^{d_i}$ denotes the corresponding bias vector, and $\sigma(.)$ represents the vector of the element-wise outcomes of the neuron-specific activation functions. The usual logistic sigmoid activation function is used in the present research. Consider a supervised training dataset $\mathcal{D} = \{(\mathbf{x}_j, \hat{\mathbf{y}}_j) | j = 1, \ldots, k\}$. Given a generic input pattern $\mathbf{x}_j \in \mathbb{R}^n$ and the corresponding target output $\hat{\mathbf{y}}_j \in \mathbb{R}^m$ drawn from $\mathcal{D}$, the state $\mathbf{h}_0 \in \mathbb{R}^n$ of the input layer of *dnet* is then defined as $\mathbf{h}_0 = \mathbf{x}_j$, while the target state $\hat{\mathbf{h}}_l \in \mathbb{R}^m$ of the output layer is $\hat{\mathbf{h}}_l = \hat{\mathbf{y}}_j$. Relying on this notation, it is seen that the function $f_i(.)$ realized by the generic i-th layer in *dnet* can be written as

$$f_i(\mathbf{h}_{i-1}) = \sigma(W_i\mathbf{h}_{i-1} + \mathbf{b}_i)$$

Therefore, the mapping $F_i : \mathbb{R}^n \to \mathbb{R}^{d_i}$ realized by the i bottommost layers of *dnet* over current input $\mathbf{x}_j$ can be expressed as the composition of i layer-specific functions as follows:

$$F_i(\mathbf{x}_j) = f_i(f_{i-1}...(f_1(\mathbf{x}_j)))$$

Eventually, the function realized by *dnet* (that is an l-layer network) is $F_l(\mathbf{x}_j)$. Bearing in mind the definition of $\mathcal{D}$, the goal of training *dnet* is having $F_l(\mathbf{x}_j) \simeq \hat{\mathbf{y}}_j$ for $j = 1, \ldots, k$. This is achieved by minimizing a point-wise loss function measured at the output layer. In this paper such a loss is the usual squared error $\mathcal{L}(\mathbf{x}_j; \theta) = (F_l(\mathbf{x}_j) - \hat{\mathbf{y}}_j)^2$ where θ represents the overall set of the parameters of *dnet*. In the traditional supervised learning framework the targets are defined only at the output layer. Nevertheless, while no explicit "loss" functions are associated to the hidden layers, the backpropagation (BP) algorithm allows the update of the hidden layers weights by back-propagating the gradients of the top-level loss $\mathcal{L}(.)$. To the contrary, target propagation consists in propagating the topmost layer targets $\hat{\mathbf{y}}_j$ to lower layers, in order to obtain explicit target states for the hidden units of the network, as well. Eventually, standard gradient-descent with no BP is applied in order to learn the layer-specific parameters as a function of the corresponding targets. In this research, at the core of the target propagation algorithm there is another, subsidiary network called the *inversion net*. Its nature and its application to target propagation are handed out in the following section.

2.1   The Inversion Net

Let us assume that the target value $\hat{\mathbf{h}}_i$ is known for a certain layer i (e.g. for the output layer, in the first place). The inversion net is then expected to estimate the targets $\hat{\mathbf{h}}_{i-1}$ for the preceding layer, that is layer $i - 1$. In this research the inversion net is a standard feed-forward neural network having a much smaller number of parameters than *dnet* has, e.g. having a single hidden layer. In principle, as in [16], the inversion net could be trained such that it learns to realize a function $g_i() : \mathbb{R}^{d_i} \to \mathbb{R}^{d_{i-1}}$ defined as

$$g_i(\hat{\mathbf{h}}_i) = \hat{\mathbf{h}}_{i-1}$$

where $\hat{\mathbf{h}}_{i-1}$ represents the estimated target at layer $i-1$. Let us assume that such inversion nets were trained properly to realize $g_i(.)$ for $i = l, \ldots, 1$. Then, layer-specific targets could be defined according to the following recursive procedure. First of all (basis of the recursion), if the layer i is the output layer, i.e. $i = l$, then $\hat{\mathbf{h}}_i = \hat{\mathbf{y}}$ and $g_l(\hat{\mathbf{y}}) = \hat{\mathbf{h}}_{l-1}$. Then (recursive step) the target outputs for the subsequent layers $(l - 1, \ldots, 1)$ are obtained by applying $g_i(.)$ to the estimated targets available at the adjacent upper (i.e., i-th) layer.

The actual error-driven training procedure for the inversion net proposed herein modifies this basic framework in the following manner. Given the generic layer i for which we want to learn the inversion function $g_i(.)$, let us define a layer-specific dataset $\mathcal{D}_i = \{(\mathbf{x}'_{i,j}, \hat{\mathbf{y}}'_{i,j}) | j = 1, \ldots, k\}$ where, omitting the pattern-specific index j for notational convenience, the generic input pattern is $\mathbf{x}'_i = (\hat{\mathbf{h}}_i, \mathbf{e}_i)$ given by the concatenation of the target value at layer i (either known, if $i = l$, or pre-computed from the upper layers if $i < l$) and the corresponding layer-specific signed error $\mathbf{e}_i = \mathbf{h}_i - \hat{\mathbf{h}}_i$. Herein $\mathbf{h}_i$ is the actual state of layer i of *dnet* upon forward propagation of its input, such that $\mathbf{x}'_i \in \mathbb{R}^{2 \times d_i}$. In turn, $\hat{\mathbf{y}}'_i$ is defined to be the state of the $(i - 1)$-th layer of *dnet*, namely $\hat{\mathbf{y}}'_i = \mathbf{h}_{i-1}$. Once the supervised dataset $\mathcal{D}_i$ has been built this way, the inversion net can be trained using standard BP with an early-stopping criterion. We say that this scheme is error-driven, meaning that the inversion net learns a target-estimation mapping which relies on the knowledge of the errors $\mathbf{e}_i$ stemming from the forward-propagation process in *dnet*.

Once training of the inversion net is completed, the proper target-propagation step (from layer i to layer $i - 1$) can be accomplished as follows. The inversion network is fed with the vector $(\hat{\mathbf{h}}_i, \mathbf{e}_i)$ where we let $\mathbf{e}_i = \mathbf{0}$ in order to get $g_i(\hat{\mathbf{h}}_i) = \hat{\mathbf{h}}_{i-1} \simeq f_i^{-1}(\hat{\mathbf{h}}_i)$. In so doing, the inversion net generates layer-specific targets that, once propagated forward by *dnet*, are expected to result in a null error, as sought. The resulting training procedure is formalized in Algorithms 1 and 2 in the form of pseudo-code. The algorithms assume the availability of two procedures, namely: *feedForward*(*net*, **x**), realizing the forward propagation of an input pattern **x** through a generic neural network *net*; and, *backpropagation*(*net*, $\mathcal{D}$) that implements the training of the network *net* via BP from the generic supervised training set $\mathcal{D}$.

In practice, in order to reduce the bias intrinsic to the training algorithm, target propagation is accomplished relying on a modified strategy, as in the difference target propagation scheme [16], accounting for the bias that the layer-specific inversion nets $g_i(.)$ are likely to introduce in estimating the corresponding target outputs $\hat{\mathbf{h}}_{i-1}$. To this end we let

$$\hat{\mathbf{h}}_{i-1} = \mathbf{h}_{i-1} + g_i(\hat{\mathbf{h}}_i, \mathbf{0}) - g_i(\mathbf{h}_i, \mathbf{0}) \tag{1}$$

The rationale behind this equation is the following. First of all, $g_i(.)$ can be applied to invert the actual state $\mathbf{h}_i$ of *dnet* instead of the target state $\hat{\mathbf{h}}_i$. Ideally, if the mapping realized by the inversion net were perfect, we would have $g_i(\mathbf{h}_i, \mathbf{0}) = \mathbf{h}_{i-1}$. To the contrary, since $g_i(.)$ is the noisy outcome of an empirical learning procedure, in practice $g_i(\mathbf{h}_i, \mathbf{0}) \neq \mathbf{h}_{i-1}$ holds, i.e. an offset is observed whose magnitude is given by $|g_i(\mathbf{h}_i, \mathbf{0}) - \mathbf{h}_{i-1}|$. Equation (1) exploits this offset as a bias corrector when applying $g_i(.)$ to the computation of $\hat{\mathbf{h}}_{i-1}$, as well. Note that whenever $g_i(\mathbf{h}_i, \mathbf{0}) = \mathbf{h}_{i-1}$ (unbiased inversion net) then the equation reduces to $\hat{\mathbf{h}}_{i-1} = g_i(\hat{\mathbf{h}}_i, \mathbf{0})$, as before. The details of the present bias-correction strategy are handed out in [16].

Algorithm 1. Training of the inversion net

Procedure `train_inv_net`($invNet_i$, $dnet$, $\mathcal{D}$, i, $\hat{\mathbf{h}}_i$)
Input: initialized inversion net $invNet_i$ with $2 \times d_i$ input units and d_{i-1} output units, deep network $dnet$, training set $\mathcal{D} = \{(\mathbf{x}_j, \hat{\mathbf{y}}_j) | j = 1, \ldots, k\}$, layer i, targets $\hat{\mathbf{h}}_i$ at layer i
Output: The trained inversion net $invNet_i$ for layer i, capable of computing $\hat{\mathbf{h}}_{i-1}$ from $\hat{\mathbf{h}}_i$
  1: $\mathcal{D}_i = \varnothing$
  2: **for** $j = 1$ to k **do**
  3:     `feedForward`($dnet$, $\mathbf{x}_j$)
  4:     $\mathbf{e}_{i,j} \leftarrow \hat{\mathbf{h}}_{i,j} - \mathbf{h}_{i,j}$
  5:     $\mathbf{x}'_{i,j} \leftarrow (\hat{\mathbf{h}}_{i,j}, \mathbf{e}_{i,j})$
  6:     $\mathbf{y}'_{i,j} \leftarrow \mathbf{h}_{i-1,j}$
  7:     $\mathcal{D}_i = \mathcal{D}_i \cup \{(\mathbf{x}'_{i,j}, \mathbf{y}'_{i,j})\}$
  8: **end for**
  9: $invNet_i = $ `backpropagation`($invNet_i$, $\mathcal{D}_i$)

Algorithm 2. Target propagation

Procedure tgt_prop($invNet_i$, i, k, $\hat{\mathbf{h}}_{i,1}, \ldots, \hat{\mathbf{h}}_{i,k}$)
Input: The inversion net $invNet_i$, layer i, number of patterns k, targets to be propagated $\hat{\mathbf{h}}_{i,j}$ for $j = 1 \ldots, k$
Output: The propagated targets $\hat{\mathbf{h}}_{i-1,1}, \ldots, \hat{\mathbf{h}}_{i-1,k}$
1: **for** $j = 1$ to k **do**
2:     $\mathbf{e}_{i,j} = \mathbf{0}$
3:     $\mathbf{x}'_{i,j} = (\hat{\mathbf{h}}_{i,j}, \mathbf{e}_{i,j})$
4:     $\hat{\mathbf{h}}_{i-1,j} = $ feedForward($invNet_i$, $\mathbf{x}'_{i,j}$)
5: **end for**

Algorithm 3. Deep learning with refinement based on target propagation

Procedure network_refinement($dnet$, $\mathcal{D}$)
Input: deep network $dnet$, supervised training set $\mathcal{D} = \{(\mathbf{x}_j, \hat{\mathbf{y}}_j) | j = 1, \ldots, k\}$
Output: the refined network $dnet$
1: **for** $j = 1$ to k **do**
2:     **for** $i = l$ to 1 **do**
3:         **if** $i = l$ **then**
4:             $\hat{\mathbf{h}}_{i,j} = \hat{\mathbf{y}}_j$
5:         **end if**
6:         $\mathbf{h}_{i,j} = F_i(\mathbf{x}_j)$
7:         $\mathbf{h}_{i-1,j} = F_{i-1}(\mathbf{x}_j)$
8:     **end for**
9: **end for**
10: **for** $i = l$ to 2 **do**
11:     Initialize_Network($invNet_i$)
12:     $invNet_i - $ train_inv_net($invNet_i$, $dnet$, $\mathcal{D}$, i, $\hat{\mathbf{h}}_i$)
13:     $\{\hat{\mathbf{h}}_{i-1,1}, \ldots, \hat{\mathbf{h}}_{i-1,k}\} - $ tgt_prop($invNet_i$, i, k, $\hat{\mathbf{h}}_{i,1}, \ldots, \hat{\mathbf{h}}_{i,k}$)
14: **end for**
15: **for** $j = 1$ to k **do**
16:     $\mathbf{h}_{0,j} = \mathbf{x}_j$
17:     layer_backprop($\mathbf{h}_{0,j}$, $\hat{\mathbf{h}}_{1,j}$)
18:     **for** $i = 2$ to l **do**
19:         $\mathbf{h}_{i-1,j} = F_{i-1}(\mathbf{x}_j)$
20:         layer_backprop($\mathbf{h}_{i-1,j}$, $\hat{\mathbf{h}}_{i,j}$)
21:     **end for**
22: **end for**

2.2   Refinement of Deep Learning via Target Propagation

The algorithms presented in the previous section form the basis for building a refinement technique for pre-trained deep networks. The overall approach goes as follows. In a first phase the deep network is trained via BP, as usual. In a second phase, targets are propagated downward through the layers, as in Algorithms 1 and 2, and the network is trained layer-wise accordingly. This phase is called "refinement". Algorithm 2 provides a detailed description of this refinement strategy in terms of pseudo-code. The algorithm invokes a routine *Initialize_Network*(.) used for initializing a generic feed-forward neural net with random parameters before the actual training takes place. Finally, the routine *layer_backprop*($\mathbf{h}_{i-1,j}$, $\hat{\mathbf{h}}_{i,j}$) realizes the adaptation of the weights between layers $i - 1$ and i (for $i = 1, \ldots, l$) via online gradient-descent. This application of gradient-descent uses $\mathbf{h}_{i-1,j}$ as its input, and $\hat{\mathbf{h}}_{i,j}$ as the corresponding target output. It is seen that extensions of the procedure to batch gradient-descent and/or multi-epochs training are straightforward by working out the skeleton of pseudo-code offered by Algorithm 3.

3   Experiments

Experiments were conducted on the popular MNIST dataset[2] [14]. We used all the 70,000 MNIST patterns, representing pixel-based images of handwritten digits (10 classes overall) having a dimensionality equal to 784. A 10-fold cross-validation strategy was applied, where for each fold as much as 80% of the data were used for training, 10% for validation/model selection, and 10% for test. The most significant results on MNIST published so far, obtained with a variety of different approaches, are listed in [15]. Variants on the theme of convolutional neural nets are known to yield the highest accuracies to date [6,23], as expected given the visual nature of the dataset. Our aim here is to exploit MNIST as a significant and difficult learning task suitable to assess the effectiveness of the present approach, and to compare the proposed algorithms to established non-convolutional feed-forward networks and target propagation methods previously applied to MNIST [16,20].

The topology of each layer and the hyperparameters were selected via grid search. Gradient-based training of the main network *dnet* (the classifier) relied on the root mean square propagation (RMSProp) variant of BP [22], while for the inversion net and the layer-wise refinement of *dnet* upon target propagation (routine `layer_backprop(.)` in Algorithm 3) the Adam variant of stochastic BP [13] turned out to be best. Besides a 784-dimensional input layer with linear activation functions and a class-wise 10-dimensional output layer with softmax activations, *dnet* had 3 hidden layers having 140, 120, and 100 neurons, respectively. Logistic sigmoid activation functions were used in the hidden layers. Connection weights and bias values for the sigmoids were initialized at random from a uniform distribution over the range $(-0.5, 0.5)$. RMSProp was applied for a maximum of 10^4 steps with early stopping (based on the generalization error not improving over the last 2000 steps), using a mini-batch size of 128 and a learning rate set to 0.01. As for the inversion nets, the dimensionality of the input and output layers were fixed according to the topology of the specific, adjacent layers in *dnet* between which the output targets had to be propagated (the input layer of *InvNet* had linear activation functions, while its output layer

---

[2] Available at http://yann.lecun.com/exdb/mnist/.

used logistic sigmoids), as explained in the previous section. A single hidden layer of 200 sigmoid units was used. The Adam optimizer was applied for a maximum of 1000 steps with early stopping, using a mini-batch size of 128 and a learning rate set to 0.001. Finally, the adaptation of the layer-specific weights in *dnet* upon propagation of the corresponding targets via the trained *InvNet* (procedure layer_backprop(.) in Algorithm 3) relied on the Adam optimizer, as well, with mini-batch size of 32 and learning rate set to 0.001.

Table 1 presents the average accuracies (± the corresponding standard deviations) on the 10-fold crossvalidation for *dnet* trained with RMSProp, with the bare target propagation, and with the refinement algorithm, respectively, evaluated on the training and the test sets. It is seen that the target propagation scheme required a proper BP-based initialization in order to achieve significant accuracies. in fact, In terms of learning capabilities (evaluated on the training sets), target propagation applied to the pre-trained *dnet* according to the refinement strategy yielded a relative 32.75% average error rate reduction over RMSProp, along with a much more stable behavior (the standard deviation was reduced as much as 42%). The statistical significance of the improvement evaluated via Welch's t-test (in order to account for the different variances of the two populations) results in a confidence level that is $\geq$ 99.75%. In terms of generalization capabilities (evaluated on the test sets), when applying the refinement strategy a significant relative 8.20% error rate reduction over RMSProp was observed on average, preserving the same stability of the performance (in fact, the difference between the standard deviations yielded by the two approaches is neglectable, namely 0.002%). Welch's t-test assessed a statistical significance of the gap between the results yielded by the two algorithms which is even higher than before (due to the much smaller variance of the RMSProp results), that is a confidence level $\geq$ 99.9%.

Table 1. Accuracies on the MNIST 10-class classification task (avg. ± std. dev. on a 10-fold crossvalidation).

Algorithm	Training	Test
RMSProp	99.48 ± 0.13	98.12 ± 0.05
Target propagation	87.30 ± 0.29	86.64 ± 0.27
Refinement	99.65 ± 0.08	98.27 ± 0.06

Table 2 offers a comparison among MNIST classifiers based on non-convolutional feed-forward deep neural networks using no augmentation of the training set (see [7,17] for established results obtained using augmentation). The comparison involves the error rate as observed on the test set (average ± standard deviation on the 10-fold crossvalidation, whenever available) and the number of free (i.e., adaptive) parameters in the model, that is an index of the model complexity. The proposed technique (target propagation with refinement) is compared with the approach by [20], that is a 2-hidden layer network with

Table 2. Comparison between the proposed algorithm and the established approaches, in terms of error rate and number of adaptive parameters.

Algorithm	Test error	#Parameters
Refinement	1.73 ± 0.06	3.04×10^5
[16]	1,94	5.36×10^5
[20]	1,6	1.28×10^6

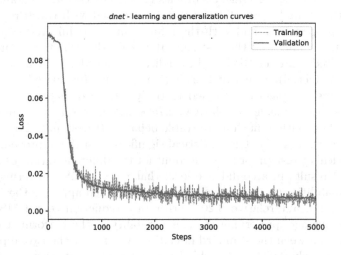

Fig. 1. Learning and generalization curves for *dnet*.

800 units per layer (resulting in a very complex machine), and by [16], that is a 7 hidden layer network having 240 neurons per layer. It is seen that the error rate achieved by the proposed refinement algorithm is in the middle between its competitors, but the complexity of the machine is dramatically smaller. A relative 11.02% error rate reduction is yielded by the present refinement approach over the difference target propagation algorithm, while a relative 7.25% reduction is still offered by [20] (credited by [11] of being the best performance yielded by a "regular" feed-forward net) over the present refinement procedure, at the expense of the number of adaptive parameters, which is one order of magnitude higher. Figure 1 presents the learning and generalization curves (mean squared error on training and validation sets, respectively) obtained running regular BP learning of *dnet* in one of the 10-folds of the present experiment. For graphical convenience, the plot is limited to the first 5000 steps (no evident changes in behavior were observed during the following steps). Note that the loss used to plot the learning curve was evaluated, from step to step, on the corresponding training mini-batch only, while the generalization curve was always evaluated on the whole validation set. This is the reason why the learning curve fluctuates locally, while the generalization curve is much smoother. The curves are compared with those corresponding to the refinement via target propagation,

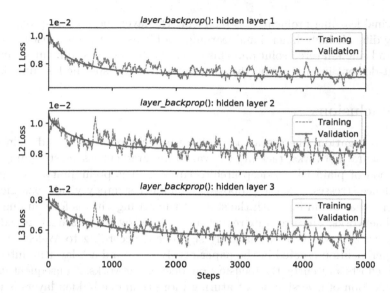

Fig. 2. Learning and generalization curves of the procedure `layer_backprop(.)` applied to the three hidden layers of *dnet*.

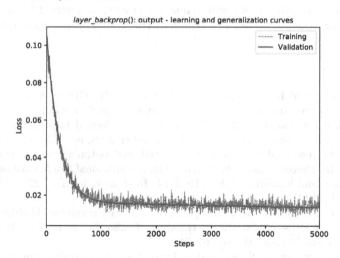

Fig. 3. Learning and generalization curves of the procedure `layer_backprop(.)` applied to the output layer of *dnet*.

namely Figs. 2 and 3. The former plots the learning and generalization curves of the layer-specific gradient-descent adaptation of the weights in the 1st, 2nd, and 3rd hidden layers of *dnet*, respectively, by means of the application of the procedure `layer_backprop(.)` to the target propagated via the inversion net. Similarly, Fig. 3 shows the curves for `layer_backprop(.)` applied to the weights in the topmost layer of *dnet*. Although eventually one is interested in solving the original learning problem, it is seen that the layer-specific sub-problems are actually difficult high-dimensional learning problems, which may just not admit any sound single-layered solution. This explains the observed difficulties met by gradient-descent in minimizing the corresponding layer-specific loss functions.

4   Conclusions

Target propagation emerges as a viable approach to learning and refinement of deep neural networks, tackling the vanishing-gradient issues stemming from application of plain BP to deep architectures. Albeit preliminary, the empirical evidence stresses that the proposed refinement strategy yields classification accuracies that are in line with the state-of-the-art algorithms for training feedforward networks. The error rate reduction observed over the bare BP-based deep learning was shown to be statistically significant according to Welch's t-tests. The experiments presented in the paper revolved around a 5-layer architecture, yet our efforts are currently focusing on deeper networks. Consequently, also the application of inversion nets featuring more than one hidden layers is under investigation. The training set for the inversion net can be enriched, as well, by synthetically generating layer-specific input-output pairs obtained from the original ones with the addition of random noise, resulting in different examples of the signed errors e_i used to drive the learning of the target-propagation relationship.

References

1. Bengio, Y., Lamblin, P., Popovici, D., Larochelle, H.: Greedy layer-wise training of deep networks. In: Advances in Neural Information Processing Systems 19, Proceedings of the Twentieth Annual Conference on Neural Information Processing Systems, Vancouver, BC, Canada, 4–7 December 2006, pp. 153–160 (2006)
2. Carreira-Perpiñán, M.Á., Wang, W.: Distributed optimization of deeply nested systems. In: Proceedings of the Seventeenth International Conference on Artificial Intelligence and Statistics, AISTATS 2014, Reykjavik, Iceland, 22–25 April 2014, pp. 10–19 (2014)
3. Castelli, I., Trentin, E.: Semi-unsupervised weighted maximum-likelihood estimation of joint densities for the co-training of adaptive activation functions. In: Schwenker and Trentin [19], pp. 62–71
4. Castelli, I., Trentin, E.: Supervised and unsupervised co-training of adaptive activation functions in neural nets. In: Schwenker and Trentin [19], pp. 52–61
5. Castelli, I., Trentin, E.: Combination of supervised and unsupervised learning for training the activation functions of neural networks. Pattern Recognit. Lett. **37**, 178–191 (2014)
6. Ciresan, D., Meier, U., Schmidhuber, J.: Multi-column deep neural networks for image classification. In: Proceedings of the 25th IEEE Conference on Computer Vision and Pattern Recognition (CVPR 2012), pp. 3642–3649. IEEE Computer Society (2012)
7. Ciresan, D.C., Meier, U., Gambardella, L.M., Schmidhuber, J.: Deep, big, simple neural nets for handwritten digit recognition. Neural Comput. **22**(12), 3207–3220 (2010)

8. Glorot, X., Bengio, Y.: Understanding the difficulty of training deep feedforward neural networks. In: Proceedings of the Thirteenth International Conference on Artificial Intelligence and Statistics, AISTATS 2010, Chia Laguna Resort, Sardinia, Italy, 13–15 May 2010, pp. 249–256 (2010)
9. Glorot, X., Bordes, A., Bengio, Y.: Deep sparse rectifier neural networks. In: Proceedings of the Fourteenth International Conference on Artificial Intelligence and Statistics, AISTATS 2011, Fort Lauderdale, USA, 11–13 April 2011, pp. 315–323 (2011)
10. Goodfellow, I., Bengio, Y., Courville, A.: Deep Learning. MIT Press, Cambridge (2016)
11. Hinton, G.E., Srivastava, N., Krizhevsky, A., Sutskever, I., Salakhutdinov, R.: Improving neural networks by preventing co-adaptation of feature detectors. CoRR abs/1207.0580 (2012). http://arxiv.org/abs/1207.0580
12. Jaderberg, M., et al.: Decoupled neural interfaces using synthetic gradients. In: Proceedings of the 34th International Conference on Machine Learning, ICML 2017, Sydney, NSW, Australia, 6–11 August 2017, pp. 1627–1635 (2017)
13. Kingma, D.P., Ba, J.: Adam: a method for stochastic optimization. CoRR abs/1412.6980 (2014)
14. Lecun, Y., Bottou, L., Bengio, Y., Haffner, P.: Gradient-based learning applied to document recognition. Proc. IEEE **86**(11), 2278–2324 (1998)
15. LeCun, Y., Cortes, C., Burges, C.: MNIST handwritten digit database (2018). http://yann.lecun.com/exdb/mnist/. Accessed 02 Feb 2018
16. Lee, D.-H., Zhang, S., Fischer, A., Bengio, Y.: Difference target propagation. In: Appice, A., Rodrigues, P.P., Santos Costa, V., Soares, C., Gama, J., Jorge, A. (eds.) ECML PKDD 2015. LNCS (LNAI), vol. 9284, pp. 498–515. Springer, Cham (2015). https://doi.org/10.1007/978-3-319-23528-8_31
17. Meier, U., Ciresan, D.C., Gambardella, L.M., Schmidhuber, J.: Better digit recognition with a committee of simple neural nets. In: Proceedings of the 2011 International Conference on Document Analysis and Recognition (ICDAR 2011), pp. 1250–1254. IEEE Computer Society, Washington, D.C. (2011)
18. Rumelhart, D.E., Hinton, G.E., Williams, R.J.: Learning internal representations by error propagation. In: Rumelhart, D.E., McClelland, J.L., Group, P.R. (eds.) Parallel Distributed Processing: Explorations in the Microstructure of Cognition, vol. 1, pp. 318–362. MIT Press, Cambridge (1986)
19. Schwenker, F., Trentin, E. (eds.): PSL 2011. LNCS (LNAI), vol. 7081. Springer, Heidelberg (2012). https://doi.org/10.1007/978-3-642-28258-4
20. Simard, P.Y., Steinkraus, D., Platt, J.C.: Best practices for convolutional neural networks applied to visual document analysis. In: 7th International Conference on Document Analysis and Recognition (ICDAR 2003), 2-Volume Set, 3–6 August 2003, Edinburgh, Scotland, UK, pp. 958–962 (2003)
21. Srivastava, N., Hinton, G., Krizhevsky, A., Sutskever, I., Salakhutdinov, R.: Dropout: a simple way to prevent neural networks from overfitting. J. Mach. Learn. Res. **15**(1), 1929–1958 (2014)
22. Tieleman, T., Hinton, G.: Lecture 6.5-rmsprop: divide the gradient by a running average of its recent magnitude. Technical report (2012)
23. Wan, L., Zeiler, M.D., Zhang, S., LeCun, Y., Fergus, R.: Regularization of neural networks using DropConnect. In: Proceedings of the 30th International Conference on Machine Learning, ICML 2013, Atlanta, GA, USA, 16–21 June 2013, pp. 1058–1066 (2013)

Combining Deep Learning and Symbolic Processing for Extracting Knowledge from Raw Text

Andrea Zugarini[1,3], Jérémy Morvan[2], Stefano Melacci[3(✉)], Stefan Knerr[4], and Marco Gori[3]

[1] DINFO, University of Florence, Florence, Italy
andrea.zugarini@unifi.it
[2] Criteo, Paris, France
morvanjeremy@gmail.com
[3] DIISM, University of Siena, Siena, Italy
{mela,marco}@diism.unisi.it
[4] CogniTalk, Nantes, France
stefan.knerr@cognitalk.com

Abstract. This paper faces the problem of extracting knowledge from raw text. We present a deep architecture in the framework of Learning from Constraints [5] that is trained to identify mentions to entities and relations belonging to a given ontology. Each input word is encoded into two latent representations with different coverage of the local context, that are exploited to predict the type of entity and of relation to which the word belongs. Our model combines an entropy-based regularizer and a set of First-Order Logic formulas that bridge the predictions on entity and relation types accordingly to the ontology structure. As a result, the system generates symbolic descriptions of the raw text that are interpretable and well-suited to attach human-level knowledge. We evaluate the model on a dataset composed of sentences about simple facts, that we make publicly available. The proposed system can efficiently learn to discover mentions with very few human supervisions and that the relation to knowledge in the form of logic constraints improves the quality of the system predictions.

Keywords: Information Extraction · Learning from Constraints
Deep Learning · Symbolic knowledge representation

1 Introduction

Information Extraction (IE) is one of the most important fields in Natural Language Processing (NLP), and it is about extracting structured knowledge from unstructured text [17]. IE encompasses a large variety of sub-problems, and, for

---

J. Morvan—Word done while at CogniTalk, Nantes, France.

© Springer Nature Switzerland AG 2018
L. Pancioni et al. (Eds.): ANNPR 2018, LNAI 11081, pp. 90–101, 2018.
https://doi.org/10.1007/978-3-319-99978-4_7

the purpose of this work, we mostly consider Named Entity Recognition (NER) and Relation Extraction (RE).

The goal of NER systems is to detect and classify proper nouns according to a predefined set of entity types, such as "Person", "Organization", "Location", and others. Many NER systems [10, 18] rely on handcrafted features and external knowledge, such as gazetteers or capitalization information. On one hand, they can help in spotting named entities, but, on the other hand, these techniques are usually tied to the considered task. Differently, Collobert et al. [4] deeply studied a neural model requiring only minor feature engineering. Their model was applied to several NLP problems, such as part-of-speech tagging, chunking, semantic role labeling, NER, and language modeling. More recent approaches make a wide use of recurrent neural networks (mostly LSTMs [6]), such as the one of Lample et al. [8], Chiu and Nichols [2] exploited similar networks, but character-level features are detected by convolutional nets, also used in [20].

Relation Extraction addresses the problem of finding and categorizing relations between entities in a given text document. This problem is even harder than NER, since relations are expressed in much more ambiguous ways than entity names. There is also a big issue related to RE, that is the lack of large collections of high quality labeled data. Relations can be implicit, they can have fuzzy boundaries, and they can also be constituted of non-contiguous words. Labeling can be hard even for humans, and it can be strongly inconsistent among different supervisors. Some approaches rely only on unsupervised models [16, 19], segmenting the word sequences ("mentions") bounded by two defined entities. Mintz et al. [13] proposed an alternative paradigm, the so called "distant supervision", that is a simple form of weak supervision. Intuitively, the distant approach is founded on the idea that sentences containing the same pair of entities are likely to express the same relation. Entities are taken from Freebase[1], and the considered relations are the ones that link the entity pair in the knowledge base. Miwa and Bansal [15] presented an end-to-end solution to extract both relations and entities from sentences. Their approach is based on stacked bidirectional tree structured LSTMs, where entities are extracted first, then relations are predicted.

This review shows that Deep Learning achieved serious improvements in NLP and IE-related applications. The renewed interest in recurrent neural networks and the introduction of distributed representations of words and sentences [1, 4, 12] allowed researchers to construct several systems that can be trained end-to-end, removing the costly efforts in feature engineering. However, these methods require large amounts of data to work properly, that in most of the cases need to be labeled. Supervisions are expensive, and, in the specific case of IE, researchers tend to focus on precise sub-tasks that are well studied and defined. Some of them (e.g. NER and RE) share several aspects, and addressing those problems jointly can be fruitful.

This work faces the problem of linking text portions to a given ontology with a known schema that is composed of entity and relation types. NER and RE can

---

[1] https://developers.google.com/freebase/.

be seen as special instances of the considered setting. The problems of recognizing and segmenting mentions to both entity and relation types are treated in a uniform way, reformulating them as problems of making predictions on a word given its context. While predicting the entity type of the mention to which the word belongs usually requires just a local context, spotting the relation type in which the word is involved needs a larger context. Following this intuition, we propose the idea that every word in the sentence can be considered from two different perspectives that we refer to as "narrow" and "broad" views. We propose a deep architecture that develops two latent representations of each word, associated to the just mentioned views. A predictor of entity types is attached to the former representation, whereas a relation type prediction operates on the latter. Our architecture is an instance of the generic framework of Learning from Constraints [5], where the unifying notion of "constraint" is used to inject knowledge coming from supervised and unsupervised data as well. In particular, an entropy-based index (that resembles the mutual information from the input views to the predictors) is maximized over all the data that is read by the system (labeled or not), while First-Order Logic (FOL) formulas are used to bridge predictions of entity and relation types. Formulas are converted into constraints by means of T-Norms. Linking the predictions on the two views allows the system to mutually improve their quality, differently from those models that treat them independently. When tested on a collection of sentences about factual knowledge, our model achieves good performances without requiring a large number of supervisions. This becomes more evident when logic constraints are introduced between the two views. We notice that this approach allows us to build neural models that provide an interpretable description of the unstructured raw text, by means of the FOL formalism. This interpretability, that is usually missing in neural architectures, offers a suitable basis to easily introduce additional information provided by an external supervisor. As a matter of fact, having a human-in-the-loop is known to be a crucial element in those models that learn and expand their internal knowledge bases in a life-long-learning setting [14].

This paper is organized as follows. Section 2 describes the proposed architecture and the logic constraints. Section 3 reports our experimental results, while Sect. 4 concludes the paper.

2   Model

We are given a data collection $\mathcal{D}$ composed of b utterances. Every utterance $u \in \mathcal{D}$ consists of $|u|$ words indicated with $w_j, \forall j = 1, \ldots, |u|$. We are also given an ontology $\mathcal{O}$, composed of k_n entity types and k_b relation types. Relations involve pairs of entities of pre-defined types, as sketched in Fig. 1. For each word w_i, the goal of our model is to learn to predict what is the entity type associated to word w_i, and what is the relation type to which w_i participates. For example, in the sentence *Paris is the capital of France*, the system should predict that *Paris* is an entity of type "city", that *France* is an entity of type "country", and that each word of the sentence is associated to the relation type "capitalof", where all the mentioned types belong to the given ontology.

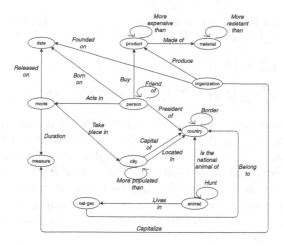

Fig. 1. Ontology. Nodes are entity types, while edges are relation types.

We follow the idea of developing two latent representations of w_i, that we refer to as "narrow" and "broad" views, respectively. We indicate such representations as $x_i^{(n)} \in \mathcal{X}^{(n)}$ and $x_i^{(b)} \in \mathcal{X}^{(b)}$, where $\mathcal{X}^{(n)}, \mathcal{X}^{(b)}$ are the generic spaces to which they belong, and n, b stand for "narrow" and "broad". Determining the entity type of w_i can be usually done by considering a local view around it, and that is what $x_i^{(n)}$ encodes. Finding the relation type of w_i usually requires to have a wider view around w_i, since mentions to relations involve larger spans of text, that is the rationale behind representation $x_i^{(b)}$.

We consider a fixed-size vocabulary of words $\mathcal{V}$, so that each w_j is a 1-hot representation of size $|\mathcal{V}|$, and those w_j that are not covered by $\mathcal{V}$ are marked with a generic symbol **unk**. Computing each $x_i^{(\cdot)}$ (being it narrow or broad) is the outcome of two main computational stages. The first stage consists in projecting the target symbol w_i into a latent (distributed) representation $e_i \in \mathbb{R}^d$, where d is the dimensionality of the embedding space. The embeddings of our vocabulary $\{e_i, \ i = 1, \dots, |\mathcal{V}|\}$, are stored (column-wise) into $\mathcal{W} \subset \mathbb{R}^{|\mathcal{V}| \times d}$, so that

$$e_i = \mathcal{W} w_i . \tag{1}$$

The second stage consists in fusing the distributed representations of the target word itself and of the other words around it, thus generating $x_i^{(\cdot)}$ by means of a Bidirectional Recurrent Neural Network (Bi-RNN). In detail, the Bi-RNN is composed by two RNNs that process two sequences of word embeddings,

$$S^{\rightarrow} = e_1, e_2, \dots, e_i$$
$$S^{\leftarrow} = e_{|u|}, e_{|u|-1}, \dots, e_i .$$

Both sequences terminate in the position of the target word, but $S^{\rightarrow}$ starts from the beginning of the sentence, while $S^{\leftarrow}$ starts from the end. Hidden states of the RNNs at the end of the sequences are concatenated, generating $x_i^{(\cdot)}$. We use Long Short Term Memories (LSTMs) to implement each RNN,

$$x_i^{(n)} = \left[LSTM^{(n\rightarrow)}(S^\rightarrow),\ LSTM^{(n\leftarrow)}(S^\leftarrow) \right]$$
$$x_i^{(b)} = \left[LSTM^{(b\rightarrow)}(S^\rightarrow),\ LSTM^{(b\leftarrow)}(S^\leftarrow) \right].$$

The global architecture is depicted in Fig. 2. While the embeddings $\{e_i\}$, are shared by the "narrow" and "broad" paths, the Bidirectional RNNs are independently developed in the two views. We implement the broad path of Fig. 2 (right side) by stacking multiple layers of Bidirectional RNNs. The last layer will embrace a larger context (due to the compositional effects of the stacking process), and it will model a higher-level/coarser representation of the input word. We notice that, in general, the broad representation could embrace multiple sentences, or even paragraphs.

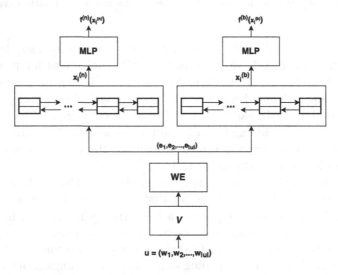

Fig. 2. Architecture of the proposed model. The utterance u is converted in a sequence of embeddings $e_1 \ldots, e_{|u|}$, feeding two bidirectional LSTMs, that compute two representations $(x_i^{(n)}, x_i^{(b)})$ of each word w_i, also referred to as "narrow" and "broad" views (left and right paths, respectively, where the right path usually includes multiple layers of LSTMs). The predictor (MLP) on the "narrow" view outputs the entity type to which w_i belong, while the MLP on the "broad" view is about the relation type of w_i.

For each word w_i, we make predictions on the entity/relation types of the ontology $\mathcal{O}$ that are more compatible with w_i. In particular, we introduce two set of functions that model the classifiers of each entity/relation type of $\mathcal{O}$,

$$f^{(n)} = [f_1^{(n)}, \ldots, f_{k_n}^{(n)}] : \mathcal{X}^n \rightarrow [0,1]^{k_n} \tag{2}$$
$$f^{(b)} = [f_1^{(b)}, \ldots, f_{k_b}^{(b)}] : \mathcal{X}^b \rightarrow [0,1]^{k_b} \tag{3}$$

where $f^{(n)}$ are about entities and $f^{(b)}$ are about relations, and every component of vectors $f^{(n)}$ and $f^{(b)}$ is in the range $[0, 1]$. Both $f^{(n)}$ and $f^{(b)}$ are multilayer perceptrons having k_n and k_b output units with sigmoidal activations, as can be observed in the upper portions of the architecture in Fig. 2.

2.1   Semantic Features

Since predictors in (2) and (3) are associated to interpretable entity/relation types of the ontology $\mathcal{O}$, we will also refer to them as "semantic features".

We expect semantic features to be learned from data by enforcing a combination of several constraints (they are enforced in a soft manner, so they are implemented as penalty functions to be minimized - we keep using the generic term "constraint" when referring to them). Each constraint guides the learning process accordingly to a specific principle that we describe in what follows. The objective function of our problem is Z, and we seek for those $f^{(n)}, f^{(b)}$ for which Z is minimal,

$$\min_{f^{(n)}, f^{(b)}} Z\left(f^{(n)}, f^{(b)}\right) = \min_{f^{(n)}, f^{(b)}} \sum_{f \in \{f^{(n)}, f^{(b)}\}} [R(f) + C(f, \mathcal{L}) + U(f)] + \Phi\left(f^{(n)}, f^{(b)}\right). (4)$$

where $R(f)$ is a regularization term (implemented with the classical weight decay approach). The term $C(f, \mathcal{L})$ is the traditional cross-entropy loss, commonly used in supervised learning, that enforces a supervision constraint on those words that have been supervised by an expert, collected (together with the supervision) in $\mathcal{L}$. Not all the words are supervised, and in the experiments we will also evaluate the case in which no-supervisions are provided at all. The term $U(f)$, is a constraint inspired by the idea of maximizing the mutual information from the space of word representations to the space of semantic features [11],

$$U(f) = \sum_{s-1}^{b} \left[\sum_{i-1}^{|u_s|} P(f(x_{si}), \theta_v) \mid P(\max_{p} f(x_{si}), \theta_h) \right] + \lambda_g \cdot G(f). \quad (5)$$

where s is the index of a sentence in $\mathcal{D}$, while i is the word index. In detail, $U(f)$ is a sum of two contributions: the one in square brackets enforces the development of only a small number of features on each word/sentence, while $G(f)$ ensures an unbiased development of the features over all the dataset $\mathcal{D}$. The sets θ_v, θ_h collect some customizable positive scalars ($U(f)$ in Eq. (4) is applied to narrow and broad features independently, so we have two independent pairs of θ_v, θ_h), while λ_g is a tuneable weight > 0. In detail, if $H(v) = -\sum_{k=1}^{|v|} v_k \log v_k$ is the Shannon entropy, we have

$$G(f) = -H\left(\frac{1}{b} \sum_{s=1}^{b} \frac{1}{|u_s|} \sum_{i=1}^{|u_s|} f(x_{si}) \right) \quad (6)$$

$$P(v, \theta = \{\lambda_1, \lambda_2, \gamma\}) = \lambda_1 \cdot H(v) + \lambda_2 \left(\sum_{k=1}^{|v|} v_k - \gamma \right)^2. \quad (7)$$

The term $G(f)$ is minimized when all the semantic features are activated uniformly over all $\mathcal{D}$, on average. The loss P is minimized when the activations provided as its first argument are "close" to 1-hot vectors, that means that we want a few-strongly-activated elements. Notice that the term P is used twice in (5). Its first instance applies to the activations of semantic features on each "word". The second instance is about the activations of semantic features on each "sentence" (pooled with the max operator). As a matter of fact, P constraints our model to develop a few, well activated features on each word, and a few well activated features on each sentence. Both the terms G and P involve the entropy function H, that is meaningful when its input is a probability distribution. For this reason, the squared term in P introduces a competition among the provided activations, that are enforced to sum to γ. If $\gamma = 1$ we have a probability distribution, while if $\gamma > 1$ (but still small) we have a sort of relaxed implementation of the probabilistic relationships. This makes the system more tolerant to multiple activations on the same input, that, from the practical point of view, turns out to be desirable. The last term of (4), $\Phi(f^{(n)}, f^{(b)})$, is a constraint coming from First-Order Logic (FOL), that introduces a link between the two views on the data, with the goal of improving the quality of both the categories of semantic features.

2.2   Logic Constraints

Narrow and broad semantic features are related to each other due to their link in the ontology $\mathcal{O}$. Consider, for example, an ontology composed of entity types "*city*", "*country*", and of the relation type "*capitalof*". In the following sentence, *Paris is the capital of France*, we can associate the entity type "*city*" to *Paris*, the type "*country*" to *France* and the relation type "*capitalof*" to each word of the sentence (since they all contribute to such relation). Our system is expected to give strong activation to the features indicate below (for the purpose of this description, we make explicit the entity/relation type to which each feature is associated),

$$\overbrace{\underbrace{Paris}_{f^{(n)}_{city}} \ is \ the \ capital \ of \ \underbrace{France}_{f^{(n)}_{country}}}^{f^{(b)}_{capitalof}}.$$

We can clearly grasp that whenever the narrow features $f^{(n)}_{city}$ and $f^{(n)}_{country}$ are active together in a sentence, it is very likely that the sentence involves the relation "*being a capital of*", i.e., that $f^{(b)}_{capitalof}$ should be active too, and viceversa[1]. Since the functions f model the activation of predicates of the ontology $\mathcal{O}$, we can implement this relationship by means of FOL formulas, such as

---

[1] In general, this could be ambiguous, since multiple relations could be associated to a city and a country. We solve this problem by introducing a special narrow feature for each broad function (to simplify the presentation, we avoid going into further details).

$$f_{city}^{(n)} \wedge f_{country}^{(n)} \Leftrightarrow f_{capitalof}^{(b)} \cdot ^1 \tag{8}$$

We repeat this process for each relation in $\mathcal{O}$, getting multiple formulas, that are then translated into real-valued constraints $\phi_r(f^{(n)}, f^{(b)})$ by means of T-Norms, as studied in [5]. For each T-Norm, there exists a unique function $\Rightarrow$ called residuum, satisfying certain properties which is the natural translation of the logic implication. In this work, we considered the residuum of the Łukasiewicz T-Norm. Łukasiewicz logic presents good properties such as the involutive negation. However, translating a large chain of $\wedge$ operations with such T-Norm requires a strong activation of all the involved components because the sum of n features should be greater than $n-1$. This could be sometimes a strong requirement to satisfy. Hence, we converted the $\wedge$ operator using the Gödel T-Norm, which instead defines such operator as the minimum among the whole predicates,

$$f_1 \wedge f_2 \wedge \ldots \wedge f_n = \min(f_1, f_2, \ldots, f_n) . \tag{9}$$

Departing from the provided example, in the ontology $\mathcal{O}$ we have a large number of relations and, for each of them, we can build a FOL formula as (8), and translate it into a real-valued penalty function. Summing up all the penalties, we get

$$\Phi(f^{(n)}, f^{(b)}) = \lambda_l \cdot \sum_{r=1}^{k_b} \phi_r(f^{(n)}, f^{(b)}), \tag{10}$$

where $\lambda_l > 0$ is a customizable scalar. We remark that whenever the activations of the premises and of the conclusions of (8) are both small (i.e., false), the corresponding constraints are automatically satisfied. The actual contribution of each $\phi_r(f^{(n)}, f^{(b)})$ becomes significant whenever there is disagreement between the semantic features computed on the broad and narrow sides.

2.3   Segmentation

In Eq. (8) we did not make explicit the arguments on which semantic features operate. While semantic features are designed to make predictions on single words, the FOL constraints can involve longer portions of text, uniformly referred to as "segments" (the previous example involved two single-word segments for narrow features - *Paris*; *France* - but, in general, we can have longer segments). In order to evaluate the FOL constraints we need to segment the input sentence and compute segment-level activations of the semantic features.

Segmentation is performed as follows: for each word, we mark as "active" only the narrow features whose activation score is beyond a decision threshold (assumed to be 0.5). If multiple nearby words share the same active feature, we collapse them into a single segment. This procedure generates multiple segmentation hypotheses for each narrow feature. We prune the hypotheses by keeping only the segment with the strongest activation (we kept also a second hypothesis for those narrow features involved twice in the same relation). In the case of broad features, for simplicity, we assume that there is only a single segment that cover the whole sentence. Finally, segment-level activations are computed by averaging the word-level scores. An example of segment activation/selection procedure is illustrated in Table 1 (the first entity is badly segmented).

Table 1. Segment generation/selection. Shaded elements are above the activation threshold (0.5), whereas bordered rectangles indicates the segments we select.

	Paris	is	the	capital	of	France	Avg Score	Activated
$f_{city}^{(n)}$	0.95	0.71	0.2	0.14	0	0.3	0.83	✓
$f_{country}^{(n)}$	0.5	0.65	0.05	0	0.1	0.9	0.9	✓
$f_{capitalof}^{(b)}$	0.35	0.7	0.9	0.8	0.8	0.1	0.6	✓

3 Experiments

We manually created (and made public) a dataset[2] $\mathcal{D}$ with sentences that are word-by-word linked to a given ontology. The dataset is a collection of 1000 sentences, where 700 are used to train our model and 300 are used as test set. Each sentence is constituted by a triple structured as *entity1-relation-entity2*. Our ontology is composed of 11 entity types, and 23 relation types, whose organization is exactly the one that we have already shown in Fig. 1, where nodes are entity types and links are relation types. We kept data simple; sentences have no co-references, quite explicit relation expressions, and the vocabulary $\mathcal{V}$ covers almost all the words. We intentionally introduced some noise in the labeling process, to make the task more challenging.

Word embeddings $\mathcal{W}$ were initialized (and kept fixed) with the ones from [3], that are vectors with 50 components. The sizes of recurrent network states (500) and the hidden layers of multilayer perceptron blocks have been chosen by cross-validation. In the case of broad features we have two layers of recurrence. The narrow and broad MLP-based predictors have a single layer with 1200 and 800 hidden units, respectively. Our cost function was optimized with ADAM [7], using mini-batches of size 32 (sentences), and we also introduced some gradient noise and gradient clipping.

The objective function in Eq. 4 requires the tuning of several hyper-parameters. However, the values of the parameter γ in the sets θ_v and θ_h can be defined exploiting prior knowledge on the sentence structure (recall that we have two independent pairs (θ_v, θ_h) for broad and narrow features). Broad features are supposed to be 1-hot in each word, and the active broad feature should be the same in the whole sentence. Thus, we set $\gamma = 1$ in both θ_v and θ_h. Likewise, we expect only one narrow feature per word, which means $\gamma = 1$ in the case of θ_v, but here the number of features per sentence is set to $\gamma = 3$ in the case of θ_h. The remaining hyper-parameters were tuned by cross-validation.

---

[2] http://sailab.diism.unisi.it/onto-sentences-dataset/.

Experiments are evaluated with different scores. A common metric used in Information Extraction is the F_1 score, but it can be only applied with labeled data, so it is not possible to measure entropy-based constraints when enforced in a completely unsupervised framework. In such situations we evaluate the mutual information between the semantic features and the ground truth. In particular, we adopted the Normalized Mutual Information (NMI) [9].

We consider a sparsely supervised setting and we compare our model against two simplified instances of it: one is trained only using constraints on the supervised examples, i.e. without entropy-based and logic constraints, another one exploits also entropy-based constraints but not logic formulas. We varied the number of labeled sentences in the training set ranging from only 1 supervised sentence per relation type, to a fully labeled case ("all"). Additionally, one of the models is also trained without considering any supervised data at all.

Figure 3 reports our results. First, we focus on the scores obtained in the case in which supervised constraints are not exploited. Since we are in a fully unsupervised case, we do not introduce logic constraints, so that only one plot is meaningful (green line, first dot of the plot). This is due to the fact that in the unsupervised case we do not have access to the symbolic elements of the ontology that are associated to the semantic features. The NMI scores in the narrow and broad cases (Fig. 3 (a, c)) show that although entropy constraints produce a significant score in the case of broad features, the result on narrow features are not encouraging. As a matter of fact, words in the borders of two entity types are sources of errors. In the case of broad features, since we output a prediction on the whole sentence, this issue is not present.

When supervised examples are introduced, Fig. 3 (a) shows that even only one supervised sentence per formula remarkably improves the NMI score of narrow features. Interestingly, the unsupervised case, despite its low performances, is still better than using a single supervision. Differently, broad features are less affected by the introduction of the first supervised example (Fig. 3 (c)), since they were already showing good performances in the fully unsupervised case. Performances of semi-supervised models (both in the case of entropy and entropy + logic) are significantly better than the model trained only with supervisions (NMI and F1, Fig. 3 (a, b, c, d)). More generally, the entropy-based terms are crucial whenever the number of supervised data is limited. Only when we go beyond 10 supervised sentences per formula ($\approx$ one third of the training set) the supervised case gets closer to the semi-supervised entropy-based case, but still does not reach the case in which logic formulas are added. Introducing logic formulas almost constantly gives improvements over the entropy-only case, confirming that bridging the predictions on broad and narrow views is important to allow a positive transfer of information between the two views.

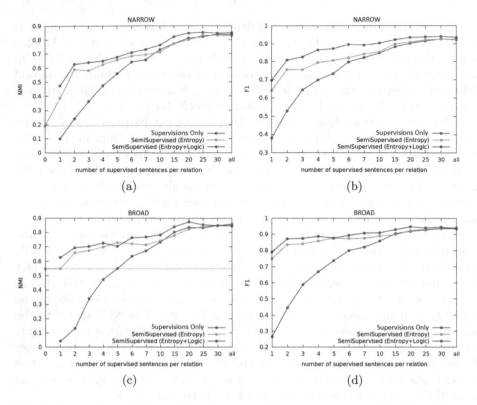

Fig. 3. Comparison of the quality of semantic features on narrow (a), (b) and broad (c), (d) views in the case of different models (test set). While (a), (c) are about the NMI score, (b), (d) report the F1 measure. The entropy-based constraints (green curve, NMI only, (a), (c)) are also evaluated in the unsupervised case (the yellow line repeats this result over all the graph, as reference). (Color figure online)

4   Conclusions

We presented a deep architecture in the framework of Learning from Constraints [5], that was designed to extract and identify mentions to entity and relation types belonging to a given ontology. Thanks to the introduction of two latent representations (views) of the input data, we implemented entity and relation detectors in a uniform way, differently from several existing systems. Our results have shown that introducing ontology-related information, represented as First-Order Logic formulas, helps the system to improve the quality of its predictions.

Our model must be extended to larger scale data and evaluated in less controlled environments. We plan to investigate more challenging settings, following the idea of life-long learning, and departing from the usual batch-mode approach toward a framework where there is an online interaction with humans. This is made possible by the interpretable representations of the raw text that are generated by our model.

References

1. Bengio, Y., Ducharme, R., Vincent, P., Jauvin, C.: A neural probabilistic language model. J. Mach. Learn. Res. **3**(Feb), 1137–1155 (2003)
2. Chiu, J., Nichols, E.: Named entity recognition with bidirectional LSTM-CNNs. Trans. Assoc. Comput. Linguist. **4**(1), 357–370 (2016)
3. Collobert, R., Weston, J.: A unified architecture for natural language processing: deep neural networks with multitask learning. In: Proceedings of the 25th International Conference on Machine Learning, pp. 160–167. ACM (2008)
4. Collobert, R., Weston, J., Bottou, L., Karlen, M., Kavukcuoglu, K., Kuksa, P.: Natural language processing (almost) from scratch. J. Mach. Learn. Res. **12**(Aug), 2493–2537 (2011)
5. Gnecco, G., Gori, M., Melacci, S., Sanguineti, M.: Foundations of support constraint machines. Neural Comput. **27**(2), 388–480 (2015)
6. Hochreiter, S., Schmidhuber, J.: Long short-term memory. Neural Comput. **9**(8), 1735–1780 (1997)
7. Kingma, D., Ba, J.: Adam: a method for stochastic optimization. arXiv preprint arXiv:1412.6980 (2014)
8. Lample, G., Ballesteros, M., Subramanian, S., Kawakami, K., Dyer, C.: Neural architectures for named entity recognition. In: Proceedings of NAACL-HLT, pp. 260–270 (2016)
9. Lancichinetti, A., Fortunato, S., Kertész, J.: Detecting the overlapping and hierarchical community structure in complex networks. New J. Phys. **11**(3), 033015 (2009)
10. Luo, G., Huang, X., Lin, C.Y., Nie, Z.: Joint named entity recognition and disambiguation. In: Proceedings EMNLP (2015)
11. Melacci, S., Gori, M.: Unsupervised learning by minimal entropy encoding. IEEE Trans. Neural Netw. Learn. Syst. **23**(12), 1849–1861 (2012)
12. Mikolov, T., Sutskever, I., Chen, K., Corrado, G.S., Dean, J.: Distributed representations of words and phrases and their compositionality. In: Advances in Neural Information Processing Systems, pp. 3111–3119 (2013)
13. Mintz, M., Bills, S., Snow, R., Jurafsky, D.: Distant supervision for relation extraction without labeled data. In: Proceedings of ACL, pp. 1003–1011 (2009)
14. Mitchell, T., Cohen, W., Hruschka, E., Talukdar, P., Yang, B., et al.: Never-ending learning. Commun. ACM **61**(5), 103–115 (2018)
15. Miwa, M., Bansal, M.: End-to-end relation extraction using LSTMs on sequences and tree structures. arXiv preprint arXiv:1601.00770 (2016)
16. Pantel, P., Bhagat, R., Coppola, B., Chklovski, T., Hovy, E.H.: ISP: learning inferential selectional preferences. In: HLT-NAACL, pp. 564–571 (2007)
17. Piskorski, J., Yangarber, R.: Information extraction: past, present and future. In: Poibeau, T., Saggion, H., Piskorski, J., Yangarber, R. (eds.) Multi-Source, Multilingual Information Extraction And Summarization. NLP, pp. 23–49. Springer, Heidelberg (2013). https://doi.org/10.1007/978-3-642-28569-1_2
18. Ratinov, L., Roth, D.: Design challenges and misconceptions in named entity recognition. In: Proceedings of the Thirteenth Conference on Computational Natural Language Learning, pp. 147–155. Association for Computational Linguistics (2009)
19. Shinyama, Y., Sekine, S.: Preemptive information extraction using unrestricted relation discovery. In: Proceedings of NAACL HLT, pp. 304–311 (2006)
20. Strubell, E., Verga, P., Belanger, D., McCallum, A.: Fast and accurate sequence labeling with iterated dilated convolutions. arXiv preprint arXiv:1702.02098 (2017)

SeNA-CNN: Overcoming Catastrophic Forgetting in Convolutional Neural Networks by Selective Network Augmentation

Abel Zacarias$^{(\boxtimes)}$ (iD) and Luís A. Alexandre$^{(\boxtimes)}$ (iD)

Instituto de Telecomunicações, Universidade da Beira Interior,
Rua Marquês d'Ávila e Bolama, 6201-001 Covilhã, Portugal
{abel.zacarias,luis.alexandre}@ubi.pt

Abstract. Lifelong learning aims to develop machine learning systems that can learn new tasks while preserving the performance on previous learned tasks. In this paper we present a method to overcome catastrophic forgetting on convolutional neural networks, that learns new tasks and preserves the performance on old tasks without accessing the data of the original model, by selective network augmentation (SeNA-CNN). The experiment results showed that SeNA-CNN, in some scenarios, outperforms the state-of-art Learning without Forgetting algorithm. Results also showed that in some situations it is better to use SeNA-CNN instead of training a neural network using isolated learning.

Keywords: Lifelong learning · Catastrophic forgetting
Convolutional neural networks · Supervised learning

1 Introduction

Deep learning is a sub-field of machine learning which uses several learning algorithms to solve real-world tasks as image recognition, facial detection, signal processing, on supervised, unsupervised and reinforcement learning using feature representations at successively higher, more abstract layers. Even with the growth and success of deep learning on many applications, some issues still remain unsolved. One of these issues is the catastrophic forgetting problem [8]. This issue can be seen as an handicap to develop truly intelligent systems.

Catastrophic forgetting arises when a neural network is not capable of preserving the past learned task when learning a new task. There are some approaches that benefit from previously learned information to improve performance of learning new information, for example fine-tuning [7] where the parameters of the old tasks are adjusted for adapting to a new task and, as was shown

This work was supported by National Founding from the FCT-Fundação para a Ciência e a Tecnologia, through the UID/EEA/50008/2013 Project. The GTX Titan X used in this research was donated by the NVIDIA Corporation.

L. Pancioni et al. (Eds.): ANNPR 2018, LNAI 11081, pp. 102–112, 2018.
https://doi.org/10.1007/978-3-319-99978-4_8

in [3], this method implies forgetting the old task while learning the new task. Other approach well known is feature extraction [6] where the parameters of the old network are unchanged and the parameters of the outputs of one or more layers are used to extract features for the new task. There is also a paradigm called joint train [4] where parameters of old and new tasks are jointly trained to minimize the loss in all tasks.

There are already some methods built to overcome the problem of catastrophic forgetting [9,11,13]. But even with these and other approaches, the problem of catastrophic forgetting is still a big challenge for the Artificial Intelligence (AI) community and according to [18] is now appropriate to the AI community to move toward algorithms that are capable of learning multiple problems over time.

In this paper we present a new method that is capable of preserving the previous learned task while learning a new tasks without requiring a training set with previous tasks data. This is achieved by selective network augmentation, where new nodes are added to an existing neural network trained on an original problem, to deal with the new tasks.

SeNA-CNN is similar to progressive neural networks proposed in [16] and in the next section we present the main differences between the two methods.

This paper is structured as follows: Sect. 2 presents related works on existing techniques to overcome the problem of catastrophic forgetting in neural networks. In Sect. 3 we describe SeNA-CNN and some implementation details. Section 4 presents the experiments and results of SeNA-CNN and on Sect. 5 we present the conclusions.

2   Related Work

The problem of catastrophic forgetting is a big issue in machine learning and artificial intelligence if the goal is to build a system that learns through time, and is able to deal with more than a single problem. According to [12], without this capability we will not be able to build truly intelligent systems, we can only create models that solve isolated problems in a specific domain. There are some recent works that tried to overcome this problem, e.g., domain adaptation that uses the knowledge learned to solve one task and transfers it to help learning another, but those two tasks have to be related. This approach was used in [10] to avoid the problem of catastrophic forgetting. They used two properties to reduce the problem of catastrophic forgetting. The first properties was to keep the decision boundary unchanged and the second was that the feature extractor from the source data by the target network should be present in a position close to the features extracted from the source data by the source network. As was shown in the experiments, by keeping the decision boundaries unchanged new classes can not be learned and it is a drawback of this approach because it can only deal with related tasks, with the same number of classes, while in our approach, we are able to deal with unrelated problems with different number of classes.

The Learning without Forgetting (LwF) algorithm proposed in [11] adds nodes to an existing network for a new task only in the fully connected layers and this approach demonstrated to preserve the performance on old tasks without accessing training data for the old tasks. We compare SeNA-CNN with LwF algorithm. The main difference is that instead of adding nodes in fully connected layers, we add convolutional and fully connected layers of the new tasks to an existing model and SeNA-CNN has a better capability of learning new problems than LwF because we train a series of convolutional and fully connected layers while LwF only trains the added nodes in the fully connected layer and hence, depends on the original task's learned feature extractors to represent the data from all problems to be learned.

Progressive Neural Networks (PNN), proposed in [16], also addressed the problem of catastrophic forgetting via lateral connection to a previous learned network. The main difference to SeNA-CNN is that the experiment was in reinforcement learning while our proposal is designed to work with supervised learning for image classification problems. This approach, as SeNA-CNN begins with one column, a CNN trained on a single problem. When adding new tasks parameters from the previous task are frozen and new columns are added and initialised from scratch. Another difference between PNN and SeNA-CNN, is that SeNA-CNN use the two first convolutional layers of the original model trained on isolated learning and by doing that SeNA-CNN can learn the new tasks faster than if all the layers had to be trained from scratch, while PNN adds an entire column each time that new tasks come and the new column is randomly initialised. In the experimental section [16] they demonstrated the proposed method with 2, 3 and 4 columns architecture on Atari Game and 3D maze game. For future work, as in our approach, the authors aims to solve the problem of adding the capability to automatically choose at which task a label belongs because during the experiment it was necessary on test time to choose which task to use for inference.

3   Proposed Method

Our proposal is a method that is able to preserve the performance on old tasks while learning new tasks, without seeing again the training data for old tasks, as is necessary in [11], using selective network augmentation.

A model that is capable of learning two or more tasks has several advantages against that which only learns one task. First is that the previous learned task can help better and faster learning the new task. Second, the model that learns multiple tasks may result in more universal knowledge and it can be used as a key to learn new task domains [17].

Initially a network is instantiated with L layers with hidden layers h_i and parameters θ_n with random initialization. The network is then trained until convergence. Figure 1(a) presents the original model for old task trained on isolated learning, Fig. 1(b) is our proposed model with two tasks. In Fig. 1(b) the blue colour represents the old task network and the orange corresponds to the new added nodes for the new task.

When a new tasks is going to be learned instead of adding nodes only in fully connected layers as is done in [11], we add layers for the new task Typically the added layers contain a structure similar to the network that we trained on isolated learning. We consider the option of not adding the first two layers, because the neurons in those layers find several simple structures, such as oriented edges as demonstrated in [15]. The remaining layers seem to be devoted to more complex objects, and hence, are more specific to each problem, and that is why we choose to create these new layers. It also resembles the idea of mini-columns in the brain [14]. We add those layers and train them initialized with weights of old tasks, keeping the old task layers frozen.

When switching to a third task, we freeze the two previous learned tasks and only train the new added layers. This process can be generalized to any number of tasks that we wish to learn.

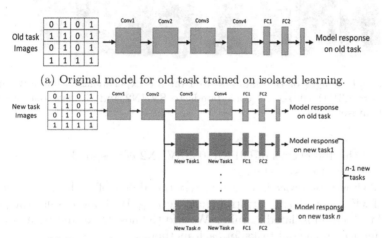

(a) Original model for old task trained on isolated learning.

(b) Proposed model: adds new layers for the second task.

Fig. 1. Original and our model used in the experiment process to avoid the catastrophic forgetting by selective network augmentation. The blue coloured boxes correspond to the old task and the orange coloured correspond to the added layers. (Color figure online)

4   Experiments

We compared our method with the algorithm LwF proposed in [11].

Our experiments evaluate if the proposed method can effectively avoid the catastrophic forgetting problem. We conducted our experiments using three well known datasets namely CIFAR10 [2], CIFAR100 [2] and SVHN2. Table 1 shows information on each dataset, and the number of images on training and test sets. CIFAR10 AND CIFAR100 are very similar. CIFAR10 has 10 classes and these

Table 1. Number of images for train and test sets.

Data set	CIFAR10	CIFAR100	SVHN2
Train	50000	50000	73257
Test	10000	10000	26032

Fig. 2. Example images of the datasets used on the experiments. First row images corresponds to CIFAR10, second corresponds to SVHN2 and the last one are from CIFAR100 dataset.

are subset of the 100 classes of CIFAR100. SVHN2 corresponds to street house numbers and has 11 classes (Fig. 2).

Figure 3 shows the procedure used to test the ability of both models (SeNA-CNN and LwF) to overcome catastrophic forgetting. Both models use the previous model trained on isolated learning. We add the new tasks and then evaluate the performance on the old tasks for each method.

4.1 Network Architecture

The neural network used on isolated learning was a standard network architecture with 4 convolutional layers, the first one is the input, 6 activation layers (one of them is the softmax), 2 maxpooling layers, 3 dropout layers, a flatten layer and 2 dense layers. For new tasks the architecture was almost the same. The difference was that for the new tasks we did not add the first two convolutinal layers, we used the first two layers of the model trained on isolated. Figure 4 shows the proposed approach when the three branches corresponding to each task are connected. This is a functional model and overall this model had 8 convolutinal layers, 6 fully-connected layers, 11 ReLUs activation layers, 4 pooling layers and 7 dropout layers. The model receives tensor input and this input is propagated to all branches and each branch produce an output. To choose the branch to predict at test time, we set all other tasks, images and targets values to zero and only show to the model the images and targets we want to predict. So

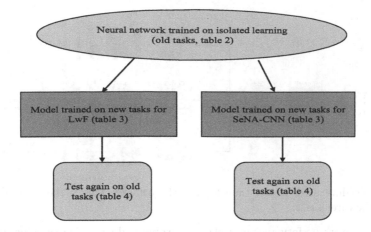

Fig. 3. Procedure used to test both evaluated models to overcome catastrophic forgetting.

far this process is done by hand and we consider for future work the automatic choice of which task to predict.

Input images are RGB and have 32×32 pixels. The first convolution layer has filters with dimensions 32×32 while the other two convolution layers have filters with 64×64. We used the keras API [5] running on tensorflow [1].

4.2 Training Methodology

Our main goal is to evaluate if the proposed model learns new tasks while preserving the performance on old tasks. During training we followed the same practice as [11], the main difference is that we first freeze all layers of the original model and only train the added nodes. Then we train all weights for convergence using back-propagation with SGD algorithm with dropout enabled. All the networks had the same architecture, and the learning rate was set to 0.01, weight decay of $1e-6$ and momentum 0.9. All networks use the same train, validation and test split for a given seed number. Table 2 shows the performance and execution time of each network after 12 training epochs. We run each experiment ten times and present results corresponding to the mean and standard deviation of these 10 repetitions. We run our experiments using a GeForce GTX TITAN X with 12 GiB.

4.3 Isolated Learning

We started by training 3 networks, one for each of the 3 data sets. Results of the experiment are shown in Table 2 where for each network we present the mean performance, its standard deviation and the execution time for train and test. These networks will be used both for SeNA-CNN and LwF in the next experiments.

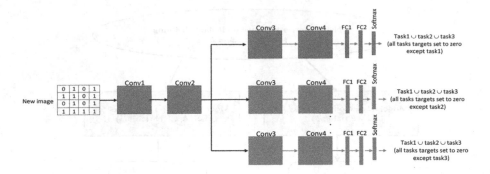

Fig. 4. Procedure used at test time for the three tasks. This is the stage when we combine the three tasks.

Table 2. Network performance on isolated learning and execution time for train and test sets.

Train	Test	Baseline [%]	Execution time [s]
CIFAR10	CIFAR10	74.10 ± 0.70	312
CIFAR100	CIFAR100	51.44 ± 0.40	423
SVHN2	SVHN2	92.27 ± 0.80	438

4.4 Adding New Tasks to the Models

As Fig. 3 shows, we used the networks trained on isolated learning to implement our method by adding layers of the new tasks in such way that the model can learn a new task without forgetting the original one. Table 3 presents the performance of the proposed method when adding new tasks and compares it with the baseline [11]. These results correspond to the performance of our model and LwF when using a model trained on cifar10 for isolated learning and we added to the model as new tasks svhn2 and cifar100. This process was repeated for the other two tasks.

Results shows that SeNA-CNN outperformed LwF algorithm almost in all scenarios, showing that selectively adding layers to an existing model can preserve the performance on the old tasks when learning a new one, also is not necessary to train again the previous model and the new task learned will not interfere on the previous learned one. Overall SeNA-CNN outperformed LwF algorithm in 2/3 of the experiments showing the effectiveness of the proposed method to learn new tasks.

We also evaluated if, when adding a new task, the knowledge previous learned was not overwritten. As shown in Fig. 3 we tested if the model was able to preserve the previous learned task. Table 4 presents the results of these experiments. The second and third columns represent results of cifar10 as old task using the others two as new tasks. Similar setups are presented in the remaining columns. Results shows that our method outperformed LwF when remembering the

Table 3. SeNA-CNN and LwF test accuracy (and standard deviation) on new tasks.

Old	New	LwF	SeNA-CNN
CIFAR10	SVHN2	**84.02**(0.47)	82.27(0.38)
CIFAR10	CIFAR100	53.10(0.55)	**55.67**(0.52)
CIFAR100	CIFAR10	75.23(0.53)	**75.69**(0.52)
CIFAR100	SVHN2	86.49(0.39)	**90.04**(0.38)
SVHN2	CIFAR10	66.42(0.62)	**67.27**(0.58)
SVHN2	CIFAR100	**49.05**(0.63)	47.15(0.45)

previous learned tasks in all cases, and once again. We also verified that in some scenarios such as cifar100↦cifar10 (for both methods), cifar100 performance increased compared to isolated learning, and it suggests using both proposed models instead of training from a random weights initialization, without interaction with other problems. These results are understandable since cifar10 and cifar100 are very similar and the two layers shared during the train of the new tasks increased the performance. Results show that by applying our method it is possible to overcome the problem of catastrophic forgetting when new tasks are added to the model.

Table 4. SeNA-CNN and LwF test accuracy (and standard deviation) showing that our method does not forget old tasks after learning the new ones and outperforms the LwF method in all cases.

New	Old	LwF	SeNA-CNN
CIFAR10	SVHN2	87.96(0.75)	**89.84**(0.68)
CIFAR10	CIFAR100	52.39(0.43)	**53.34**(0.58)
CIFAR100	CIFAR10	69.37(0.65)	**70.59**(0.59)
CIFAR100	SVHN2	89.01(0.39)	**89.53**(0.57)
SVHN2	CIFAR10	65.80(0.47)	**67.83**(0.59)
SVHN2	CIFAR100	48.11(0.41)	**49.40**(0.72)

4.5   Three Tasks Scenario

To demonstrate that SeNA-CNN is able to deal with several different problems, we experiment by learning three tasks. In this case we used the three datasets previously presented and we combine them two by two as old and one as new task. In Table 5 we presents results when adding a new task to a model that had already learned two tasks. From this scenario clearly in all cases SeNA-CNN outperformed LwF when learning a new task, and also the performance for cifar100 continue increasing for both methods and consolidating what we previously said.

Table 5. Three tasks SeNA-CNN and LwF test accuracy (and standard deviation) on new tasks.

Old	New	LwF	SeNA-CNN
SVHN2, CIFAR10	CIFAR100	46.96(0.29)	**47.15**(0.48)
CIFAR10, CIFAR100	SVHN2	87.21(0.30)	**87.87**(0.50)
CIFAR100, SVHN2	CIFAR10	74.71(0.50)	**75.69**(0.14)
CIFAR10, SVHN2	CIFAR100	54.24(0.37)	**54.87**(0.63)
SVHN2, CIFAR100	CIFAR10	65.99(0.47)	**66.00**(0.48)
CIFAR100, CIFAR10	SVHN2	87.68(0.43)	**89.08**(0.37)

In this scenario we also evaluated the ability to preserve the performance of the two old learned tasks. Table 6 present results of both methods when they have to recall the old tasks. Comparing results, both algorithms typically had the same percentage of performance, meaning that in some scenarios SeNA-CNN performed better than LwF and vice-versa. Once again these results shows the ability to overcome the catastrophic forgetting problem in convolutional neural networks by selectively network augmentation.

Table 6. Three tasks SeNA-CNN and LwF test accuracy (and standard deviation) on old tasks.

New	Old	LwF	SeNA-CNN
CIFAR100	SVHN2, CIFAR10	**89.23**(0.70), 75.14(0.14)	89.01(0.44), **76.81**(0.64)
SVHN2	CIFAR10, CIFAR100	**73.99**(0.12), **56.78**(0.37)	71.11(0.37), 56.20(0.58)
CIFAR10	CIFAR100, SVHN2	**52.41**(0.26), 87.10(0.22)	49.14(0.58), **89.17**(0.57)
CIFAR100	CIFAR10, SVHN2	74.28(0.25), **90.04**(0.39)	**75.58**(0.52), 88.07(0.94)
CIFAR10	SVHN2, CIFAR100	90.13(0.59), **48.11**(0.27)	**90.19**(0.64), 46.96(0.51)
SVHN2	CIFAR100, CIFAR10	47.20(0.40), 74.95(0.43)	**47.87**(0.63), **75.24**(0.39)

5   Conclusion

In this paper we presented a new method, SeNA-CNN to avoid the problem of catastrophic forgetting by selective network augmentation and the proposed method demonstrated to preserve the previous learned tasks without accessing the old task's data after the original training had been done. We demonstrated the effectiveness of SeNA-CNN to avoid catastrophic forgetting for image classification by running it on three different datasets and compared it with the baseline LwF algorithm.

It has the advantage of being able to learn better new tasks than LwF since we train a series of convolutional and fully connected layers for each new task, whereas LwF only adds nodes to the fully connected layers and hence, depends on the original task's learned feature extractors to represent the data from all problems to be learned.

We also showed that in some scenarios SeNA-CNN and LWF increases the performance when compared to isolated training for classification problems with some similarity. This is understandable since by reusing partial information from previous tasks, we are somehow doing fine-tuning on the new task.

As future work we consider adapting SeNA-CNN for on-line learning and make it automatically choose which task is to be classified.

References

1. Abadi, M., et al.: TensorFlow: large-scale machine learning on heterogeneous systems, software available from TensorFlow.org (2015). https://www.tensorflow.org/
2. Acemoglu, D., Cao, D.: MIT and CIFAR (2010)
3. Aljundi, R., Babiloni, F., Elhoseiny, M., Rohrbach, M., Tuytelaars, T.: Memory aware synapses: learning what (not) to forget. CoRR abs/1711.09601 (2017). http://arxiv.org/abs/abs/1711.09601
4. Caruana, R.: Multitask learning. Mach. Learn. **28**(1), 41–75 (1997). https://doi.org/10.1023/A:1007379606734
5. Chollet, F., et al.: Keras (2015)
6. Donahue, J., et al.: Decaf: a deep convolutional activation feature for generic visual recognition. In: Xing, E.P., Jebara, T. (eds.) Proceedings of the 31st International Conference on Machine Learning. Proceedings of Machine Learning Research, vol. 32, pp. 647–655. PMLR, Bejing, 22–24 June 2014. http://proceedings.mlr.press/v32/donahue14.html
7. Girshick, R., Donahue, J., Darrell, T., Malik, J.: Rich feature hierarchies for accurate object detection and semantic segmentation. In: 2014 IEEE Conference on Computer Vision and Pattern Recognition, pp. 580–587, June 2014. https://doi.org/10.1109/CVPR.2014.81
8. Goodfellow, I.J., Mirza, M., Xiao, D., Courville, A., Bengio, Y.: An Empirical Investigation of Catastrophic Forgetting in Gradient-Based Neural Networks. arXiv e-prints, December 2013
9. Gutstein, S., Stump, E.: Reduction of catastrophic forgetting with transfer learning and ternary output codes. In: 2015 International Joint Conference on Neural Networks (IJCNN), pp. 1–8, July 2015. https://doi.org/10.1109/IJCNN.2015.7280416
10. Jung, H., Ju, J., Jung, M., Kim, J.: Less-forgetting learning in deep neural networks. CoRR abs/1607.00122 (2016). http://arxiv.org/abs/abs/1607.00122
11. Li, Z., Hoiem, D.: Learning without forgetting. CoRR abs/1606.09282 (2016). http://arxiv.org/abs/abs/1606.09282
12. Liu, B.: Lifelong machine learning: a paradigm for continuous learning. Front. Comput. Sci. **11**, 1–3 (2016). https://doi.org/10.1007/s11704-016-6903-6
13. van Merriënboer, B., et al.: Blocks and Fuel: frameworks for deep learning. CoRR abs/1506.00619 (2015). http://arxiv.org/abs/abs/1506.00619
14. Mountcastle, V.B.: Modality and topographic properties of single neurons of cat's somatic sensory cortex. J. Neurophysiol. **20**(4), 408–434 (1957). https://doi.org/10.1152/jn.1957.20.4.408. pMID: 13439410
15. Rafegas, I., Vanrell, M., Alexandre, L.A.: Understanding trained CNNs by indexing neuron selectivity. CoRR abs/1702.00382 (2017). http://arxiv.org/abs/abs/1702.00382

16. Rusu, A.A., et al.: Progressive neural networks. arXiv preprint arXiv:1606.04671 (2016)
17. Shin, H., Lee, J.K., Kim, J., Kim, J.: Continual learning with deep generative replay. CoRR abs/1705.08690 (2017). http://arxiv.org/abs/abs/1705.08690
18. Silver, D.L., Yang, Q., Li, L.: Lifelong machine learning systems: beyond learning algorithms. In: AAAI Spring Symposium Series (2013)

Classification Uncertainty of Deep Neural Networks Based on Gradient Information

Philipp Oberdiek[1]([✉]), Matthias Rottmann[2], and Hanno Gottschalk[2]

[1] Technische Universität Dortmund, 44227 Dortmund, Germany
philipp.oberdiek@udo.edu
[2] Bergische Universität Wuppertal, 42119 Wuppertal, Germany
{hanno.gottschalk,rottmann}@uni-wuppertal.de

Abstract. We study the quantification of uncertainty of Convolutional Neural Networks (CNNs) based on gradient metrics. Unlike the classical softmax entropy, such metrics gather information from all layers of the CNN. We show for the EMNIST digits data set that for several such metrics we achieve the same meta classification accuracy – i.e. the task of classifying predictions as correct or incorrect without knowing the actual label – as for entropy thresholding. We apply meta classification to unknown concepts (out-of-distribution samples) – EMNIST/Omniglot letters, CIFAR10 and noise – and demonstrate that meta classification rates for unknown concepts can be increased when using entropy together with several gradient based metrics as input quantities for a meta classifier. Meta classifiers only trained on the uncertainty metrics of known concepts, i.e. EMNIST digits, usually do not perform equally well for all unknown concepts. If we however allow the meta classifier to be trained on uncertainty metrics for some out-of-distribution samples, meta classification for concepts remote from EMNIST digits (then termed known unknowns) can be improved considerably.

Keywords: Deep learning · Uncertainty quantification
Meta classification

1 Introduction

In recent years deep learning has outperformed other classes of predictive models in many applications. In some of these, e.g. autonomous driving or diagnostics in medicine, the reliability of a prediction is of highest interest. In classification tasks, the thresholding on the highest softmax probability or thresholding on the entropy of the classification distributions (softmax output) are commonly used metrics to quantify classification uncertainty of neural networks, see e.g. [11]. However, misclassification is oftentimes not detected by these metrics and it is also well known that these metrics can be fooled easily. Many works demonstrated how an input can be designed to fool a neural network such that it incorrectly classifies the input with high confidence (termed adversarial examples, see e.g. [9,13,18,19]). This underlines the need for measures of uncertainty.

© Springer Nature Switzerland AG 2018
L. Pancioni et al. (Eds.): ANNPR 2018, LNAI 11081, pp. 113–125, 2018.
https://doi.org/10.1007/978-3-319-99978-4_9

A basic statistical study of the performance of softmax probability thresholding on several datasets was developed in [11]. This work also assigns proper out-of-distribution candidate datasets to many common datasets. For instance a network trained on MNIST is applied to images of handwritten letters, scaled gray scale images from CIFAR10, and different types of noise. This represents a baseline for comparisons.

Using classical approaches from uncertainty quantification for modeling input uncertainty and/or model uncertainty, the detection rate of misclassifications can be improved. Using the baseline in [11], an approach named ODIN, which is based on input uncertainty, was published in [15]. This approach shows improved results compared to pure softmax probability thresholding. Uncertainty in the weights of a neural network can be modeled using Bayesian neural networks. A practically feasible approximation to Bayesian neural networks was introduced in [8], known as Monte-Carlo dropout, which also improves over classical softmax probability thresholding.

Since the softmax removes one dimension from its input by normalization, some works also perform outlier detection on the softmax input (the penultimate layer) and outperform softmax probability thresholding as well, see [2].

In this work we propose a different approach to measure uncertainty of a neural network based on gradient information. Technically, we compute the gradient of the negative log-likelihood of a single sample during inference where the class argument in the log-likelihood is the predicted class. We then extract compressed representations of the gradients, e.g., the norm of a gradient for a chosen layer. E.g., a large norm of the gradient is interpreted as a sign that, if the prediction would be true, major re-learning would be necessary for the CNN. We interpret this 're-learning-stress' as uncertainty and study the performance of different gradient metrics used in two meta classification tasks: separating correct and incorrect predictions and detecting in- and out-of-distribution samples.

The closest approaches to ours are probably [2,11] as they also establish a self evaluation procedure for neural networks. However they only incorporate (non-gradient) metrics for particular layers close to the networks output while we consider gradient metrics extracted from all the layers. Just as [2,11] our approach does not make use of input or model uncertainty. However these approaches, as well as our approach, are somewhat orthogonal to classical uncertainty quantification and should be potentially combinable with input uncertainty and model uncertainty, as used in [8,15], respectively.

The remainder of this work is structured as follows: First, in Sect. 2 we introduce (gradient) metrics, the concept of meta classification and threshold independent performance measures for meta classification, AUROC and AUPR, that are used in the experiments. In Sect. 3 we introduce the network architecture and the experiment setup containing the choice of data sets. We use EMNIST [6] digits as a known concept on which the CNN is trained and EMNIST letters, CIFAR10 images as well as different types of noise as unknown/unlearned concepts. Then we statistically investigate the separation performance of our metrics for correct vs. incorrect classifications provided by CNNs. This is followed by a performance study for the detection of in- and out-of-distribution samples (detection of unlearned concepts) in Sect. 4. Therefore we also combine available metrics for training and comparing different meta classifiers. In this section meta classifiers are trained only using known concepts, i.e., EMNIST digits. Afterwards, in Sect. 5, we insert unlearned concepts (which therefore become known unknowns) into the training of the meta classifiers. While the softmax baseline achieves an AUROC value of 95.83% our approach gains 0.81% in terms of AUROC and even more in terms of AUPR values.

2   Entropy, Softmax Baseline and Gradient Metrics

Given an input $x \in \mathbb{R}^n$, weights $w \in \mathbb{R}^p$ and class labels $y \in \mathcal{C} = \{1, \ldots, q\}$, we denote the output of a neural network by $f(y|x, w) \in [0, 1]$. The entropy of the estimated class distribution conditioned on the input (also called Shannon information, [16])

$$E(x, w) = -\frac{1}{\log(q)} \sum_{y \in \mathcal{C}} f(y|x, w) \log(f(y|x, w)), \tag{1}$$

is a well known dispersion measure and widely used for quantifying classification uncertainty of neural networks. In the following we will use the term entropy in the sense explained above. Note that this should not be confused with the entropy underlying the (not estimated and joint) statistical distribution of inputs and labels. The softmax baseline proposed by [11] is calculated as

$$S(x, w) = \max_{y \in \mathcal{C}} f(y|x, w). \tag{2}$$

Using the maximum a posteriori principle (MAP), the predicted class is defined by

$$\hat{y}(x, w) := \arg \max_{y \in \mathcal{C}} f(y|x, w) \tag{3}$$

according to the Bayes decision rule [3], or as one hot encoded label $\hat{g}(x, w) \in \{0, 1\}^q$ with

$$\hat{g}_k(x, w) = \begin{cases} 1, & \hat{y}(x, w) = k \\ 0, & \text{else} \end{cases} \tag{4}$$

for $k = 1, \ldots, q$. Given an input sample x^i with one hot label y^i, predicted class label $\hat{g}^i$ (from Eq. (4)) and a loss function $L = L(f(y|x^i, w), y^i)$, we can calculate the gradient of the loss function with respect to the weights $\nabla_w L = \nabla_w L(f(y|x^i, w), \hat{g}^i)$. In our experiments we use the gradient of the negative log-likelihood at the predicted class label, which means

$$L = L(f(y|x^i, w), \hat{g}^i) = -\sum_{y \in \mathcal{C}} \hat{g}^i_y \log\left(f(y|x^i, w)\right) = -\log\left(f(\hat{y}|x^i, w)\right). \tag{5}$$

We apply the following metrics to this gradient:

- Absolute norm ($\|\nabla_w L\|_1$)
- Euclidean norm ($\|\nabla_w L\|_2$)
- Minimum ($\min(\nabla_w L)$)
- Maximum ($\max(\nabla_w L)$)
- Mean ($\text{mean}(\nabla_w L)$)
- Skewness ($\text{skew}(\nabla_w L)$)
- Kurtosis ($\text{kurt}(\nabla_w L)$)

These metrics can either be applied layerwise by restricting the gradient to those weights belonging to a single layer in the neural network or to the whole gradient on all layers.

The metrics can be sampled over the input X and conditioned to the event of either correct or incorrect classification. Let $T(w)$ and $F(w)$ denote the subset of correctly and incorrectly classified samples for the network $f(y|x,w)$, respectively. Given a metric M (e.g. the entropy E or any gradient based one), the two conditioned distributions $M(X,w)|_{T(w)}$ and $M(X,w)|_{F(w)}$ are further investigated. For a threshold t, we measure $P(M(X,w) < t \,|\, T(w))$ and $P(M(X,w) \geq t \,|\, F(w))$ by sampling X. If both probabilities are high, t gives a good separation between correctly and incorrectly classified samples. This concept can be transfered to the detection of out-of-distribution samples by defining these as incorrectly classified. We term this procedure (classifying $M(X,w) < t$ vs. $M(X,w) \geq t$) *meta classification*.

Since there are many possible ways to compute thresholds t, we compute our results threshold independent by using **A**rea **U**nder the **R**eceiver **O**perating **C**urve (AUROC) and **A**rea **U**nder the **P**recision **R**ecall curve (AUPR). For any chosen threshold t we define

$$TP = \#\{\text{correctly predicted positive cases}\},$$
$$TN = \#\{\text{correctly predicted negative cases}\},$$
$$FP = \#\{\text{incorrectly predicted positive cases}\},$$
$$FN = \#\{\text{incorrectly predicted negative cases}\}.$$

and can compute the quantities

$$R = TPR = \frac{TP}{TP + FN} \qquad \text{(True positive rate or Recall)},$$
$$FPR = \frac{FP}{FP + TN} \qquad \text{(False positive rate)},$$
$$P = \frac{TP}{TP + FP} \qquad \text{(Precision)}.$$

When dealing with threshold dependent classification techniques, one calculates TPR (R), FPR and P for many different thresholds in the value range of the variable. The AUROC is the area under the receiver operating curve, which has the FPR as ordinate and the TPR as abscissa. The AUPR is the area under the precision recall curve, which has the recall as the ordinate and the precision as abscissa. For more information on these performance measures see [7].

The AUPR is in general more informative for datasets with a strong imbalance in positive and negative cases and is sensitive to which class is defined as the positive case. Because of that we are computing the AUPR-In and AUPR-Out, for which the definition of a positive case is reversed. In addition the values of one variable are multiplied by -1 to switch between AUPR-In and AUPR-Out as in [11].

3   Meta Classification – A Benchmark Between Maximum Softmax Probability and Gradient Metrics

We perform all our statistical experiments on the EMNIST data set [6], which contains 28×28 gray scale images of 280 000 handwritten digits (0–9) and 411 302 handwritten letters (a–z, A–Z). We train the CNNs only on the digits, in order to test their behavior on untrained concepts. We split the EMNIST data set (after a random permutation) as follows:

- 60,000 digits (0–9) for training
- 20,000 digits (0–9) for validation
- 200,000 digits (0–9) for testing
- 20,000 letters (a–z, A–Z) as untrained concepts

Additionally we included the CIFAR10 library [12], shrinked and converted to gray scale, as well as 20,000 images generated from random uniform noise. All concepts can be seen in Fig. 1.

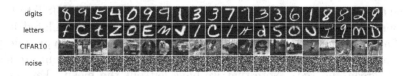

Fig. 1. Different concepts used for our statistical experiments

The architecture of the CNNs consists of three convolutional (conv) layers with 16 filters of size 3×3 each, with a stride of 1, as well as a dense layer with a 10-way softmax output. Each of the first two conv layers are equipped with leaky ReLU activations

$$LeakyReLU(x) = \begin{cases} x, & x > 0 \\ 0.1x, & x < 0 \end{cases} \tag{6}$$

118   P. Oberdiek et al.

and followed by 2×2 max pooling. We employ L^2 regularization with a regularization parameter of 10^{-3}. Additionally, dropout [17] is applied after the first and third conv layer. The dropout rate is 33%.

The models are trained using stochastic gradient descent with a batch size of 256, momentum of 0.9 and categorical cross entropy as cost function. The initial learning rate is 0.1 and is reduced by a factor of 10 every time the average validation accuracy stagnates, until a lower limit for the learning rate of 0.001 is reached. All models were trained and evaluated using Keras [5] with Tensorflow backend [1]. Note, that the parameters where chosen from experience and not tuned to any extent. The goal is not to achieve a high accuracy, but to detect the uncertainty of a neural network reliably.

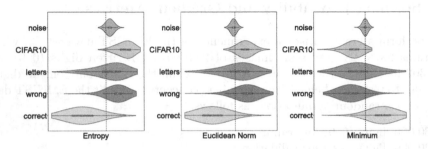

Fig. 2. Empirical distribution for entropy, euclidean norm and minimum applied to correctly predicted and incorrectly predicted digits from the test data (green and red) of one CNN. Further distributions are generated from EMNIST samples with unlearned letters (blue), CIFAR10 images (gray) and uniform noise images (purple). (Color figure online)

In this section, we study the performance of gradient metrics, the softmax baseline and the entropy in terms of AUROC and AUPR for EMNIST test data, thus considering the *error and success prediction* problem, formulated in [11]. First of all we demonstrate that gradient metrics are indeed able to provide good separations. Results for the entropy, euclidean norm and minimum are shown in Fig. 2 (green and red). Note that we have left out the mean, skewness and kurtosis metric, as their violin plots showed, that they are not suitable for a threshold meta classifier.

In what follows we define EMNISTc as the set containing all correctly classified samples of the EMNIST test set and EMNISTw as the set containing all incorrectly classified ones. From now on we resample the data splitting and use ensembles of CNNs. More precisely, the random splitting of the 280,000 digit images in training, validation and test data is repeated 10 times and we train one CNN for each splitting. In this way we train 10 CNNs that differ with respect to initial weights, training, validation and test data. We then repeat the above meta classification for each of the CNNs. With this non parametric bootstrap, we try to get as close as possible to a true sampling of the statistical law underlying the EMNIST ensemble of data and obtain results with statistic validity.

Table 1. AUROC, AUPR-In (EMNISTc as positive case) and AUPR-Out (EMNISTc as negative case) values for the threshold classification on the softmax baseline, entropy as well as selected gradient metrics. All values are in percentage and averaged over 10 differently initialized CNNs with distinct splittings of the training data. Values in brackets are the standard deviation of the mean in percentage. To get the standard deviation within the sample, multiply by $\sqrt{10}$.

Metric	EMNISTc/ EMNISTw	EMNISTc/ EMNIST letters	EMNISTc/ CIFAR10	EMNISTc/ uniform noise
AUROC				
Softmax baseline	**97.82** (0.03)	87.62 (0.20)	99.13 (0.11)	92.95 (1.86)
Entropy	97.74 (0.04)	**88.44** (0.21)	**99.24** (0.03)	**93.52** (1.83)
Absolute norm	97.77 (0.03)	87.22 (0.16)	98.19 (0.03)	90.66 (1.96)
Euclidean norm	97.78 (0.03)	87.27 (0.17)	98.38 (0.02)	91.05 (1.92)
Minimum	97.78 (0.03)	87.30 (0.20)	98.40 (0.03)	90.50 (2.16)
Maximum	97.70 (0.03)	86.92 (0.20)	98.31 (0.04)	87.05 (2.70)
Standard deviation	97.78 (0.03)	87.26 (0.17)	98.38 (0.02)	90.98 (1.93)
AUPR-In				
Softmax baseline	**99.97** (0.00)	98.39 (0.03)	**99.98** (0.00)	99.31 (0.19)
Entropy	99.97 (0.00)	98.38 (0.04)	99.95 (0.00)	**99.36** (0.19)
Absolute norm	99.97 (0.00)	**98.42** (0.02)	99.89 (0.00)	99.07 (0.21)
Euclidean norm	99.97 (0.00)	98.42 (0.02)	99.90 (0.00)	99.11 (0.21)
Minimum	99.97 (0.00)	95.20 (5.40)	99.90 (0.00)	99.05 (0.23)
Maximum	99.97 (0.00)	95.03 (5.65)	99.89 (0.00)	98.67 (0.31)
Standard deviation	99.97 (0.00)	95.04 (5.70)	99.90 (0.00)	99.11 (0.21)
AUPR-Out				
Softmax baseline	39.96 (0.57)	59.04 (0.37)	77.10 (1.90)	40.10 (4.87)
Entropy	**95.56** (0.05)	**60.36** (0.42)	**91.27** (0.39)	**42.46** (5.38)
Absolute norm	95.28 (0.06)	58.39 (0.37)	66.62 (0.35)	33.08 (3.58)
Euclidean norm	95.30 (0.06)	58.27 (0.38)	70.81 (0.52)	34.03 (3.68)
Minimum	95.36 (0.05)	58.76 (0.42)	72.72 (0.36)	33.00 (3.83)
Maximum	95.32 (0.06)	55.01 (0.41)	74.59 (0.71)	26.84 (3.02)
Standard deviation	95.30 (0.06)	58.26 (0.38)	70.75 (0.52)	33.88 (3.66)

Table 1 shows that the softmax baseline as well as some selected gradient metrics exhibit comparable performance on the test set in the error and success prediction task. Column one corresponds to the empirical distributions depicted in Fig. 2 for 200,000 test images.

In a next step we aggregate entropy and all gradient based metrics (evaluated on the gradient of each layer in the CNN) in a more sophisticated classification technique. Therefore we choose a variety of regularized and unregularized logistic regression techniques, namely a **G**eneralized **L**inear **M**odel (GLM) equipped with the logit link function, the **L**east **A**bsolute **S**hrinkage and **S**election **O**perator (LASSO) with a L^1 regularization term and a regularization parameter $\lambda_1 = 1$, the ridge regression with a L^2 regularization term and a regularization parameter of $\lambda_2 = 1$ and finally the Elastic net with one half L^1 and one half L^2 regularization, which means $\lambda_1 = \lambda_2 = 0.5$. For details about these methods, cf. [10].

To include a non linear classifier we train a feed forward NN with one hidden layer containing 15 rectified linear units (ReLUs) with L^2 weight decay of 10^{-3} and 2-way softmax output. The neural network is trained in the same fashion as the CNNs with stochastic gradient descent. Both groups of classifiers are trained on the EMNIST validation set. Results for the logistic regression techniques can be seen in Table 2 (column one) and those for the neural network in Table 3 (first row of each evaluation metric). For comparison we also include the entropy and softmax baseline in each table. The regression techniques perform equally well or better compared to the softmax baseline. This is however not true for the NN. For the logistic regression types including more features from early layers did not improve the performance, the neural network however showed improved results. This means the additional information in those layers can only be utilized by a non linear classifier.

4   Recognition of Unlearned Concepts

A (C)NN, being a statistical classifier, classifies inside the prescribed label space. In this section, we empirically test the hypothesis that test samples out of the label space will be all misclassified, however at a statistically different level of entropy or gradient metric, respectively. We test this hypothesis for three cases: First we feed the CNN with images from the EMNIST letter set and determine the entropy as well as the values for all gradient metrics for each of it. Secondly we follow the same procedure, however the inputs are gray scale CIFAR10 images coarsened to 28×28 pixels. Finally, we use uncorrelated noise that is uniformly distributed in the gray scales with the same resolution. Roughly speaking, we test empirical distributions for unlearned data that is close to the learned concept as in the case of EMNIST letters, data that represents a somewhat remote concept as in the case of CIFAR10 or, as in the case of noise, do not represent any concept at all.

We are classifying the output of a CNN on such input as incorrect label, this way we solve the *in- and out-of-distribution detection problem* from [11], but are still detecting misclassifications in the prescribed label space. The empirical distributions of unlearned concepts can be seen in Fig. 2. As we can observe, the distributions for incorrectly classified samples are in a statistical sense significantly different from those for correctly classified ones. The gradient metrics however are not able to separate the noise samples very well, but also resulting in an overall good separation of the other concepts, as for the entropy. The threshold classification evaluation metrics can be seen in Table 1. For the logistic regression results in Table 2 one can see that the GLM is inferior to the other methods. Regression techniques with a regularization term like LASSO, Ridge and Elastic net are performing best. We get similar AUROC values as for the threshold classification with single metrics, but can improve between 5% and 14.08% over the softmax baseline in terms of AUPR-Out values for unknown concepts, showing a better generalization.

Table 2. Average AUROC, AUPR-In and AUPR-Out values for different regression types trained on the validation set and all metric features including the entropy but excluding the softmax baseline. The values are averaged over 10 CNNs and displayed in percentage. The values in brackets are the standard deviations of the mean in percentage. To get the standard deviation within the sample, multiply by $\sqrt{10}$.

Metric/Regression technique	EMNISTc/ EMNISTw	EMNISTc/ EMNIST letters	EMNISTc/ CIFAR10 noise	EMNISTc/ uniform
AUROC				
Softmax baseline	**97.82** (0.03)	87.62 (0.20)	99.13 (0.11)	92.95 (1.86)
Entropy	97.74 (0.04)	88.44 (0.21)	**99.42** (0.10)	93.52 (1.83)
GLM	94.76 (0.70)	85.94 (0.46)	80.26 (5.46)	89.41 (2.90)
LASSO	97.75 (0.03)	**89.34** (0.17)	99.23 (0.03)	93.86 (1.04)
Ridge	97.59 (0.03)	88.63 (0.11)	98.93 (0.02)	**94.08** (0.67)
Elastic net	97.79 (0.06)	89.27 (0.24)	98.82 (0.06)	93.47 (0.67)
AUPR-In				
Softmax baseline	**99.97** (0.00)	**98.39** (0.03)	99.98 (0.00)	99.31 (0.19)
Entropy	99.97 (0.00)	98.38 (0.04)	**99.99** (0.00)	**99.36** (0.19)
GLM	99.81 (0.05)	96.80 (0.21)	95.51 (1.15)	97.81 (0.84)
LASSO	99.97 (0.00)	98.30 (0.06)	99.95 (0.00)	99.33 (0.12)
Ridge	99.97 (0.00)	97.86 (0.04)	99.93 (0.00)	99.36 (0.08)
Elastic net	99.97 (0.00)	98.26 (0.09)	99.92 (0.00)	99.29 (0.08)
AUPR-Out				
Softmax baseline	39.96 (0.57)	59.04 (0.37)	77.10 (1.90)	40.10 (4.87)
Entropy	**95.56** (0.05)	60.36 (0.42)	86.07 (1.76)	42.46 (5.38)
GLM	31.27 (0.79)	57.72 (0.74)	62.90 (6.77)	46.43 (5.24)
LASSO	36.27 (0.32)	**64.04** (0.26)	**91.18** (0.44)	**48.38** (3.12)
Ridge	38.17 (0.34)	61.92 (0.18)	82.95 (0.61)	47.30 (2.20)
Elastic net	38.71 (0.65)	63.43 (0.62)	79.56 (1.76)	45.03 (1.92)

5   Meta Classification with Known Unknowns

In the previous section we trained the meta classifier on the training or validation data only. This means it has no knowledge of entropy or metric distributions for unlearned concepts, hence we followed a puristic approach treating out of distribution cases as unknown unknowns. The classification accuracy could be improved, by extending the training set of the meta classifier with the entropy and gradient metric values of a few unlearned concepts and labeling them as false, i.e., incorrectly predicted. As in the previous sections we then train meta classifiers on the metrics. For this we use the same data sets as [11], namely the omniglot handwritten characters set [14], the notMNIST dataset [4] consisting of letters from different fonts, the CIFAR10 dataset [12] coarsened and converted to gray scale as well as normal and uniform noise. In order to investigate the influence of unknown concepts in the training set of the meta classifier, we used the LASSO regression and the NN introduced in Sect. 3 and supplied them with different training sets, consisting of

- – EMNIST validation set
- – EMNIST validation set and 200 uniform noise images
- – EMNIST validation set, 200 uniform noise images and 200 CIFAR10 images
- – EMNIST validation set, 200 uniform noise images, 200 CIFAR10 images and 200 omniglot images

We are omitting the results for the LASSO here, since they are inferior to those of the NN. Including known unknowns into the training set, the NN has far better performance on the unknown concepts, even though the amount of additional training data is small. Noteworthily the validation set together with only 200 uniform noise images increases the results on the AUPR-Out values for all unknown concepts already significantly by 13.74%, even comparable to using all concepts. Together with the fact, that noise is virtually available at no cost, it is a very promising candidate for improving the generalization of the meta classifier without the need of generating labels for more datasets. The in-distribution detection rate of correct and wrong predictions is also increased when using additional training concepts, making it only beneficial to include noise into the training set of the meta classifier. Our experiments show however that normal noise does not have such a high influence on the performance as uniform noise and is even decreasing the in-distribution meta classification performance. All in all we reach a 3.48% higher performance on the out of distribution examples compared to the softmax baseline in AUPR-Out and 0.81% in AUROC, whereas the increase in AUPR-In is marginal (0.12%).

Table 3. AUROC, AUPR-In (EMNISTc is positive case) and AUPR-Out (EMNISTc is negative case) values for a NN meta classifier. "All" contains omniglot, notMNIST, CIFAR10, normal noise and uniform noise. We used 200 samples of each concept that was additionally included into the training set. The supplied features are all gradient based metrics as well as the entropy. The displayed values are averages over 5 differently initialized NN meta classifiers for each of the 10 CNNs trained on the EMNIST dataset. All values are in percentage and the values in brackets are the standard deviations of the mean in percentage. To get the mean within the sample multiply by $\sqrt{10}$.

Wrong datasets	Entropy	Softmax baseline [11]	Training set for the neural network meta classifier			
			EMNIST validation	EMNIST validation + uniform noise	EMNIST validation + uniform noise + CIFAR10	EMNIST validation + uniform noise + CIFAR10 + omniglot
AUROC						
EMNISTw	97.74 (0.02)	**97.84** (0.02)	94.59 (0.17)	96.51 (0.08)	96.69 (0.08)	96.68 (0.07)
Omniglot	**98.05** (0.03)	97.84 (0.03)	94.38 (0.15)	97.29 (0.12)	97.44 (0.10)	97.84 (0.06)
notMNIST	**95.41** (0.15)	95.24 (0.15)	85.90 (0.49)	93.22 (0.46)	94.49 (0.28)	94.86 (0.22)
CIFAR10	**99.24** (0.03)	99.03 (0.04)	81.19 (1.40)	96.27 (0.63)	99.12 (0.03)	99.09 (0.03)
Normal noise	94.36 (0.54)	94.49 (0.50)	56.09 (1.56)	**98.37** (0.08)	98.34 (0.09)	98.17 (0.10)
Uniform noise	94.31 (0.84)	93.87 (0.85)	86.77 (1.16)	94.22 (0.54)	93.87 (0.71)	**94.42** (0.70)
All	96.04 (0.19)	95.83 (0.19)	80.55 (0.49)	95.49 (0.29)	96.36 (0.19)	**96.64** (0.16)
AUPR-In						
EMNISTw	99.97 (0.02)	**99.97** (0.02)	99.89 (0.17)	99.95 (0.08)	99.96 (0.08)	99.96 (0.07)
Omniglot	**99.84** (0.03)	99.82 (0.03)	99.04 (0.15)	99.73 (0.12)	99.75 (0.10)	99.80 (0.06)
notMNIST	**99.45** (0.15)	99.43 (0.15)	95.86 (0.49)	98.83 (0.46)	99.19 (0.28)	99.29 (0.22)
CIFAR10	**99.95** (0.03)	99.94 (0.04)	95.47 (1.40)	99.41 (0.63)	99.94 (0.03)	99.93 (0.03)
Normal noise	99.59 (0.54)	99.60 (0.50)	92.72 (1.56)	99.89 (0.08)	**99.89** (0.09)	99.88 (0.10)
Uniform noise	**99.65** (0.84)	99.62 (0.85)	98.05 (1.16)	99.56 (0.54)	99.53 (0.71)	99.57 (0.70)
All	98.66 (0.19)	98.59 (0.19)	84.98 (0.49)	97.72 (0.29)	98.53 (0.19)	**98.71** (0.16)
AUPR-Out						
EMNISTw	35.83 (0.30)	**39.94** (0.32)	32.95 (0.28)	36.02 (0.39)	35.98 (0.42)	35.33 (0.40)
Omniglot	**83.48** (0.21)	80.45 (0.22)	74.17 (0.38)	80.36 (0.71)	81.46 (0.60)	83.40 (0.39)
notMNIST	74.86 (0.38)	14.59 (0.06)	64.57 (0.42)	71.53 (0.79)	74.91 (0.61)	**75.13** (0.49)
CIFAR10	**91.27** (0.39)	87.38 (0.46)	54.45 (1.17)	73.93 (2.39)	90.84 (0.55)	89.82 (0.64)
Normal noise	54.98 (2.16)	57.32 (1.79)	18.57 (0.76)	**68.73** (1.73)	67.89 (1.77)	65.12 (1.74)
Uniform noise	37.97 (2.50)	36.63 (2.23)	56.66 (1.90)	58.56 (2.70)	56.59 (2.82)	**59.53** (2.94)
All	89.17 (0.40)	88.07 (0.39)	75.64 (0.48)	89.38 (0.47)	91.23 (0.35)	**91.55** (0.32)

6   Conclusion and Outlook

We introduced a new set of metrics that measures the uncertainty of deep CNNs. These metrics have a comparable performance with the widely used entropy and maximum softmax probability to meta-classify whether a certain classification proposed by the underlying CNN is presumably correct or incorrect. Here the performance is measured by AUROC, AUPR-In and AUPR-Out. Entropy and softmax probability perform equally well or slightly better than any single member of the new gradient based metrics for the detection of unknown concepts like EMNIST letters, gray scale converted CIFAR10 images and uniform noise where simple thresholding criteria are applied. But still, our new metrics allow contributions of different layers and weights to the total uncertainty. Combining the gradient metrics together with entropy in a more complex meta classifier increases the ability to identify out-of-distribution examples, so that in some cases these meta classifiers outperform the baseline. Additional calibration by including a few samples of unknown concepts increases the performance significantly. Uniform noise proved to raise the overall performance, without the need of more labels. Overall the results for the classification of correct or incorrect predictions increased when the meta classifier was supplied with more distinct concepts in the training set. It seems that the higher number of uncertainty metrics helps to better hedge the correctly classified samples from the variety of out of sample classes, which would be difficult, if only one metric is available. Note that this increase in meta classification is particularly valuable, if one does not want to deteriorate the classification performance of the underlying classifier by additional classes for the known unknowns.

As future work we want to evaluate the performance and robustness of such gradient metrics on different tasks in pattern recognition. Further features could be generated by applying the metrics to activations rather than gradients. One could also investigate the possibility of generating artificial samples, labeled as incorrect, for the training set of the meta classifier in order to further improve the results.

Acknowledgement. We thank Fabian Hüger and Peter Schlicht from Volkswagen Group Research for discussion and remarks on this work. We also thank the referees for their comments and criticism helping us to improve the paper.

References

1. Abadi, M., Agarwal, A., Barham, P., et al.: TensorFlow: large-scale machine learning on heterogeneous systems (2015). http://tensorflow.org/
2. Bendale, A., Boult, T.E.: Towards open set deep networks. CoRR abs/1511.06233 (2015). http://arxiv.org/abs/1511.06233
3. Berger, J.O.: Statistical Decision Theory and Bayesian Analysis. Springer, New York (1980). https://doi.org/10.1007/978-1-4757-4286-2
4. Bulatov, Y.: NotMNIST dataset (2011). http://yaroslavvb.blogspot.de/2011/09/notmnist-dataset.html
5. Chollet, F., et al.: Keras (2015). https://github.com/fchollet/keras
6. Cohen, G., Afshar, S., Tapson, J., van Schaik, A.: EMNIST: an extension of MNIST to handwritten letters (2017). http://arxiv.org/abs/1702.05373
7. Davis, J., Goodrich, M.: The relationship between precision-recall and ROC curves. In: Machine Learning, Proceedings of the Twenty-Third International Conference (ICML 2006), Pittsburgh, Pennsylvania, USA, 25–29 June 2006, pp. 233–240 (2006). https://doi.org/10.1145/1143844.1143874
8. Gal, Y., Ghahramani, Z.: Dropout as a Bayesian approximation: representing model uncertainty in deep learning. In: Proceedings of the 33rd International Conference on International Conference on Machine Learning, ICML 2016, vol. 48, pp. 1050–1059. JMLR.org (2016). http://dl.acm.org/citation.cfm?id=3045390.3045502

9. Goodfellow, I.J., Shlens, J., Szegedy, C.: Explaining and harnessing adversarial examples. arXiv preprint (2014). http://arxiv.org/abs/1412.6572
10. Hastie, T., Tibshirani, R., Friedman, J.: The Elements of Statistical Learning. Springer Series in Statistics. Springer, New York (2001). https://doi.org/10.1007/978-0-387-84858-7
11. Hendrycks, D., Gimpel, K.: A baseline for detecting misclassified and out-of-distribution examples in neural networks. CoRR abs/1610.02136 (2016). http://arxiv.org/abs/1610.02136
12. Krizhevsky, A.: CIFAR-10 (2009). https://www.cs.toronto.edu/~kriz/cifar.html
13. Kurakin, A., Goodfellow, I.J., Bengio, S.: Adversarial examples in the physical world. CoRR abs/1607.02533 (2016). http://arxiv.org/abs/1607.02533
14. Lake, B.M., Salakhutdinov, R., Tenenbaum, J.B.: Human-level concept learning through probabilistic program induction (2015)
15. Liang, S., Li, Y., Srikant, R.: Principled detection of out-of-distribution examples in neural networks. CoRR abs/1706.02690 (2017). http://arxiv.org/abs/1706.02690
16. Shannon, C.E.: A mathematical theory of communication. Bell Syst. Tech. J. **27**, 379–423, 623–656 (1948). http://math.harvard.edu/~ctm/home/text/others/shannon/entropy/entropy.pdf
17. Srivastava, N., Hinton, G., Krizhevsky, A., Sutskever, I., Salakhutdinov, R.: Dropout: a simple way to prevent neural networks from overfitting. J. Mach. Learn. Res. **15**, 1929–1958 (2014). http://jmlr.org/papers/v15/srivastava14a.html
18. Szegedy, C., et al.: Intriguing properties of neural networks. CoRR abs/1312.6199 (2013). http://arxiv.org/abs/1312.6199
19. Yuan, X., He, P., Zhu, Q., Bhat, R.R., Li, X.: Adversarial examples: attacks and defenses for deep learning. CoRR abs/1712.07107 (2017)

Learning Neural Models for End-to-End Clustering

Benjamin Bruno Meier[1,2]($\boxtimes$), Ismail Elezi[1,3], Mohammadreza Amirian[1,4],
Oliver Dürr[1,5], and Thilo Stadelmann[1]

[1] ZHAW Datalab & School of Engineering, Winterthur, Switzerland
benjamin.meier70@gmail.com
[2] ARGUS DATA INSIGHTS Schweiz AG, Zurich, Switzerland
[3] Ca' Foscari University of Venice, Venice, Italy
[4] Institute of Neural Information Processing, Ulm University, Ulm, Germany
[5] Institute for Optical Systems, HTWG Konstanz, Konstanz, Germany

Abstract. We propose a novel end-to-end neural network architecture
that, once trained, directly outputs a probabilistic clustering of a batch
of input examples in one pass. It estimates a distribution over the number of clusters k, and for each $1 \leq k \leq k_{\max}$, a distribution over
the individual cluster assignment for each data point. The network is
trained in advance in a supervised fashion on separate data to learn
grouping by any perceptual similarity criterion based on pairwise labels
(same/different group). It can then be applied to different data containing different groups. We demonstrate promising performance on high-
dimensional data like images (COIL-100) and speech (TIMIT). We call
this "learning to cluster" and show its conceptual difference to deep metric learning, semi-supervise clustering and other related approaches while
having the advantage of performing learnable clustering fully end-to-end.

Keywords: Perceptual grouping · Learning to cluster
Speech & image clustering

1 Introduction

Consider the illustrative task of grouping images of cats and dogs by *perceived*
similarity: depending on the intention of the user behind the task, the similarity
could be defined by animal type (foreground object), environmental nativeness
(background landscape, cp. Fig. 1) etc. This is characteristic of clustering perceptual, high-dimensional data like images [15] or sound [24]: a user typically has
some similarity criterion in mind when thinking about naturally arising groups
(e.g., pictures by holiday destination, or persons appearing; songs by mood, or
use of solo instrument). As defining such a similarity for every case is difficult,
it is desirable to learn it. At the same time, the learned model will in many
cases not be a classifier—the task will not be solved by classification—since the
number and specific type of groups present at application time are not known

© Springer Nature Switzerland AG 2018
L. Pancioni et al. (Eds.): ANNPR 2018, LNAI 11081, pp. 126–138, 2018.
https://doi.org/10.1007/978-3-319-99978-4_10

in advance (e.g., speakers in TV recordings; persons in front of a surveillance camera; object types in the picture gallery of a large web shop).

Grouping objects with machine learning is usually approached with clustering algorithms [16]. Typical ones like K-means [25], EM [14], hierarchical clustering [29] with chosen distance measure, or DBSCAN [8] each have a specific inductive bias towards certain similarity structures present in the data (e.g., K-means: Euclidean distance from a central point; DBSCAN: common point density). Hence, to be applicable to above-mentioned tasks, they need high-level features that already encode the aspired similarity measure. This may be solved by learning salient embeddings [28] with a deep metric learning approach [12], followed by an off-line clustering phase using one of the above-mentioned algorithm.

However, it is desirable to combine these distinct phases (learning salient features, and subsequent clustering) into an end-to-end approach that can be trained globally [19]: it has the advantage of each phase being perfectly adjusted to the other by optimizing a global criterion, and removes the need of manually fitting parts of the pipeline. Numerous examples have demonstrated the success of neural networks for end-to-end approaches on such diverse tasks as speech recognition [2], robot control [21], scene text recognition [34], or music transcription [35].

Fig. 1. Images of cats (top) and dogs (bottom) in urban (left) and natural (right) environments.

In this paper, we present a conceptually novel approach that we call *"learning to cluster"* in the above-mentioned sense of grouping high-dimensional data by some perceptually motivated similarity criterion. For this purpose, we define a novel neural network architecture with the following properties: (a) during training, it receives pairs of similar or dissimilar examples to learn the intended similarity function implicitly or explicitly; (b) during application, it is able to group objects of groups never encountered before; (c) it is trained end-to-end in a supervised way to produce a tailor-made clustering model and (d) is applied like a clustering algorithm to find both the number of clusters as well as the cluster membership of test-time objects in a fully probabilistic way.

Our approach builds upon ideas from *deep metric embedding*, namely to learn an embedding of the data into a representational space that allows for specific perceptual similarity evaluation via simple distance computation on feature vectors. However, it goes beyond this by adding the actual clustering step—grouping by similarity—directly to the same model, making it trainable end-to-end. Our approach is also different from *semi-supervised clustering* [4], which uses labels for some of the data points in the inference phase to guide the creation of groups. In contrast, our method uses absolutely no labels during inference, and moreover doesn't expect to have seen any of the groups it encounters during inference already during training (cp. Fig. 2). Its training stage may be compared to creating K-means, DBSCAN etc. in the first place: it creates a specific clustering model, applicable to data with certain similarity structure, and once created/trained, the model performs "unsupervised learning" in the sense of finding groups. Finally, our approach differs from traditional cluster *analysis* [16] in how the clustering algorithm is applied: instead of looking for patterns in the data in an unbiased and exploratory way, as is typically the case in unsupervised learning, our approach is geared towards the use case where users know perceptually what they are looking for, and can make this explicit using examples. We then learn appropriate features and the similarity function simultaneously, taking full advantage of end-to-end learning.

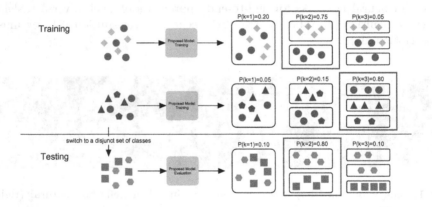

Fig. 2. Training vs. testing: cluster types encountered during application/inference are never seen in training. Exemplary outputs (right-hand side) contain a partition for each k (1–3 here) and a corresponding probability (best highlighted blue). (Color figure online)

Our main contribution in this paper is the creation of a neural network architecture that learns to *group* data, i.e., that outputs the same "label" for "similar" objects regardless of (a) it has ever seen this group before; (b) regardless of the actual value of the label (it is hence not a "class"); and (c) regardless of the number of groups it will encounter during a single application run, up to a predefined maximum. This is novel in its concept and generality (i.e., learn to cluster previously unseen groups end-to-end for arbitrary, high-dimensional input without any optimization on test data). Due to this novelty in approach, we focus here on the general idea and experimental demonstration of the principal workings, and leave comprehensive hyperparameter studies and optimizations for future work.

In Sect. 2, we compare our approach to related work, before presenting the model and training procedure in detail in Sect. 3. We evaluate our approach on different datasets in Sect. 4, showing promising performance and a high degree of generality for data types ranging from 2D points to audio snippets and images, and discuss these results with conclusions for future work in Sect. 5.

2   Related Work

Learning to cluster based on neural networks has been approached mostly as a supervised learning problem to extract embeddings for a subsequent off-line clustering phase. The core of all deep metric embedding models is the choice of the loss function. Motivated by the fact that the softmax-cross entropy loss function has been designed as a classification loss and is not suitable for the clustering problem per se, *Chopra et al.* [7] developed a "Siamese" architecture, where the loss function is optimized in a way to generate similar features for objects belonging to the same class, and dissimilar features for objects belonging to different classes. A closely related loss function called "triplet loss" has been used by *Schroff et al.* [32] to get state-of-the-art accuracy in face detection. The main difference from the Siamese architecture is that in the latter case, the network sees same and different class objects with every example. It is then optimized to jointly learn their feature representation. A problem of both approaches is that they are typically difficult to train compared to a standard cross entropy loss.

Song et al. [37] developed an algorithm for taking full advantage of all the information available in training batches. They later refined the work [36] by proposing a new metric learning scheme based on structured prediction, which is designed to optimize a clustering quality metric (normalized mutual information [27]). Even better results were achieved by *Wong et al.* [38], where the authors proposed a novel angular loss, and achieved state-of-the-art results on the challenging real-world datasets *Stanford Cars* [17] and *Caltech Birds* [5]. On the other hand, *Lukic et al.* [23] showed that for certain problems, a carefully chosen deep neural network can simply be trained with softmax-cross entropy loss and still achieve state-of-the-art performance in challenging problems like speaker clustering. Alternatively, *Wu et al.* [26] showed that state-of-the-art results can be achieved simply by using a traditional margin loss function and being careful on how sampling is performed during the creation of mini-batches.

On the other hand, attempts have been made recently that are more similar to ours in spirit, using deep neural networks only and performing clustering end-to-end [1]. They are trained in a fully unsupervised fashion, hence solve a different task then the one we motivated above (that is inspired by speaker- or image clustering based on some human notion of similarity). Perhaps first to group objects together in an unsupervised deep learning based manner where *Le et al.* [18], detecting high-level concepts like cats or humans. *Xie et al.* [40] used an autoencoder architecture to do clustering, but experimental evaluated it only simplistic datasets like *MNIST*. CNN-based approaches followed, e.g. by *Yang et al.* [42], where clustering and feature representation are optimized together. *Greff et al.* [10] performed perceptual grouping (of pixels within an image into the objects constituting the complete image, hence a different task than ours) fully unsupervised using a neural expectation maximization algorithm. Our work differs from above-mentioned works in several respects: it has no assumption on the type of data, and solves the different task of grouping whole input objects.

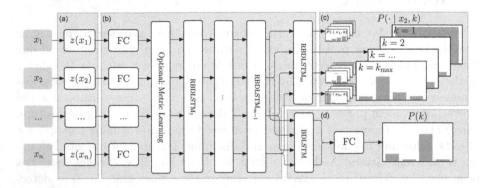

Fig. 3. Our complete model, consisting of (a) the embedding network, (b) clustering network (including an optional metric learning part, see Sect. 3.3), (c) cluster-assignment network and (d) cluster-count estimating network.

3   A Model for End-to-End Clustering of Arbitrary Data

Our method learns to cluster end-to-end purely ab initio, without the need to explicitly specify a notion of similarity, only providing the information whether two examples belong together. It uses as input $n \geq 2$ examples x_i, where n may be different during training and application and constitutes the number of objects that can be clustered at a time, i.e. the maximum number of objects in a partition. The network's output is two-fold: a probability distribution $P(k)$ over the cluster count $1 \leq k \leq k_{\max}$; and probability distributions $P(\cdot \mid x_i, k)$ over all possible cluster indexes for each input example x_i and for each k.

3.1   Network Architecture

The network architecture (see Fig. 3) allows the flexible use of different input types, e.g. images, audio or 2D points. An input x_i is first processed by an embedding network (a) that produces a lower-dimensional representation $z_i = z(x_i)$. The dimension of z_i may vary depending on the data type. For example, 2D points do not require any embedding network. A fully connected layer (FC) with LeakyReLU activation at the beginning of the clustering network (b) is then used to bring all embeddings to the same size. This approach allows to use the identical subnetworks (b)–(d) and only change the subnet (a) for any data type. The goal of the subnet (b) is to compare each input $z(x_i)$ with all other $z(x_{j \neq i})$, in order to learn an abstract grouping which is then concretized into an estimation of the number of clusters (subnet (d)) and a cluster assignment (subnet (c)).

To be able to process a non-fixed number of examples n as input, we use a recurrent neural network. Specifically, we use stacked residual bi-directional LSTM-layers (RBDLSTM), which are similar to the cells described in [39] and visualized in Fig. 4. The residual connections allow a much more effective gradient flow during training [11] and avoid vanishing gradients. Additionally, the network can learn to use or bypass certain layers using the residual connections, thus reducing the architectural decision on the number of recurrent layers to the simpler one of finding a reasonable upper bound.

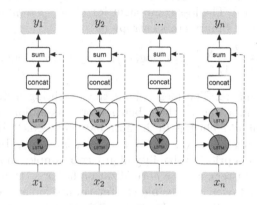

Fig. 4. RBDLSTM-layer: A BDLSTM with residual connections (dashed lines). The variables x_i and y_i are named independently from the notation in Fig. 3.

The first of overall two outputs is modeled by the cluster assignment network (c). It contains a softmax-layer to produce $P(\ell \mid x_i, k)$, which assigns a cluster index ℓ to each input x_i, given k clusters (i.e., we get a distribution over possible cluster assignments for each input and every possible number of clusters). The second output, produced by the cluster-count estimating network (d), is built from another BDLSTM-layer. Due to the bi-directionality of the network, we concatenate its first and the last output vector into a fully connected layer of twice as many units using again LeakyReLUs. The subsequent softmax-activation finally models the distribution $P(k)$ for $1 \leq k \leq k_{\max}$. The next subsection shows how this neural network learns to approximate these two complicated probability distributions [20] purely from pairwise constraints on data that is completely separate from any dataset to be clustered. No labels for clustering are needed.

3.2 Training and Loss

In order to define a suitable loss-function, we first define an approximation (assuming independence) of the probability that x_i and x_j are assigned to the same cluster for a given k as

$$P_{ij}(k) = \sum_{\ell=1}^{k} P(\ell \mid x_i, k) P(\ell \mid x_j, k).$$

By marginalizing over k, we obtain P_{ij}, the probability that x_i and x_j belong to the same cluster:

$$P_{ij} = \sum_{k=1}^{k_{\max}} P(k) \sum_{\ell=1}^{k} P(\ell \mid x_i, k) P(\ell \mid x_j, k).$$

Let $y_{ij} = 1$ if x_i and x_j are from the same cluster (e.g., have the same group label) and 0 otherwise. The loss component for *cluster assignments*, L_{ca}, is then given by the weighted binary cross entropy as

$$L_{\text{ca}} = \frac{-2}{n(n-1)} \sum_{i<j} (\varphi_1 y_{ij} \log(P_{ij}) + \varphi_2 (1 - y_{ij}) \log(1 - P_{ij}))$$

with weights φ_1 and φ_2. The idea behind the weighting is to account for the imbalance in the data due to there being more dissimilar than similar pairs (x_i, x_j) as the number of clusters in the mini batch exceeds 2. Hence, the weighting is computed using $\varphi_1 = c\sqrt{1 - \varphi}$ and $\varphi_2 = c\sqrt{\varphi}$, with φ being the expected value of y_{ij} (i.e., the a priori probability of any two samples in a mini batch coming from the same cluster), and c a normalization factor so that $\varphi_1 + \varphi_2 = 2$. The value φ is computed over all possible cluster counts for a fixed input example count n, as during training, the cluster count is randomly chosen for each mini batch according to a uniform distribution. The weighting of the cross entropy given by φ is then used to make sure that the network does not converge to a sub-optimal and trivial minimum. Intuitively, we thus account for permutations in the sequence of examples by checking rather for pairwise correctness (probability of same/different cluster) than specific indices.

The second loss term, L_{cc}, penalizes a wrong *number of clusters* and is given by the categorical cross entropy of $P(k)$ for the true number of clusters k in the current mini batch:

$$L_{\text{cc}} = -\log(P(k)).$$

The complete loss is given by $L_{\text{tot}} = L_{\text{cc}} + \lambda L_{\text{ca}}$. During training, we prepare each mini batch with N sets of n input examples, each set with $k = 1 \ldots k_{\max}$ clusters chosen uniformly. Note that this training procedure requires only the knowledge of y_{ij} and is thus also possible for weakly labeled data. All input examples are randomly shuffled for training and testing to avoid that the network learns a bias w.r.t. the input order. To demonstrate that the network really learns an intra-class distance and not just classifies objects of a fixed set of classes, it is applied on totally different clusters at evaluation time than seen during training.

3.3 Implicit vs. Explicit Distance Learning

To elucidate the importance and validity of the implicit learning of distances in our subnetwork (b), we also provide a modified version of our network architecture for comparison, in which the calculation of the distances is done explicitly. Therefore, we add an extra component to the network before the RBDLSTM layers, as can be seen in Fig. 3: the optional metric learning block receives the fixed-size embeddings from the fully connected layer after the embedding network (a) as input and outputs the pairwise distances of the data points. The recurrent layers in block (b) then subsequently cluster the data points based on this pairwise distance information [3,6] provided by the metric learning block.

We construct a novel metric learning block inspired by the work of *Xing et al.* [41]. In contrast to their work, we optimize it end-to-end with backpropagation. This has been proposed in [33] for classification alone; we do it here for a clustering task, for the whole covariance matrix, and jointly with the rest of our network. We construct the non-symmetric, non-negative dissimilarity measure d_A^2 between two data points x_i and x_j as

$$d_A^2(x_i, x_j) = (x_i - x_j)^T A(x_i - x_j)$$

and let the neural network training optimize A through L_{tot} without intermediate losses. The matrix A as used in d_A^2 can be thought of as a trainable distance metric. In every training step, it is projected into the space of positive semidefinite matrices.

4 Experimental Results

To assess the quality of our model, we perform clustering on three different datasets: for a proof of concept, we test on a set of generated *2D points* with a high variety of shapes, coming from different distributions. For speaker clustering, we use the *TIMIT* [9] corpus, a dataset of studio-quality speech recordings frequently used for pure speaker clustering in related work. For image clustering, we test on the *COIL-100* [30] dataset, a collection of different isolated objects in various orientations. To compare to related work, we measure the performance with the standard evaluation scores misclassification rate (MR) [22] and normalized mutual information (NMI) [27]. Architecturally, we choose $m = 14$ BDLSTM layers and 288 units in the FC layer of subnetwork (b), 128 units for the BDLSTM in subnetwork (d), and $\alpha = 0.3$ for all LeakyReLUs in the experiments below. All hyperparameters where chosen based on preliminary experiments to achieve reasonable performance, but not tested nor tweaked extensively. The code and further material and experiments are available online[1].

We set $k_{\max} = 5$ and $\lambda = 5$ for all experiments. For the 2D point data, we use $n = 72$ inputs and a batch-size of $N = 200$ (We used the batch size of $N = 50$ for metric learning with 2D points). For TIMIT, the network input consists of

---

[1] See https://github.com/kutoga/learning2cluster.

$n = 20$ audio snippets with a length of 1.28 s, encoded as mel-spectrograms with 128×128 pixels (identical to [24]). For COIL-100, we use $n = 20$ inputs with a dimension of $128 \times 128 \times 3$. For TIMIT and COIL-100, a simple CNN with 3 conv/max-pooling layers is used as subnetwork (a). For TIMIT, we use 430 of the 630 available speakers for training (and 100 of the remaining ones each for validation and evaluation). For COIL-100, we train on 80 of the 100 classes (10 for validation, 10 for evaluation). For all runs, we optimize using Adadelta [43] with a learning rate of 5.0. Example clusterings are shown in Fig. 5. For all configurations, the used hardware set the limit on parameter values: we used the maximum possible batch size and values for n and k_{max} that allow reasonable training times. However, values of $n \geq 1000$ where tested and lead to a large decrease in model accuracy. This is a major issue for future work.

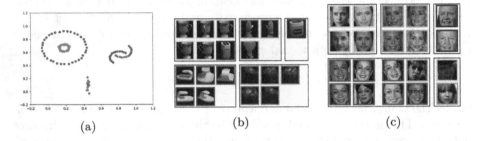

(a)   (b)   (c)

Fig. 5. Clustering results for (a) 2D point data, (b) COIL-100 objects, and (c) faces from FaceScrub (for illustrative purposes). The color of points/colored borders of images depict true cluster membership. (Color figure online)

Table 1. NMI $\in [0,1]$ and MR $\in [0,1]$ averaged over 300 evaluations of a trained network. We abbreviate our "learning to cluster" method as "L2C".

	2D points (self generated)		TIMIT		COIL-100	
	MR	NMI	MR	NMI	MR	NMI
L2C (=our method)	0.004	0.993	0.060	0.928	0.116	0.867
L2C + Euclidean	0.177	0.730	0.093	0.883	0.123	0.884
L2C + Mahalanobis	0.185	0.725	0.104	0.882	0.093	0.890
L2C + Metric Learning	0.165	0.740	0.101	0.880	0.100	0.880
Random cluster assignment	0.485	0.232	0.435	0.346	0.435	0.346
Baselines (related work)	k-Means: MR = 0.178, NMI = 0.796 DBSCAN: MR = 0.265, NMI = 0.676		[24]: MR = 0		[42]: NMI = 0.985	

The results on 2D data as presented in Fig. 5a demonstrate that our method is able to learn specific and diverse characteristics of intuitive groupings. This is superior to any single traditional method, which only detects a certain class of cluster structure (e.g., defined by distance from a central point). Although [24] reach moderately better scores for the speaker clustering task and [42] reach a superior NMI for COIL-100, our method finds reasonable clusterings, is more flexible through end-to-end training and is not tuned to a specific kind of data. Hence, we assume, backed by the additional experiments to be found online, that our model works well also for other data types and datasets, given a suitable embedding network. Table 1 gives the numerical results for said datasets in the row called "L2C" without using the explicit metric learning block. Extensive preliminary experiments on other public datasets like e.g. FaceScrub [31] confirm these results: learning to cluster reaches promising performance while not yet being on par with tailor-made state-of-the-art approaches.

We compare the performance of our implicit distance metric learning method to versions enhanced by different explicit schemes for pairwise similarity computation prior to clustering. Specifically, three implementations of the optional metric learning block in subnetwork (b) are evaluated: using a fixed diagonal matrix A (resembling the Euclidean distance), training a diagonal A (resembling Mahalanobis distance), and learning the entire coefficients of the distance matrix A. Since we argue above that our approach combines *implicit* deep metric embedding with clustering in an end-to-end architecture, one would not expect that adding *explicit* metric computation changes the results by a large extend. This assumption is largely confirmed by the results in the "L2C+..." rows in Table 1: for COIL-100, Euclidean gives slightly worse, and the other two slightly better results than L2C alone; for TIMIT, all results are worse but still reasonable. We attribute the considerable performance drop on 2D points using all three explicit schemes to the fact that in this case much more instances are to be compared with each other (as each instance is smaller than e.g. an image, n is larger). This might have needed further adaptations like e.g. larger batch sizes (reduced here to $N = 50$ for computational reasons) and longer training times.

5   Discussion and Conclusions

We have presented a novel approach to learn neural models that directly output a probabilistic clustering on previously unseen groups of data; this includes a solution to the problem of outputting similar but unspecific "labels" for similar objects of unseen "classes". A trained model is able to cluster different data types with promising results. This is a complete end-to-end approach to clustering that learns both the relevant features and the "algorithm" by which to produce the clustering itself. It outputs probabilities for cluster membership of all inputs as well as the number of clusters in test data. The learning phase only requires pairwise labels between examples from a separate training set, and no explicit similarity measure needs to be provided. This is especially useful for high-dimensional, perceptual data like images and audio, where similarity is usually semantically defined by humans. Our experiments confirm that our algorithm is able to implicitly learn a metric and directly use it for the included clustering.

This is similar in spirit to the very recent work of *Hsu et al.* [13], but does not need and optimization on the test (clustering) set and finds k autonomously. It is a novel approach to *learn to cluster*, introducing a novel architecture and loss design.

We observe that the clustering accuracy depends on the availability of a large number of different classes during training. We attribute this to the fact that the network needs to learn intra-class distances, a task inherently more difficult than just to distinguish between objects of a fixed amount of classes like in classification problems. We understand the presented work as an early investigation into the new paradigm of learning to cluster by perceptual similarity specified through examples. It is inspired by our work on speaker clustering with deep neural networks, where we increasingly observe the need to go beyond surrogate tasks for learning, training end-to-end specifically for clustering to close a performance leak. While this works satisfactory for initial results, points for improvement revolve around scaling the approach to practical applicability, which foremost means to get rid of the dependency on n for the partition size.

The number n of input examples to assess simultaneously is very relevant in practice: if an input data set has thousands of examples, incoherent single clusterings of subsets of n points would be required to be merged to produce a clustering of the whole dataset based on our model. As the (RBD) LSTM layers responsible for assessing points simultaneously in principle have a long, but still local (short-term) horizon, they are not apt to grasp similarities of thousands of objects. Several ideas exist to change the architecture, including to replace recurrent layers with temporal convolutions, or using our approach to seed some sort of differentiable K-means or EM layer on top of it. Preliminary results on this exist. Increasing n is a prerequisite to also increase the maximum number of clusters k, as $k \ll n$. For practical applicability, k needs to be increased by an order of magnitude; we plan to do this in the future. This might open up novel applications of our model in the area of transfer learning and domain adaptation.

Acknowledgements. We thank the anonymous reviewers for helpful feedback.

References

1. Aljalbout, E., Golkov, V., Siddiqui, Y., Cremers, D.: Clustering with deep learning: taxonomy and new methods. arXiv preprint arXiv:1801.07648 (2018)
2. Amodei, D., et al.: Deep speech 2: end-to-end speech recognition in English and Mandarin. In: ICML, pp. 173–182 (2016)
3. Arias-Castro, E.: Clustering based on pairwise distances when the data is of mixed dimensions. IEEE Trans. Inf. Theory **57**, 1692–1706 (2011)
4. Basu, S., Banerjee, A., Mooney, R.: Semi-supervised clustering by seeding. In: ICML, pp. 19–26 (2002)
5. Branson, S., Horn, G.V., Wah, C., Perona, P., Belongie, S.J.: The ignorant led by the blind: a hybrid human-machine vision system for fine-grained categorization. IJCV **108**, 3–29 (2014)
6. Chin, C.F., Shih, A.C.C., Fan, K.C.: A novel spectral clustering method based on pairwise distance matrix. J. Inf. Sci. Eng. **26**, 649–658 (2010)

7. Chopra, S., Hadsell, R., LeCun, Y.: Learning a similarity metric discriminatively, with application to face verification. In: CVPR, vol. 1, pp. 539–546 (2005)
8. Ester, M., Kriegel, H.P., Sander, J., Xu, X.: A density-based algorithm for discovering clusters in large spatial databases with noise. In: KDD, pp. 226–231 (1996)
9. Garofolo, J.S., Lamel, L.F., Fisher, W.M., Fiscus, J.G., Pallett, D.S., Dahlgren, N.L.: DARPA TIMIT acoustic phonetic continuous speech corpus CDROM (1993)
10. Greff, K., van Steenkiste, S., Schmidhuber, J.: Neural expectation maximization. In: NIPS, pp. 6694–6704 (2017)
11. He, K., Zhang, X., Ren, S., Sun, J.: Deep residual learning for image recognition. In: CVPR, pp. 770–778 (2016)
12. Hoffer, E., Ailon, N.: Deep metric learning using triplet network. In: Feragen, A., Pelillo, M., Loog, M. (eds.) SIMBAD 2015. LNCS, vol. 9370, pp. 84–92. Springer, Cham (2015). https://doi.org/10.1007/978-3-319-24261-3_7
13. Hsu, Y., Lv, Z., Kira, Z.: Learning to cluster in order to transfer across domains and tasks. In: ICLR (2018, accepted)
14. Jin, X., Han, J.: Expectation maximization clustering. In: Sammut, C., Webb, G.I. (eds.) Encyclopedia of Machine Learning, pp. 382–383. Springer, Boston (2011). https://doi.org/10.1007/978-0-387-30164-8
15. Kampffmeyer, M., Løkse, S., Bianchi, F.M., Livi, L., Salberg, A.B., Robert, J.: Deep divergence-based clustering. In: IEEE International Workshop on Machine Learning for Signal Processing (MLSP) (2017)
16. Kaufman, L., Rousseeuw, P.J.: Finding Groups in Data: An Introduction to Cluster Analysis. Wiley, Hoboken (1990)
17. Krause, J., Stark, M., Deng, J., Li, F.F.: 3D object representations for fine-grained categorization. In: Workshop on 3D Representation and Recognition at ICCV (2013)
18. Le, Q.V., et al.: Building high-level features using large scale unsupervised learning. In: ICML, pp. 8595–8598 (2012)
19. LeCun, Y., Bottou, L., Bengio, Y., Haffner, P.: Gradient-based learning applied to document recognition. Proc. IEEE **86**, 2278–2324 (1998)
20. Lee, H., Ge, R., Ma, T., Risteski, A., Arora, S.: On the ability of neural nets to express distributions. In: COLT, pp. 1271–1296 (2017)
21. Levine, S., Finn, C., Darrell, T., Abbeel, P.: End-to-end training of deep visuomotor policies. JMLR **17**(1), 1334–1373 (2016)
22. Liu, D., Kubala, F.: Online speaker clustering. In: ICASSP, vol. 1, pp. I-333-6 (2003)
23. Lukic, Y., Vogt, C., Dürr, O., Stadelmann, T.: Speaker identification and clustering using convolutional neural networks. In: IEEE International Workshop on Machine Learning for Signal Processing (MLSP) (2016)
24. Lukic, Y., Vogt, C., Dürr, O., Stadelmann, T.: Learning embeddings for speaker clustering based on voice equality. In: 2017 IEEE 27th International Workshop on Machine Learning for Signal Processing (MLSP) (2017)
25. MacQueen, J.: Some methods for classification and analysis of multivariate observations. In: 5th Berkeley Symposium on Mathematical Statistics and Probability, pp. 281–297 (1967)
26. Manmatha, R., Wu, C., Smola, A.J., Krähenbühl, P.: Sampling matters in deep embedding learning. In: ICCV, pp. 2840–2848 (2017)
27. McDaid, A.F., Greene, D., Hurley, N.: Normalized mutual information to evaluate overlapping community finding algorithms. arXiv preprint arXiv:1110.2515 (2011)
28. Mikolov, T., Chen, K., Corrado, G., Dean, J.: Efficient estimation of word representations in vector space. arXiv preprint arXiv:1301.3781 (2013)

29. Murtagh, F.: A survey of recent advances in hierarchical clustering algorithms. Comput. J. **26**, 354–359 (1983)
30. Nayar, S., Nene, S., Murase, H.: Columbia object image library (COIL 100). Department of Computer Science, Columbia University, Technical report, CUCS-006-96 (1996)
31. Ng, H.W., Winkler, S.: A data-driven approach to cleaning large face datasets. In: ICIP, pp. 343–347 (2014)
32. Schroff, F., Kalenichenko, D., Philbin, J.: FaceNet: A unified embedding for face recognition and clustering. In: CVPR, pp. 815–823 (2015)
33. Schwenker, F., Kestler, H.A., Palm, G.: Three learning phases for radial-basis-function networks. Neural Netw. **14**(4–5), 439–458 (2001)
34. Shi, B., Bai, X., Yao, C.: An end-to-end trainable neural network for image-based sequence recognition and its application to scene text recognition. PAMI **39**, 2298–2304 (2017)
35. Sigtia, S., Benetos, E., Dixon, S.: An end-to-end neural network for polyphonic piano music transcription. IEEE/ACM TASLP **24**(5), 927–939 (2016)
36. Song, H.O., Jegelka, S., Rathod, V., Murphy, K.: Deep metric learning via facility location. In: CVPR, pp. 5382–5390 (2017)
37. Song, H.O., Xiang, Y., Jegelka, S., Savarese, S.: Deep metric learning via lifted structured feature embedding. In: CVPR, pp. 4004–4012 (2016)
38. Wang, J., Zhou, F., Wen, S., Liu, X., Lin, Y.: Deep metric learning with angular loss. In: ICCV, pp. 2593–2601 (2017)
39. Wu, Y., et al.: Google's neural machine translation system: bridging the gap between human and machine translation. arXiv preprint arXiv:1609.08144 (2016)
40. Xie, J., Girshick, R., Farhadi, A.: Unsupervised deep embedding for clustering analysis. In: ICML, pp. 478–487 (2016)
41. Xing, E.P., Jordan, M.I., Russell, S.J., Ng, A.Y.: Distance metric learning with application to clustering with side-information. In: NIPS, pp. 521–528 (2003)
42. Yang, J., Parikh, D., Batra, D.: Joint unsupervised learning of deep representations and image clusters. In: CVPR, pp. 5147–5156 (2016)
43. Zeiler, M.D.: ADADELTA: an adaptive learning rate method. arXiv preprint arXiv:1212.5701 (2012)

A k-Nearest Neighbor Based Algorithm for Multi-Instance Multi-Label Active Learning

Adrian T. Ruiz$^{(\boxtimes)}$, Patrick Thiam, Friedhelm Schwenker, and Günther Palm

Institute of Neural Information Processing, Ulm University,
James-Franck-Ring, 89081 Ulm, Germany
adrian.ruiz@uni-ulm.de

Abstract. Multi-instance multi-label learning (MIML) is a framework in machine learning in which each object is represented by multiple instances and associated with multiple labels. This relatively new approach has achieved success in various applications, particularly those involving learning from complex objects. Because of the complexity of MIML, the cost of data labeling increases drastically along with the improvement of the model performance. In this paper, we introduce a MIML active learning approach to reduce the labeling costs of MIML data without compromising the model performance. Based on a query strategy, we select and request from the Oracle the label set of the most informative object. Our approach is formulated in a pool-based scenario and uses MIML-kNN as the base classifier. This classifier for MIML is based on the k-Nearest Neighbor algorithm and has achieved superior performance in different data domains. We proposed novel query strategies and also implemented previously used query strategies for MIML learning. Finally, we conducted an experimental evaluation on various benchmark datasets. We demonstrate that these approaches can achieve significantly improved results than without active selection for all datasets on various evaluation criteria.

Keywords: Multi-instance · Multi-label · Active learning
k nearest neighbors · Partially supervised learning
Acoustic classification of birds · Text categorization
Scene classification

1 Introduction

In standard supervised learning, an object consists of a single instance, represented by a feature vector, and is associated with a single class label. This framework is known as single-instance single-label (SISL) learning. The goal of SISL learning is to train a classifier model which learns from training instances how to assign a class label to any feature vector. However, in many real applications, such a learning framework is less convenient to model complex objects,

© Springer Nature Switzerland AG 2018
L. Pancioni et al. (Eds.): ANNPR 2018, LNAI 11081, pp. 139–151, 2018.
https://doi.org/10.1007/978-3-319-99978-4_11

which intrinsic representation is a collection of instances. Likewise, these complex objects may also be associated simultaneously with multiple class labels. For example, a scene image may comprise images of mountains, lakes, and trees, and we may associate it with the labels *Landscape* and *Summer* at the same time. If we extract a single instance to represent it, some useful information may get lost. In another approach, we can segment the image into multiple regions and extract one instance from each region of interest. Another example could be in text categorization tasks where a document may be annotated with multiple labels. To fully exploit the content with multiple topics, it would be more advantageous if we represent each paragraph with one instance. Zhou and Zhang [22] introduced multi-instance multi-label (MIML) learning, where each object is represented by a *bag of multiple instances* (feature vectors with fixed-length), and each object is associated with a *set of class labels*. Several algorithms for MIML have been proposed and achieved better performance in image and text classification, in comparison to conventional methods adapted for MIML classification. Other successful applications include genome protein function prediction [18], gene expression patterns annotation [20], relationship extraction [15], video understanding [19], classification of bird species [1,2], and predicting tags for web pages [14].

In most cases of supervised learning, it is necessary to use large amounts of training examples to obtain accurate models. Nevertheless, it is a typical situation that the costs of manually labeled data are expensive or time-consuming. Active learning is an approach of a partially-supervised learning algorithm [3,4,10] that reduces the required amount of training data without compromising the model performance. This goal is accomplished by selecting the most informative examples from the unlabeled examples and query their label from an oracle (expert). Pool-based sampling is the most common scenario in active learning in which queries are drawn from a static or closed pool of unlabeled examples. Many active learning strategies have been proposed to estimate the informativeness of unlabeled samples [13,17]. These query strategies are based on different measures, e.g., uncertainty, expected error reduction and information density. A comprehensive literature survey on query strategies is provided by Settles [12].

For MIML datasets, the cost of labeled data depends on the maximum amount of possible labels for a bag of instances. In some applications, MIML provides a major advantage because it is easier or less costly to obtain labels at the bag-level than at instance-level. Nevertheless, because of their multiplicity in the input and output spaces, the required amount of training data to improve the accuracy model increases dramatically. For this reason, it is of great interest to implement active learning algorithms in a MIML framework. Currently, few studies have proposed active learning methods for MIML. Retz and Schwenker [9] use MIMLSVM [23] as the base classifier in which the MIML data is reduced to a bag-level output vector. This representation is later used to formulate an active learning strategy. Another proposed method uses MIMLFAST as base classifiers and the approach actively queries the most valuable information by exploiting diversity and uncertainty in both the input and output spaces [5].

The efficiency of an active learning algorithm relies not only on the query strategy design but also on the selection of the base classifier. Two of the most commonly used classifiers are MIMLBOOST and MIMLSVM [22,23]. Nevertheless, MIMLBOOST can handle only small datasets and does not yield good performance in general [6]. MIMLSVM reaches a satisfying classification accuracy for text and image, but usually not for other types of data sets [1,6]. A better alternative is MIML-kNN[21] (Multi-Instance Multi-Label k-Nearest Neighbor) which combines the well-known k-Nearest Neighbor technique with MIML. Given a test example, MIML-kNN not only considers its κ neighbors but also considers its κ' citers, i.e., examples that consider the test example within their κ' nearest neighbors. The identification of neighbors and citers relies on the Hausdorff distance which is an estimation of the distances between bags. One advantage of using MIML-kNN with pool-based sampling is that the distance between all bags (i.e., labeled and unlabeled bags) can be precomputed and stored for later use in any model learning or prediction. Beside this, MIML-kNN classifiers have achieve a superior performance than the MIMLSVM and MIMLBOOST for different types of data such as text [11,21], image [21,22], and bio-acoustic data [1].

In this paper, we introduce an active multi-instance multi-label learning approach within a pool-based scenario and use MIML-kNN as the base classifier. This method aims to reduce the amount of training MIML data needed to achieve the highest possible classification performance. This paper presents two major contributions to active learning and MIML learning. First, we motivate and introduce several new query strategies within the MIML framework. Later we conduct an empirical study of our proposed active learning methods on a variety of benchmark MIML data.

The remainder of this paper is organized as follows. Section 2 describes in detail the proposed approach. Section 3 describes the experiments and presents their results, followed by conclusions in Sect. 4.

2 Method

2.1 MIML Framework

In a MIML framework, an example X consists of a *bag* of instances $X = \{\mathbf{x}_j\}_{j=1}^m$ where m is the number of instances and each instance $\mathbf{x}_j = [x_1, \ldots, x_D]$ is a D-dimensional feature vector. The number of instances m can variate among bags. In this framework, each bag X can be associated to one or more labels and they are represented by a *label set* $Y = \{y_k\}$ where $k \in \{1, \ldots, K\}$. For our purposes, Y is represented by a *label indicator vector* $\mathbf{I} = [I_1, \ldots, I_K]$ where the entry $I_k = 1$ if $y_k \in Y$ and $I_k = 0$ otherwise. Given a fully labeled training set $\mathcal{L} = \{(X_l, Y_l)\}_{l=1}^L$, the learning task in a MIML framework is to train a classification model which is a function $h : 2^{\mathcal{X}} \to 2^{\mathcal{Y}}$ that maps a set of instances $X \in \mathcal{X}$ to a set of labels $Y \in \mathcal{Y}$.

MIML algorithms such as MIMLSVM, MIMLRBF and MIML-kNN reduce the MIML problem to a single-instance multi-label problem by associating each bag X with a bag-level feature vector $\mathbf{z}(X) \in \mathbb{R}^K$ which combines information from the instances in the bag. Each algorithm uses different approaches to compute a bag-level feature vector. Nevertheless all these methods heavily depend on the use of some form of bag-level distance measure. The most common choice is the Hausdorff distance $D_H(X, X')$. Retz and Schwenker [9] examined several variations of this distance. For this paper we consider the maximum D_H^{max}, median D_H^{med} and average D_H^{avg} Hausdorff distances defined as:

$$D_H^{max}(X, X') = \max \left\{ \max_{\mathbf{x} \in X} \min_{\mathbf{x}' \in X'} d(\mathbf{x}, \mathbf{x}'), \max_{\mathbf{x}' \in X'} \min_{\mathbf{x} \in X} d(\mathbf{x}, \mathbf{x}') \right\} \tag{1a}$$

$$D_H^{med}(X, X') = \frac{1}{2} \left(\operatorname*{median}_{\mathbf{x} \in X} \min_{\mathbf{x}' \in X'} d(\mathbf{x}, \mathbf{x}'), \operatorname*{median}_{\mathbf{x}' \in X'} \min_{\mathbf{x} \in X} d(\mathbf{x}, \mathbf{x}') \right) \tag{1b}$$

$$D_H^{avg}(X, X') = \frac{1}{|X| + |X'|} \left(\sum_{\mathbf{x} \in X} \min_{\mathbf{x}' \in X'} d(\mathbf{x}, \mathbf{x}') + \sum_{\mathbf{x}' \in X'} \min_{\mathbf{x} \in X} d(\mathbf{x}, \mathbf{x}') \right) \tag{1c}$$

where $d(\mathbf{x}, \mathbf{x}') = \|\mathbf{x} - \mathbf{x}'\|$ is the Euclidean distance between instances.

2.2 MIML-kNN

In the following we describe MIML-kNN algorithm [21]. Given an example bag X and a training set $\mathcal{L} = \{(X_l, Y_l)\}$, first we identify in the training bags $\mathcal{X}_L = \{X_l\}$, the κ *nearest neighbors*, and the κ' *citers* of X by employing the Hausdorff metric $D_H(X, X')$. This means that we have to identify the neighbors set $\mathcal{N}_\kappa(X)$ and the citers set $\mathcal{C}_{\kappa'}(X)$. These sets are defined as follows

$$\mathcal{N}_\kappa(X) = \{A | A \text{ is one of } X\text{'s } \kappa \text{ nearest neighbors in } \mathcal{X}_\mathcal{L}\} \tag{2a}$$

$$\mathcal{C}_{\kappa'}(X) = \{B | X \text{ is one of } B\text{'s } \kappa' \text{ nearest neighbors in } \mathcal{X}_\mathcal{L} \cup \{X\}\} \tag{2b}$$

The citers bags are the bags that consider X to be one of their κ' nearest neighbors. After the computation of $\mathcal{N}_\kappa(X)$ and $\mathcal{C}_{\kappa'}(X)$, we defined a *labeling counter vector* $\mathbf{z}(X) = [z_1(X), \ldots, z_K(X)]$ where the entry $z_k(X)$ is the number of bags in $\mathcal{Z}(X) = \mathcal{N}_\kappa(X) \cup \mathcal{C}_{\kappa'}(X)$ that include label y_k in their label set. Using the binary label vector $\mathbf{I}(X)$, $\mathbf{z}(X)$ is defined as

$$\mathbf{z}(X) = \sum_{X' \in \mathcal{Z}(X)} \mathbf{I}(X') \tag{3}$$

Later, the information contained in $\mathbf{z}(X)$ is used to obtain the *predicted label set* $\hat{Y}$ associated to X by employing a *prediction function* $\mathbf{f}(X) = [f_1(X), \ldots, f_K(X)]$ such that

$$f_k(X) = \mathbf{w}_k^\top \cdot \mathbf{z}(X) \tag{4}$$

where $\mathbf{w}_k^\top$ is the kth transposed column of the *weight matrix* $\mathbf{W} = [\mathbf{w}_1, \ldots, \mathbf{w}_K]$. The classification rule is that the label $\hat{y}_k$ belongs to the *predicted label set*

$\hat{Y}(X) = \{\hat{y}_k\}$ only if $f_k(X) > 0$. Hence, for the *predicted indicator vector* $\hat{\mathbf{I}}(X) = \left[\hat{I}_1, \ldots, \hat{I}_K\right]$ the entry $\hat{I}_k = 1$ if $f_k(X) > 0$ and $\hat{I}_k = 0$ otherwise. The values of $\mathbf{W}$ are computed using a linear classification approach by minimizing the following sum-of-squares error function

$$E = \frac{1}{2} \sum_{l=1}^{L} \sum_{k=1}^{K} \left(w_K^\top \cdot \mathbf{z}(X_l) - y_k(X_l)\right)^2 \tag{5}$$

This error minimization implies to solve the weight matrix $\mathbf{W}$ as in a least sum-of-squares problem of the form $\left(\mathbf{Z}^\top \mathbf{Z}\right) \mathbf{W} = \mathbf{Z}^\top \mathbf{Y}$. In this case, the matrix $\mathbf{W}$ is computed using a linear matrix inversion technique of singular value decomposition.

2.3   Active Learning

In this part, we present the strategies of active learning for a multi-instance multi-label data set using MIML-kNN as the base classifier. Initially we have a set of *labeled data* $\mathcal{L} = \{(X_l, Y_l)\}$ with L labeled bags and a set of *unlabeled data* $\mathcal{U} = \{X_u\}$ with U unlabeled bags. In an active learning scenario, usually the amount of unlabeled data is much larger than the amount of labeled data, i.e. $U \gg L$. The main task of an active learning algorithm is to select the *most informative bag* X^* according to some *query strategy* $\phi(X)$, which is a function evaluated on each example X from the pool $\mathcal{U}$. In this work, the selection of the bag X^* is done according to

$$X^* = \operatorname*{argmax}_{X \in \mathcal{U}} \phi(X) \tag{6}$$

Algorithm 1 describes the pool-based active learning algorithm for training a MIML-kNN model. One advantage of using MIML-kNN with pool-based sampling, is that, the distance between all bags (i.e. labeled and unlabeled bags) can be precomputed and stored for later use in any model learning or prediction task. As in Algorithm 1, first we calculated the *bag distance matrix* $\mathbf{D}$ such that $d_{ij} = D_H(X_i, X_j)$ for all bags X_i, X_j. Then from this matrix we can extract the distance submatrix $\mathbf{D}_{\mathcal{L}}$ of the labeled bags and use it in the training of a MIML-kNN model (see Eq. 5). For classification of the bag X, we have to feed the trained MIML-kNN model with the subtracted matrix $\mathbf{D}_{\mathcal{L} \cup \{X\}}$ (see Eq. 2). In the following, we describe in detail the query strategies we proposed which will be later compared in an empirical study.

Uncertainty Sampling (Unc). This approach is one of the most common in SISL framework. Here a learner queries the instance that is most uncertain how to label. For a muti-label problem we define the uncertainty as $\phi(X) = 1 - P(\hat{Y}|X)$ where $P(\hat{Y}|X)$ is the *bag posterior probability* for the predicted label set $\hat{Y}$ given the bag X. We calculate $P(\hat{Y}|X)$ as the probability given the

Algorithm 1. Active kMIML

___

   input:
     $\mathcal{L}$: Labeled data set $\{(X_l, Y_l)\}$
     $\mathcal{U}$: Unlabeled data set $\{X_u\}$
     κ : Neighbors parameter
     κ': Citers parameter

   output:
     h : MIML-kNN model

1 **begin**
2   Calculate the distance matrix $\mathbf{D}$ using $D_H (X_i, X_j)$ for all bags $X_i, X_j \in \{X_u, X_l\}$
3   Train a MIML-kNN model h on $\mathcal{L}$ using κ, κ' and $\mathbf{D}_{\mathcal{L}}$

4 **repeat**
5   Classify each bag $X \in \mathcal{U}$ with trained MIML-kNN model h using $\kappa, \kappa', \mathbf{D}_{\mathcal{L} \cup \{X\}}$
6   Calculate $\phi(X)$ for all X
7   Select the *most informative* bag X^* with arg max $\phi(X)$
8   Request the label set Y^* for X^*
9   Remove X^* from $\mathcal{U}$
10   Add (X^*, Y^*) to $\mathcal{L}$
11   Train a MIML-kNN model on $\mathcal{L}$ using κ, κ' and $\mathbf{D}_{\mathcal{L}}$
12 **until** *stop criterion reached*

___

combination of labels $\hat{y}_k$ founded in $\hat{Y}(X)$. For this we use a *single-label posterior probability* $P(\hat{y}_k|X)$ to estimate the uncertainty $\phi(X)$ as

$$\phi(X) = 1 - \prod_{\hat{y}_k \in \hat{Y}} P(\hat{y}_k|X) \tag{7}$$

The MIML-kNN classifier output for the kth label is a prediction function $f_k(X)$. This function outputs higher positive or lower negative values for very certain positive or negative predictions respectively. Considering Eq. 4, this means that when $|f_k(X)| \gg 0$ the vectors $\mathbf{w}_k^\top$ and $\mathbf{z}$ are linearly codependent. For the most uncertain label prediction then $|f_k(X)| \approx 0$ which means that $\mathbf{w}_k^\top$ and $\mathbf{z}$ are linearly independent. Based on this, we estimate $P(\hat{y}_k|X)$ using a normalization on $f_k(X)$ using the Cauchy–Schwarz inequality as follows

$$P(\hat{y}_k|X) = \frac{1}{2} \left(\frac{\mathbf{w}_k^\top \cdot \mathbf{z}(X)}{\|\mathbf{w}_k^\top\| \|\mathbf{z}(X)\|} + 1 \right) \tag{8}$$

Diversity (Div). This method is based on the multi-label active learning method proposed by Huang et al. [5,6]. This method considers that the most informative bags are those where the number of predictions are inconsistent with the average of predicted labels in the training set. Using the indicator vector $\hat{I}(X)$, $\phi(X)$ is formulated as follows

$$\phi(X) = \left| \frac{1}{K} \sum_{k=1}^{K} \hat{I}_k(X) - \rho_{\mathcal{L}} \right| \tag{9}$$

where

$$\rho_{\mathcal{L}} = \frac{1}{LK} \sum_{l=1}^{L} \sum_{k=1}^{K} I_k(X_l) \tag{10}$$

Margin (Mrg). A high positive (or low negative) value of $f_k(X)$ means that the model has a high certainty that X positively (or negatively) belongs to the kth class. Meanwhile lower absolute values in $f_k(X)$ indicate a high uncertainty. This strategy chooses the bag which average output values are the nearest to zero. This means

$$\phi(X) = -\frac{1}{K} \sum_{k=1}^{K} |f_k(X)| \tag{11}$$

Range (Rng). This method is similar to the *margin* query strategy. In this case is considered that lower range of output values $f_k(X)$ indicates higher uncertainty. This strategy is defined as

$$\phi(X) = -\left(\max_k f_k(X) - \min_k f_k(X) \right) \tag{12}$$

Percentile (Prc). This approach is related to ExtMidSelect used by Retz und Schwenker [9]. This method measures the distance between the upper and lower values of $\mathbf{f}(X) = [f_1, \ldots, f_K]$ delimited by the percentile value $F_p(X) = \text{percentile}(\mathbf{f}(X), p)$ at the percentage $p = 100(1 - \rho_{\mathcal{L}})\%$, see Eq. 10. The strategy is defined as

$$\phi(X) = -|F_\uparrow(X) - F_\downarrow(X)| \tag{13}$$

where $F_\uparrow(X)$ and $F_\downarrow(X)$ are respectively the conditional means of the upper and lower values, this means $F_\uparrow(X) = E[\mathbf{f}(X)|f_k \geq F_p]$ and $F_\downarrow(X) = E[\mathbf{f}(X)|f_k < F_p]$.

Information Density (IDC & IDH). It has been suggested that uncertainty based strategies for SISL are prone to querying outliers. To address this problem, Settles et al. [13] proposed a strategy that favors uncertain samples nearest to clusters of unlabeled samples. This strategy uses a similarity measure $S(X)$ and an uncertainty sampling $\phi_u(X)$ such that

$$\phi(X) = \phi_u(X) \cdot S(X) \tag{14}$$

Table 1. Statistics on data sets used in experiments

Dataset	Bags	Labels	Inst.	Feat.	Instances per bag			Labels per bag		
					min	*max*	*mean ± std.*	*min*	*max*	*mean ± std.*
Birds	548	13	10,232	38	2	43	8.7 ± 7.9	1	5	2.1 ± 1.0
Scene	2,000	5	18,000	15	9	9	9.0 ± 0.0	1	3	1.2 ± 0.4
Reuters	2,000	7	7,119	243	2	26	3.6 ± 2.7	1	3	1.2 ± 0.4
CK+	430	79	7,915	4,391	4	66	18.4 ± 7.6	2	9	4.0 ± 1.5
UnitPro(G.s.)	379	340	1,250	216	2	8	3.1 ± 1.2	1	69	4.0 ± 7.0

The uncertainty factor $\phi_u(X)$ is formulated as in Eq. 7. We defined two types of similarity measures. The first approach (IDC) is based on a *cosine distance* using the formula

$$\cos(X, X') = \frac{\tilde{\mathbf{x}} \cdot \tilde{\mathbf{x}}'}{\|\tilde{\mathbf{x}}\| \|\tilde{\mathbf{x}}'\|} \tag{15}$$

where $\tilde{\mathbf{x}}$ is a bag-level vector that is the mean of features over all instances $\mathbf{x}_j \in X$, this is $\tilde{\mathbf{x}} = (1/m)\sum_{j=1}^{m} \mathbf{x}_j$ where $m = |X|$. The similarity measure based on cosine distance is defined as

$$S(X) = \frac{1}{U} \sum_{X' \in \mathcal{U}} \cos(X, X') \tag{16}$$

The second approach (IDH) is based on the *Hausdorff distance* from Eq. 1. The similarity measure is defined as

$$S(X) = 1 - \frac{\exp\left(\bar{D}_U(X)\right)}{\sum_{X' \in \mathcal{U}} \exp\left(\bar{D}_U(X')\right)} \tag{17}$$

where $\bar{D}_U(X)$ is the mean distance between the bag X and the unlabeled bags, this is $\bar{D}_U(X) = (1/U)\sum_{u=1}^{U} D_H(X, X_u)$. In order to have comparable measures we applied on $\bar{D}_U(X)$ a softmax averaging.

3   Experiments

We conduct a series of experiments to compare the performance of each of the query strategies presented in this work. We employed five MIML benchmark datasets including *Birds* [1,2], *Reuters* [11], *Scene* [22], *CK+* [7,8] and *Unit-Pro(G.s.)* [16,18]. A summary of the datasets is presented in Table 1. All data sets are publicly available and prepared as MIML datasets except for the *CK+* dataset. We extracted this last one from the *Cohn-Kanade* dataset and the labels correspond to action units categories. A bag represents an image sequence and

Table 2. MIML-kNN parameters

Dataset	Parameters			Performance							
	D_H	κ	κ'	$h.l. \downarrow$	$r.l. \downarrow$	$o.e. \downarrow$	$co. \downarrow$	$a.a. \uparrow$	$a.p. \uparrow$	$a.r. \uparrow$	$a.f_1 \uparrow$
Birds	*med*	5	15	0.100	0.080	0.138	2.633	0.431	0.764	0.780	0.781
Scene	*med*	1	9	0.171	0.182	0.340	0.975	0.463	0.620	0.575	0.597
Reuters	*max*	5	17	0.037	0.031	0.078	0.355	0.820	0.895	0.910	0.903
CK+	*max*	43	19	0.034	0.124	0.198	28.14	0.163	0.757	0.544	0.633
UnitPro(G.s.)	*avg*	43	11	0.025	0.356	0.653	175.9	0.267	0.237	0.297	0.263

we extracted appearance based (local binary patterns) and shape based (histogram of oriented gradients) features at each image. *UnitPro(G.s.)* dataset is a complete proteome of the bacteria *Geobacter sulfurreducens* downloaded from the UniProt databank [16].

For each dataset, we randomly sample 20% of bags as the test data, and the rest as the unlabeled pool for active learning. Before the active learning tasks, 5% of the unlabeled pool is randomly labeled to train an initial MIML-kNN model. After each query, we train a MIML-kNN model with the extended labeled data and we test the performance of this model on the test set. Additionally, we run an experiment with a bag random sampling and use it as a reference. We run each experiment until we label 50% of the original unlabeled pool. In the experiments, a simulated Oracle provides the labels requested. We repeat the experiment 30 times for each of the datasets. The performance of the MIML-kNN models using active learning was estimated with eight measures: *hamming loss, ranking loss, coverage, one error, average accuracy, average precision, average recall* and *average f_1-measure* (see [1,22,23]). These measures are common performance metrics for evaluation in MIML framework. Lower values for *hamming loss, ranking loss, coverage* and *one error* imply a better performance and vice-versa for the other four measures.

For each data set we tuned the number of neighbors κ, the number of citers κ' and the type of Hausdorff distance D_H to obtain a maximum model performance. We perform a cross-validation test over all combinations of $(\kappa, \kappa') \in \{1, 3, 5, \ldots, 75\}^2$ with $D_H \in \{D_H^{max}, D_H^{avg}, D_H^{med}\}$. For each combination we tested 30 replicas with 20% and 80% of the data randomly selected as testing and training set respectively. At last, we selected the parameters setting that maximizes the *average f_1-measure*. The results of the parameter tuning are reported in Table 2.

The results of the performance experiments are shown in Table 3. The black dot ($\bullet$) indicates that the performance is significantly better than the bag random sampling (Rnd). The white dot ($\circ$) indicates the opposite case. Regarding the query strategy, we observe that among all datasets several strategies have superior performance than Rnd. The information density based approaches (IDD & IDH) in *UnitPro(G.s.)* and *Scene* have significantly worse performance. In contrast, these strategies performed better using the *CK+* and *Birds* dataset. The

Table 3. Comparison of query strategies at 50% of data labeled. ↑ (↓) indicate that higher (lower) values imply a better performance. • (o) indicate that the query strategy is significantly better (worse) than a random bag sampling (Rnd) based on a paired t-test at the 5% significance level ($p < 0.05$).

	Rnd	Unc	Div	Mrg	Prc	Rng	IDC	IDH
Birds								
h.l. ↓	0.116	0.111•	0.107•	**0.097•**	0.100•	0.101•	0.106•	0.116
r.l. ↓	0.099	0.093•	0.089•	**0.077•**	0.077•	0.079•	0.086•	0.091•
o.e. ↓	0.188	0.183	0.173•	0.163•	**0.157•**	0.158•	0.178	0.189
co. ↓	2.889	2.804	2.752•	2.559•	**2.552•**	2.584•	2.702•	2.761•
a.a. ↑	0.730	0.720	0.724	0.718	0.767•	**0.768•**	0.738	0.731
a.p. ↑	0.821	0.826	0.835•	0.850•	**0.852•**	0.848•	0.835•	0.822
a.r. ↑	0.730	0.720	0.724	0.718	0.767•	**0.768•**	0.738	0.731
a.f₁ ↑	0.773	0.769	0.775	0.778	**0.807•**	0.806•	0.783•	0.774
Scene								
h.l. ↓	0.196	0.204o	0.200o	**0.187•**	0.190•	0.191•	0.205o	0.209o
r.l. ↓	0.210	0.221o	0.213	**0.191•**	0.193•	0.195•	0.221o	0.226o
o.e. ↓	0.380	0.396o	0.383	**0.352•**	0.362•	0.363•	0.396o	0.404o
co. ↓	1.100	1.140o	1.110	**1.036•**	1.039•	1.046•	1.140o	1.160o
a.a. ↑	0.493	0.492	0.496	0.470o	0.496	**0.506•**	0.487	0.494
a.p. ↑	0.754	0.744o	0.752	**0.771•**	0.767•	0.766•	0.744o	0.739o
a.r. ↑	0.493	0.492	0.496	0.470o	0.496	**0.506•**	0.487	0.494
a.f₁ ↑	0.596	0.592	0.597	0.584o	0.603	**0.609•**	0.588	0.592
Reuters								
h.l. ↓	0.045	0.042•	**0.041•**	0.050o	0.048o	0.051o	0.104	0.104
r.l. ↓	0.039	0.035•	0.034•	0.044o	**0.033•**	0.039	0.121	0.121
o.e. ↓	0.100	0.087•	**0.085•**	0.120o	0.090•	0.106o	0.274	0.274
co. ↓	0.409	0.387•	0.381•	0.436o	**0.374•**	0.407	0.916	0.916
a.a. ↑	0.872	0.901•	0.896•	0.826o	**0.905•**	0.883•	0.675	0.675
a.p. ↑	0.934	0.941•	**0.943•**	0.923o	0.941•	0.931	0.816	0.816
a.r. ↑	0.872	0.901•	0.896•	0.826o	**0.905•**	0.883•	0.675	0.675
a.f₁ ↑	0.902	0.921•	0.919•	0.871o	**0.922•**	0.906	0.738	0.738
CK+								
h.l. ↓	0.041	0.040	0.040	0.040	0.043o	0.042o	**0.039•**	0.040•
r.l. ↓	0.163	0.152•	0.150•	0.157	0.157	0.157	**0.146•**	0.149•
o.e. ↓	0.270	0.246•	0.247•	0.268	0.264	0.263	0.250	**0.240•**
co. ↓	32.84	31.98	31.00•	32.19	32.31	32.08	**30.54•**	30.97•
a.a. ↑	0.492	0.514•	0.520•	0.514•	0.524•	**0.526•**	0.511•	0.500
a.p. ↑	0.599	0.615•	0.617•	0.609•	0.605	0.607	0.622•	**0.622•**
a.r. ↑	0.492	0.514•	0.520•	0.514•	0.524•	**0.526•**	0.511•	0.500
a.f₁ ↑	0.540	0.560•	**0.564•**	0.557•	0.561•	0.563•	0.561•	0.554•
UnitPro(G.s.)								
h.l. ↓	0.040	0.043	0.032•	**0.027•**	0.064o	0.061o	0.076o	0.086o
r.l. ↓	0.503	0.496	**0.494**	0.498	0.501	0.514	0.531o	0.519
o.e. ↓	0.834	0.826	0.819	**0.811•**	0.824	0.828	0.865o	0.866o
co. ↓	196.9	192.3	192.5	**187.6•**	189.9•	192.6	212.5o	201.7
a.a. ↑	0.180	0.202•	0.181	0.170	**0.221•**	0.202•	0.185	0.206•
a.p. ↑	0.141	0.148	0.154	**0.168•**	0.158•	0.153	0.101o	0.108o
a.r. ↑	0.180	0.202•	0.181	0.170	**0.221•**	0.202•	0.185	0.206•
a.f₁ ↑	0.157	0.170	0.166	0.168	**0.183•**	0.173•	0.129o	0.141o

best performance among all datasets is achieved by the percentile strategy (Prc) followed by margin (Mrg) and diversity (Div) strategies. Regarding the dataset, in the *Reuters* and *UnitPro(G.s.)* dataset we observe in general a remarkable performance of the strategies. In the *Reuters* dataset, uncertainty (Unc) and diversity (Div) strategies are significantly better for all metrics.

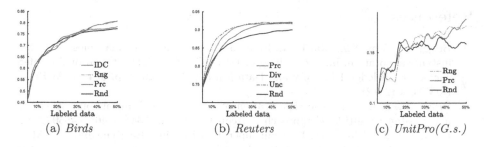

Fig. 1. Example of query strategies performance based on the *average f₁-measure*

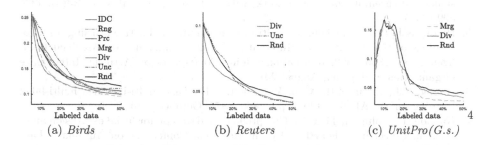

Fig. 2. Example of query strategies performance based on the *hamming loss*

Figures 1 and 2 shows the performance curves as the number of labeled data increases until the stop criterion is reached (50% labeled). We show a selection of the most representative curves based on the *avg. f₁-measure* and *hamming loss* metrics. We observe in Fig. 1b that the MIML-*k*NN model can reach its best performance with much less labeled data (∼25%) using uncertainty (Unc) or percentile (Prc) query strategies. A similar situation can be observed in Fig. 2c where the MIML-*k*NN reaches nearly the lowest *hamming loss* at approx. 35% of labeled data using the margin (Mrg) query strategy.

4   Conclusion

In this paper we proposed an active learning approach to reduce the labeling cost of the MIML dataset using MIML-*k*NN as base classifier. We introduced novel query strategies and also implemented previously used query strategies for MIML learning. Finally, we conducted an experimental evaluation on various benchmark datasets. We demonstrated that these approaches can achieve significantly improved results than no active selection for all datasets on various evaluation criteria.

References

1. Briggs, F., Fern, X.Z., Raich, R.: Rank-loss support instance machines for MIML instance annotation. In: Proceedings of 18th ACM SIGKDD International Conference on Knowledge Discovery and Data Mining, KDD 2012, p. 534. ACM Press, New York (2012)
2. Briggs, F., et al.: Acoustic classification of multiple simultaneous bird species: a multi-instance multi-label approach. J. Acoust. Soc. Am. **131**(6), 4640–4650 (2012)
3. Hady, M.F.A., Schwenker, F.: Semi-supervised learning. In: Bianchini, M., Maggini, M., Jain, L. (eds.) Handbook on Neural Information Processing, pp. 215–239. Springer, Heidelberg (2013). https://doi.org/10.1007/978-3-642-36657-4_7
4. Hady, M.F.A., Schwenker, F., Palm, G.: Semi-supervised learning for tree-structured ensembles of RBF networks with co-training. Neural Netw. **23**(4), 497–509 (2010)
5. Huang, S.J., Gao, N., Chen, S.: Multi-instance multi-label active learning. In: Proceedings of the Twenty-Sixth International Joint Conference on Artificial Intelligence, pp. 1886–1892. International Joint Conferences on Artificial Intelligence Organization, California, August 2017
6. Huang, S.J., Zhou, Z.H., Gao, W., Zhou, Z.H.: Fast multi-instance multi-label learning. In: AAAI (61321491), pp. 1868–1874, October 2014
7. Kanade, T., Cohn, J., Tian, Y.: Comprehensive database for facial expression analysis. In: Proceedings Fourth IEEE International Conference on Automatic Face and Gesture Recognition (Cat. No. PR00580), pp. 46–53. IEEE Computer Society (2000)
8. Lucey, P., Cohn, J.F., Kanade, T., Saragih, J., Ambadar, Z., Matthews, I.: The extended Cohn-Kanade dataset (CK+): a complete dataset for action unit and emotion-specified expression. In: 2010 IEEE Computer Society Conference on Computer Vision and Pattern Recognition Workshops, pp. 94–101. IEEE, June 2010
9. Retz, R., Schwenker, F.: Active multi-instance multi-label learning. In: Wilhelm, A.F.X., Kestler, H.A. (eds.) Analysis of Large and Complex Data. SCDAKO, pp. 91–101. Springer, Cham (2016). https://doi.org/10.1007/978-3-319-25226-1_8
10. Schwenker, F., Trentin, E.: Pattern classification and clustering: a review of partially supervised learning approaches. Pattern Recogn. Lett. **37**(1), 4–14 (2014)
11. Sebastiani, F.: Machine learning in automated text categorization. ACM Comput. Surv. **34**(1), 1–47 (2002)
12. Settles, B.: Active learning literature survey. Mach. Learn. **15**(2), 201–221 (1994)
13. Settles, B., Craven, M.: An analysis of active learning strategies for sequence labeling tasks. In: Proceedings of the Conference on Empirical Methods in Natural Language Processing, pp. 1070–1079. Association for Computational Linguistics (2008)
14. Shen, C., Jiao, J., Yang, Y., Wang, B.: Multi-instance multi-label learning for automatic tag recommendation. In: IEEE International Conference on Systems, Man, and Cybernetics, pp. 4910–4914. IEEE, October 2009
15. Surdeanu, M., Tibshirani, J., Nallapati, R., Manning, C.D.: Multi-instance multi-label learning for relation extraction. In: Proceedings of the 2012 Joint Conference on Empirical Methods in Natural Language Processing and Computational Natural Language Learning, pp. 455–465. Association for Computational Linguistics (2012)
16. The UniProt Consortium: UniProt: the universal protein knowledgebase. Nucleic Acids Res. **45**(D1), D158–D169 (2017)

17. Thiam, P., Meudt, S., Palm, G., Schwenker, F.: A temporal dependency based multi-modal active learning approach for audiovisual event detection. Neural Process. Lett. 1–24 (2017)
18. Wu, J.S., Huang, S.J., Zhou, Z.H.: Genome-wide protein function prediction through multi-instance multi-label learning. IEEE/ACM Trans. Comput. Biol. Bioinforma. **11**(5), 891–902 (2014)
19. Xu, X.S., Xue, X., Zhou, Z.H.: Ensemble multi-instance multi-label learning approach for video annotation task. In: Proceedings of the 19th ACM International Conference on Multimedia, MM 2011, p. 1153. ACM Press, New York (2011)
20. Li, Y.-X., Ji, S., Kumar, S., Ye, J., Zhou, Z.-H.: Drosophila gene expression pattern annotation through multi-instance multi-label learning. IEEE/ACM Trans. Comput. Biol. Bioinforma. **9**(1), 98–112 (2012)
21. Zhang, M.L.: A k-nearest neighbor based multi-instance multi-label learning algorithm. In: 2010 22nd IEEE International Conference on Tools with Artificial Intelligence, pp. 207–212. IEEE, October 2010
22. Zhou, Z.H., Zhang, M.l.: Multi-instance multi-label learning with application to scene classification. In: Advances in Neural Information Processing Systems, pp. 1609–1616 (2007)
23. Zhou, Z.H., Zhang, M.L., Huang, S.J., Li, Y.F.: Multi-instance multi-label learning. Artif. Intell. **176**(1), 2291–2320 (2012)

Manifold Learning Regression
with Non-stationary Kernels

Alexander Kuleshov, Alexander Bernstein, and Evgeny Burnaev[✉]

Skolkovo Institute of Science and Technology, Skolkovo Innovation Center,
3 Nobel Street, Moscow 121205, Russia
{A.Kuleshov,A.Bernstein,E.Burnaev}@skoltech.ru,
http://adase.group

Abstract. Nonlinear multi-output regression problem is to construct a predictive function which estimates an unknown smooth mapping from q-dimensional inputs to m-dimensional outputs based on a training data set consisting of given "input-output" pairs. In order to solve this problem, regression models based on stationary kernels are often used. However, such approaches are not efficient for functions with strongly varying gradients. There exist some attempts to introduce non-stationary kernels to account for possible non-regularities, although even the most efficient one called Manifold Learning Regression (MLR), which estimates the unknown function as well its Jacobian matrix, is too computationally expensive. The main problem is that the MLR is based on a computationally intensive manifold learning technique. In this paper we propose a modified version of the MLR with significantly less computational complexity while preserving its accuracy.

Keywords: Nonlinear multi-output regression
Manifold learning regression · Non-stationary kernel

1 Introduction

1.1 Nonlinear Multi-output Regression

We formulate a nonlinear multi-output regression task [1–3]: let $\mathbf{f}$ be an unknown smooth mapping from an input space $\mathbf{X} \subset \mathbb{R}^q$ to m-dimensional output space $\mathbb{R}^m$. Given a training data set

$$\mathbf{Z}_{(n)} = \{\mathbf{Z}_i = (\mathbf{x}_i, \mathbf{y}_i = \mathbf{f}(\mathbf{x}_i)), i = 1, 2, \ldots, n\}, \tag{1}$$

The research, presented in Sect. 1 of this paper, was partially supported by the Russian Foundation for Basic Research grants 16-01-00576 A and 16-29-09649 ofi m. The research, presented in other sections, was supported solely by the Ministry of Education and Science of Russian Federation, grant No. 14.606.21.0004, grant code: RFMEFI60617X0004.

© Springer Nature Switzerland AG 2018
L. Pancioni et al. (Eds.): ANNPR 2018, LNAI 11081, pp. 152–164, 2018.
https://doi.org/10.1007/978-3-319-99978-4_12

consisting of input-output pairs, the task is to construct the function $\mathbf{y}^* = \mathbf{f}^*(\mathbf{x}) = \mathbf{f}^*(\mathbf{x}|\mathbf{Z}_{(n)})$ to predict the true output $\mathbf{y} = \mathbf{f}(\mathbf{x})$ for an arbitrary Out-of-Sample (OoS) input $\mathbf{x} \in \mathbf{X}$ with small predictive error $|\mathbf{y}^* - \mathbf{y}|$. In engineering applications $\mathbf{f}^*(\mathbf{x})$ is usually used as a surrogate of some target function [4]. Most of optimization algorithms use gradient of the optimized function; in this case, the regression method also should allow estimating $m \times q$ Jacobian matrix $\mathbf{J}_f(\mathbf{x}) = \nabla_x \mathbf{f}(\mathbf{x})$ of the mapping $\mathbf{f}(\mathbf{x})$ at an arbitrary input point $\mathbf{x} \in \mathbf{X}$.

There exist various regression methods such as least squares (LS) techniques (linear and nonlinear), artificial neural networks, kernel nonparametric regression, Gaussian process regression, kriging regression, etc. [1–3,5–16]. A classical approach is based on Kernel Nonparametric Regression (KNR) [7]: we select the kernel function $K(\mathbf{x}, \mathbf{x}')$ (see [17]) and construct the KNR-estimator

$$\mathbf{f}_{KNR}(\mathbf{x}) = \frac{1}{K(\mathbf{x})} \sum_{j=1}^{n} K(\mathbf{x}, \mathbf{x}_j) \cdot y_j, \ K(\mathbf{x}) = \sum_{j=1}^{n} K(\mathbf{x}, \mathbf{x}_j), \tag{2}$$

which minimizes (over $\hat{\mathbf{y}}$) the residual $\sum_{j=1}^{n} K(\mathbf{x}, \mathbf{x}_j) |\hat{\mathbf{y}} - \mathbf{y}_j|^2$.

The symmetric non-negative definite function $K(\mathbf{x}, \mathbf{x}')$ can be interpreted as a covariance function of some random field $\mathbf{y}(\mathbf{x})$; thus, the unknown function $\mathbf{f}(\mathbf{x})$ can be interpreted as a realization of the random field $\mathbf{y}(\mathbf{x})$ and $K(\mathbf{x}, \mathbf{x}') = \mathrm{cov}(\mathbf{f}(\mathbf{x}), \mathbf{f}(\mathbf{x}'))$. If we consider only the first and second moments of this random field, then without loss of generality we can assume that this field is Gaussian and as a result obtain so-called Gaussian Process Regression [5,6,18,19].

One of the most popular kernel estimators is kriging, first developed by Krige [20] and popularized by Sacks [21]. Kriging provides both global predictions and their uncertainty. Kriging-based surrogate models are widely used in engineering modeling and optimization [4,22–24].

Kriging regression combines both linear LS and KNR approaches: the deviation of the unknown function $\mathbf{f}(\mathbf{x})$ from its LS estimator, constructed on basis of some functional dictionary, is modeled by a zero mean Gaussian random field with the covariance function $K(\mathbf{x}, \mathbf{x}')$. Thus we can estimate the deviation at the point $\mathbf{x}$ using some filtration procedure and known deviations at the sample points $\{\mathbf{x}_i\}$. Usually stationary covariance functions $K(\mathbf{x}, \mathbf{x}')$ are used that depend on their arguments $\mathbf{x}$ and $\mathbf{x}'$ only through the difference $(\mathbf{x} - \mathbf{x}')$.

1.2 Learning with Non-stationary Kernels

Many methods use kernels that are stationary. However, as indicated e.g. in [2,3,5,6], such methods have serious drawbacks in case of functions with strongly varying gradients. Traditional kriging "is stationary in nature" and has low accuracy in case of functions with "non-stationary responses" (significant changes in "smoothness") [25,26]. Figure 1 illustrates this phenomenon by the Xiong function $\mathbf{f}(\mathbf{x}) = \sin(30(\mathbf{x} - 0.9)^4) \cdot \cos(2(\mathbf{x} - 0.9)) + (\mathbf{x} - 0.9)/2$, $\mathbf{x} \in [0, 1]$, and its kriging estimator with a stationary kernel [25]. Therefore, non-stationary kernels with adaptive kernel width are used to estimate non-regular functions. There are strategies for constructing the non-stationary kernels [26].

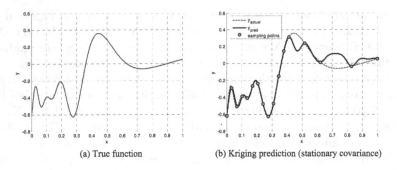

(a) True function　　　　　　(b) Kriging prediction (stationary covariance)

Fig. 1. Example of Kriging prediction with a stationary covariance [25].

The interpretable nonlinear map approach from [27] uses the one-to-one reparameterization function $\mathbf{u} = \varphi(\mathbf{x})$ with the inverse $\mathbf{x} = \psi(\mathbf{u})$ to map the Input space $\mathbf{X}$ to $\mathbf{U} = \varphi(\mathbf{X})$, such that the covariance function $k(\mathbf{u}, \mathbf{u}') = K(\psi(\mathbf{u}), \psi(\mathbf{u}')) = \mathrm{cov}(\mathbf{f}(\psi(\mathbf{u})), \mathbf{f}(\psi(\mathbf{u}')))$ becomes approximately stationary. This approach was studied for years in geostatistics in case of relatively low dimensions ($q = 2, 3$), and the general case has been considered in [25] with the reparameterization function $\varphi(\mathbf{x}) = \mathbf{x}_0 + \int_{x_0^{(1)}}^{x^{(1)}} \int_{x_0^{(2)}}^{x^{(2)}} \cdots \int_{x_0^{(q)}}^{x^{(q)}} s(\mathbf{x}) d\mathbf{x}$, where $\mathbf{x} = (x^{(1)}, x^{(2)}, \ldots, x^{(q)})$ and $s(\mathbf{x})$ is a density function, modelled by a linear combination of some "dictionary" functions with optimized coefficients. A simple one-dimensional illustration of such map is provided in Fig. 2.

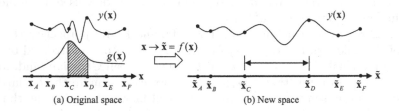

(a) Original space　　　　　　(b) New space

Fig. 2. A conceptual illustration of the nonlinear reparameterization function [25].

After such reparameterization, KNR-estimator (2) $\mathbf{g}_{KNR}(\mathbf{u})$ for the function $\mathbf{g}(\mathbf{u}) = \mathbf{f}(\psi(\mathbf{u}))$ with the stationary kernel $k(\mathbf{u}, \mathbf{u}')$ is constructed, and the function $\mathbf{f}^*(\mathbf{x}) = \mathbf{g}_{KNR}(\varphi(\mathbf{x}))$ is used as an estimator for $\mathbf{f}(\mathbf{x})$.

1.3 Manifold Learning Regression

A fundamentally different geometrical approach to KNR called Manifold Learning Regression (MLR) was proposed in [10, 11]; MLR also constructs the reparameterization function $\mathbf{u} = \varphi(\mathbf{x})$ and estimates the Jacobian matrix $\mathbf{J}_f(\mathbf{x})$.

MLR compares favourably with many conventional regression methods. In Fig. 3 (see [10]) we depict the KNR-estimator $\mathbf{f}_{KNR}$ (2) with a stationary kernel and the MLR-estimator $\mathbf{f}_{MLR}$ for the Xiong function $\mathbf{f}(\mathbf{x})$. The input values in the set $\mathbf{Z}_{(n)}$, $n = 100$ were uniformly randomly distributed on the interval $[0, 1]$.

We see that the MLR method provides the essentially smoother estimate. The mean squared errors $\mathrm{MSE}_{KNR} = 0.0024$ and $\mathrm{MSE}_{MLR} = 0.0014$ were calculated using the test sample with $n = 1001$ uniform grid points in the interval $[0, 1]$.

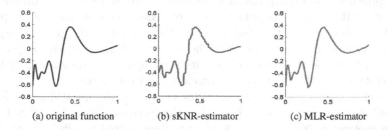

| (a) original function | (b) sKNR-estimator | (c) MLR-estimator |

Fig. 3. Reconstruction of the Xiong function (a) by KNR with stationary kernel (b) and MLR (c).

MLR is based on a Manifold Learning approach. Let us represent in the input-output space $\mathbb{R}^p$, $p = q + m$, the graph of the function $\mathbf{f}$ by the smooth q-dimensional manifold (Regression Manifold, RM)

$$\mathbf{M}(\mathbf{f}) = \{\mathbf{Z} = \mathbf{F}(\mathbf{x}) \in \mathbb{R}^p : \mathbf{x} \in \mathbf{X} \subset \mathbb{R}^q\} \subset \mathbb{R}^p, \qquad (3)$$

embedded in the ambient space $\mathbb{R}^p$ and parameterized by the single chart

$$\mathbf{F} : \mathbf{x} \in \mathbf{X} \subset \mathbb{R}^q \to \mathbf{Z} = \mathbf{F}(\mathbf{x}) = (\mathbf{x}, \mathbf{f}(\mathbf{x})) \in \mathbb{R}^p. \qquad (4)$$

Arbitrary function $\mathbf{f}^* : \mathbf{X} \to \mathbb{R}^m$ also determines the manifold $\mathbf{M}(\mathbf{f}^*)$ (substitute $\mathbf{f}^*(\mathbf{x})$ and $\mathbf{F}^*(\mathbf{x})$ instead of $\mathbf{f}(\mathbf{x})$ and $\mathbf{F}(\mathbf{x})$ in (3) and (4)).

In order to apply MLR, we estimate RM $\mathbf{M}(\mathbf{f})$ using the training data $\mathbf{Z}_{(n)}$ (1) by the Grassmann & Stiefel Eigenmaps (GSE) algorithm [28]. The constructed estimator $\mathbf{M}_{GSE} = \mathbf{M}_{GSE}(\mathbf{Z}_{(n)})$, being also a q-dimensional manifold embedded in $\mathbb{R}^p$, provides small Hausdorff distance $d_H(\mathbf{M}_{GSE}, \mathbf{M}(\mathbf{f}))$ between these manifolds. In addition, the tangent spaces $\mathbf{L}(\mathbf{Z})$ to RM $\mathbf{M}(\mathbf{f})$ at the manifold points $\mathbf{Z} \in \mathbf{M}(\mathbf{f})$ are estimated by the linear spaces $\mathbf{L}_{GSE}(\mathbf{Z})$ with "aligned" bases smoothly depending on $\mathbf{Z}$. GSE also constructs the low-dimensional parameterization $h(\mathbf{Z})$ of the manifold points $\mathbf{Z}$ and the recovery mapping $\mathbf{g}(h)$, which accurately reconstructs $\mathbf{Z}$ from $h(\mathbf{Z})$.

To get the estimator $\mathbf{f}_{MLR}(\mathbf{x})$ of the unknown function $\mathbf{f}$, we solve the equation $\mathbf{M}(\mathbf{f}_{MLR}) = \mathbf{M}_{GSE}$. Using the estimator $\mathbf{L}_{GSE}(\mathbf{F}(\mathbf{x}))$, we also construct $m \times q$ matrix $\mathbf{G}_{MLR}(\mathbf{x})$, which estimates the $m \times q$ Jacobian matrix $\mathbf{J}_f(\mathbf{x}) = \nabla_x \mathbf{f}(\mathbf{x})$ of $\mathbf{f}(\mathbf{x})$ at the arbitrary point $\mathbf{x} \in \mathbf{X}$. Here as the reparameterization function $\mathbf{u} = \varphi(\mathbf{x})$ we use approximation of the unknown function $h(\mathbf{F}(\mathbf{x}))$ (it depends on $\mathbf{f}(\mathbf{x})$, which is unknown at the OoS points $\mathbf{x} \in \mathbf{X}$).

1.4   Paper Contribution

The GSE algorithm contains several very computationally expensive steps such as construction of the aligned bases in the estimated tangent spaces, the embedding mapping and the recovery mappings, the reparameterization mapping, etc. Although the incremental version of the GSE algorithm [29] reduces its complexity, still it remains computationally expensive.

The paper proposes a new modified version of the MLR algorithm (mMLR) with significantly less computational complexity. We developed a simplified version of the MLR algorithm, which does not require computationally expensive steps, listed above, so that we can construct the estimators $(\mathbf{f}_{MLR}(\mathbf{x}), \mathbf{G}_{MLR}(\mathbf{x}))$ while preserving the same accuracy. Then instead of using the KNR procedure with a stationary kernel we developed its version with a non-stationary kernel, which is defined on basis of the constructed MLR estimators.

Note that in this paper we consider the case when the input domain $\mathbf{X} \subset \mathbb{R}^q$ is a "full-dimensional" subset of $\mathbb{R}^q$ (i.e., the intrinsic dimension of $\mathbf{X}$ is equal to q) in contrast to [6,16], where $\mathbf{X}$ is a low-dimensional manifold in $\mathbb{R}^q$. In [30] they reviewed approaches to the regression with manifold valued inputs.

The paper is organized as follows. Section 2 describes some details of the GSE/MLR algorithms; the proposed mMLR algorithm is described in Sect. 3.

2   Manifold Learning Regression

2.1   Tangent Bundle Manifold Estimation Problem

The MLR algorithm is based on the solution of the Tangent bundle manifold estimation problem [31,32]: estimate RM $\mathbf{M}(\mathbf{f})$ (3) from the dataset $\mathbf{Z}_{(n)}$ (1), sampled from $\mathbf{M}(\mathbf{f})$. The manifold estimation problem is to construct:

- the embedding mapping h from RM $\mathbf{M}(\mathbf{f})$ to the q-dimensional Feature Space (FS) $\mathbf{T}_h = h(\mathbf{M}(\mathbf{f}))$, which provides low-dimensional parameterization (coordinates) $h(\mathbf{Z})$ of the manifold points $\mathbf{Z} \in \mathbf{M}(\mathbf{f})$,
- the recovery mapping $\mathbf{g}(t)$ from FS $\mathbf{T}_h$ to $\mathbb{R}^p$, which recovers the manifold points $\mathbf{Z} = \mathbf{g}(t)$ from their low-dimensional coordinates $t = h(\mathbf{Z})$,

such that the recovered value $r_{h,g}(\mathbf{Z}) = \mathbf{g}(h(\mathbf{Z}))$ is close to the initial vector $\mathbf{Z}$:

$$\mathbf{g}(h(\mathbf{Z})) \approx \mathbf{Z}, \qquad (5)$$

i.e. the recovery error $\delta_{h,g}(\mathbf{Z}) = |r_{h,g}(\mathbf{Z}) - \mathbf{Z}|$ is small. These mappings determine the q-dimensional Recovered Regression manifold (RRM)

$$\mathbf{M}_{h,g} = r_{h,g}(\mathbf{M}(\mathbf{f})) = \{r_{h,g}(\mathbf{Z}) \in \mathbb{R}^p : \mathbf{Z} \in \mathbf{M}(\mathbf{f})\}$$
$$= \{\mathbf{Z} = \mathbf{g}(t) \in \mathbb{R}^p : t \in \mathbf{T}_h = h(\mathbf{M}(\mathbf{f})) \subset \mathbb{R}^q\}, \qquad (6)$$

which is embedded in the ambient space $\mathbb{R}^p$, covered by the single chart $\mathbf{g}$, and consists of all recovered values $r_{h,g}(\mathbf{Z})$ of the manifold points $\mathbf{Z}$. Thanks to (5) we get proximity of the manifolds $\mathbf{M}_{h,g} \approx \mathbf{M}(\mathbf{f})$, i.e. the Hausdorff distance $d_H(\mathbf{M}_{h,g}, \mathbf{M}(\mathbf{f}))$ between RM $\mathbf{M}(\mathbf{f})$ and RRM $\mathbf{M}_{h,g}$ (6) is small due the inequality $d_H(\mathbf{M}_{h,g}, \mathbf{M}(\mathbf{f})) \leq \sup_{\mathbf{Z} \in \mathbf{M}(\mathbf{f})} \delta_{h,g}(\mathbf{Z})$.

The manifold proximity (5) at the OoS point $\mathbf{Z} \in \mathbf{M}(\mathbf{f}) \setminus \mathbf{Z}_{(n)}$ characterizes the generalization ability of the solution $(h, \mathbf{g})$ at the specific point $\mathbf{Z}$. Good generalization ability requires [32] that the pair $(h, \mathbf{g})$ should provide the tangent proximities $L_{h,g}(\mathbf{Z}) \approx L(\mathbf{Z})$ between the tangent spaces $L(\mathbf{Z})$ to RM $\mathbf{M}(\mathbf{f})$ at points $\mathbf{Z} \in \mathbf{M}(\mathbf{f})$ and the tangent spaces $L_{h,g}(\mathbf{Z}) = \mathrm{Span}(\mathbf{J}_g(h(\mathbf{Z})))$ (spanned by columns of the Jacobian matrix $\mathbf{J}_g(t)$ of the mapping $\mathbf{g}$ at the point $t = h(\mathbf{Z})$) to RRM $\mathbf{M}_{h,g}$ at the recovered points $r_{h,g}(\mathbf{Z}) \in \mathbf{M}_{h,g}$. Note that the tangent proximity is defined in terms of a chosen distance between these tangent spaces considered as elements of the Grassmann manifold $\mathrm{Grass}(p, q)$, consisting of all q-dimensional linear subspaces in $\mathbb{R}^p$.

The set of manifold points equipped with the tangent spaces at these points is called the Tangent bundle of the manifold [33], and therefore we refer to the manifold estimation problem with the tangent proximity requirement as the Tangent bundle manifold learning problem [31]. The GSE algorithm, briefly described in the next section, provides the solution to this problem.

2.2 Grassmann and Stiefel Eigenmaps Algorithm

The GSE algorithm consists of the three successively performed steps: tangent manifold learning, manifold embedding, and manifold recovery.

Tangent Manifold Learning. We construct the sample-based $p \times q$ matrices $\mathbf{H}(\mathbf{Z})$ with columns $\{\mathbf{H}^{(k)}(\mathbf{Z}) \in \mathbb{R}^p, 1 \leq k \leq q\}$, smoothly depending on $\mathbf{Z}$, to meet the relations $\mathrm{Span}(\mathbf{H}(\mathbf{Z})) \approx L(\mathbf{Z})$ and $\nabla_{\mathbf{H}^{(i)}(\mathbf{Z})} \mathbf{H}^{(j)}(\mathbf{Z}) = \nabla_{\mathbf{H}^{(j)}(\mathbf{Z})} \mathbf{H}^{(i)}(\mathbf{Z})$ (covariant differentiation is used here), $1 \leq i < j \leq q$, for all points $\mathbf{Z} \in \mathbf{M}(\mathbf{f})$.

The latter condition provides that these columns are coordinate tangent fields on RM $\mathbf{M}(\mathbf{f})$ and, thus, $\mathbf{H}(\mathbf{Z})$ is the Jacobian matrix of some mapping [33]. Thus the mappings h and $\mathbf{g}$ are constructed in such a way that

$$\mathbf{J}_g(h(\mathbf{Z})) = \mathbf{H}(\mathbf{Z}). \tag{7}$$

Using Principal Component Analysis (PCA), we estimate the tangent space $L(\mathbf{Z})$ at the sample point $\mathbf{Z} \in \mathbf{Z}_{(n)}$ [34] by the q-dimensional linear space $L_{PCA}(\mathbf{Z})$, spanned by the eigenvectors of the local sample covariance matrix

$$\Sigma(\mathbf{Z}|K_p) = \frac{1}{K_p(\mathbf{Z})} \sum_{j=1}^{n} K_p(\mathbf{Z}, \mathbf{Z}_j) \cdot [(\mathbf{Z}_j - \mathbf{Z}) \cdot (\mathbf{Z}_j - \mathbf{Z})^T], \tag{8}$$

corresponding to the q largest eigenvalues; here $K_p(\mathbf{Z}) = \sum_{j=1}^{n} K_p(\mathbf{Z}, \mathbf{Z}_j)$ and $K_p(\mathbf{Z}, \mathbf{Z}')$ is a stationary kernel in $\mathbb{R}^p$ (e.g., the indicator kernel $\mathrm{I}\{|\mathbf{Z} - \mathbf{Z}'| < \varepsilon\}$ or the heat kernel [35] $K_{p,\varepsilon,\rho}(\mathbf{Z}, \mathbf{Z}') = \mathrm{I}\{|\mathbf{Z} - \mathbf{Z}'| \leq \varepsilon\} \cdot \exp\{-\rho \cdot |\mathbf{Z} - \mathbf{Z}'|^2\}$ with the parameters ε and ρ).

We construct the matrices $\mathbf{H}(\mathbf{Z})$ to meet the relations

$$\mathrm{Span}(\mathbf{H}(\mathbf{Z})) = L_{PCA}(\mathbf{Z}), \tag{9}$$

therefore, the required proximity $\mathrm{Span}(\mathbf{H}(\mathbf{Z})) \approx L(\mathbf{Z})$ follows automatically from the approximate equalities $L_{PCA}(\mathbf{Z}) \approx L(\mathbf{Z})$, which are satisfied when RM $\mathbf{M}(\mathbf{f})$ is "well sampled" and the parameter ε is small enough [36].

The principal components form the orthogonal basis in the linear space $L_{PCA}(\mathbf{Z})$. Let us denote the $p \times q$ matrix with the principal components as columns by $Q_{PCA}(\mathbf{Z})$. However, for different $\mathbf{Z}$ these bases are not agreed with each other and can be very different even in neighboring points. While preserving the requirements (9), the GSE algorithm constructs other bases in these linear spaces, determined by the $p \times q$ matrices

$$\mathbf{H}_{GSE}(\mathbf{Z}) = Q_{PCA}(\mathbf{Z}) \cdot v(\mathbf{Z}). \tag{10}$$

Here $q \times q$ nonsingular matrices $v(\mathbf{Z})$ should provide smooth dependency of $\mathbf{H}(\mathbf{Z})$ on $\mathbf{Z}$ and coordinateness of the tangent fields $\{\mathbf{H}^{(k)}(\mathbf{Z}) \in \mathbb{R}^p, 1 \leq k \leq q\}$.

At the sample points the matrices $\mathbf{H}_i = \mathbf{H}_{GSE}(\mathbf{Z}_i)$ (10) are constructed to minimize the quadratic form $\sum_{i,j=1}^{n} K_p(\mathbf{Z}_i, \mathbf{Z}_j) \cdot \|\mathbf{H}_i - \mathbf{H}_j\|_F^2$ under the coordinateness constraint and certain normalizing condition, required to avoid a degenerate solution; here $\| \cdot \|_F$ is the Frobenius matrix norm. The exact solution of this problem is obtained in the explicit form; at the OoS points $\mathbf{Z}$, the matrices $\mathbf{H}_{GSE}(\mathbf{Z})$ are constructed using certain interpolation procedure.

Manifold Embedding. After we construct the matrices $\mathbf{H}_{GSE}(\mathbf{Z})$ and assuming that the conditions (5) and (9) are satisfied, we use the Taylor series expansion of the mapping $\mathbf{g}(t)$, $t = h(\mathbf{Z})$ to get the relation $\mathbf{Z}' - \mathbf{Z} \approx \mathbf{H}_{GSE}(\mathbf{Z}) \cdot (h(\mathbf{Z}') - h(\mathbf{Z}))$ for the neighboring points $\mathbf{Z}, \mathbf{Z}' \in \mathbf{M}(\mathbf{f})$. These relations, considered further as regression equations, allow constructing the embedding mapping $h_{GSE}(\mathbf{Z})$ and FS $\mathbf{T}_h = h(\mathbf{M}(\mathbf{f}))$.

Manifold Recovery. After we construct the matrices $\mathbf{H}_{GSE}(\mathbf{Z})$ and the mapping h_{GSE}, using known values $\{\mathbf{g}(t_i) \approx \mathbf{Z}_i\}$ (5) and $\{\mathbf{J}_g(t_i) = \mathbf{H}_i\}$ (9), $t_i = h_{GSE}(\mathbf{Z}_i)$, we construct the mapping $\mathbf{g}_{GSE}(t)$ and the estimator $\mathbf{G}_{GSE}(t)$ for its covariance matrix $\mathbf{J}_g(t)$.

2.3   Manifold Learning Regression Algorithm

We split the p-dimensional vector $\mathbf{Z} = \begin{pmatrix} \mathbf{Z}_{in} \\ \mathbf{Z}_{out} \end{pmatrix}$, $p = q+m$, into the q-dimensional vector $\mathbf{Z}_{in}$ and the m-dimensional vector $\mathbf{Z}_{out}$ and obtain the corresponding partitions

$$\mathbf{H}_{GSE}(\mathbf{Z}) = \begin{pmatrix} \mathbf{H}_{GSE,in}(\mathbf{Z}) \\ \mathbf{H}_{GSE,out}(\mathbf{Z}) \end{pmatrix}, \ Q_{PCA}(\mathbf{Z}) = \begin{pmatrix} Q_{PCA,in}(\mathbf{Z}) \\ Q_{PCA,out}(\mathbf{Z}) \end{pmatrix},$$

$$\mathbf{g}_{GSE}(t) = \begin{pmatrix} \mathbf{g}_{GSE,in}(t) \\ \mathbf{g}_{GSE,out}(t) \end{pmatrix}, \ \mathbf{G}_{GSE}(t) = \begin{pmatrix} \mathbf{G}_{GSE,in}(t) \\ \mathbf{G}_{GSE,out}(t) \end{pmatrix} \tag{11}$$

of the $p \times q$ matrices $\mathbf{H}_{GSE}(\mathbf{Z})$ and $Q_{PCA}(\mathbf{Z})$, the p-dimensional vector $\mathbf{g}_{GSE}(t)$, and the $p \times q$ matrix $\mathbf{G}_{GSE}(t)$; note that the $q \times q$ matrix $\mathbf{G}_{GSE,in}(t)$ and the

$m \times q$ matrix $\mathbf{G}_{GSE,out}(t)$ are the Jacobian matrices of the mappings $\mathbf{g}_{GSE,in}(t)$ and $\mathbf{g}_{GSE,out}(t)$, respectively.

It follows from the proximities (5), (9) and the partition (11) with $\mathbf{Z} = \mathbf{F}(\mathbf{x})$ (4) that

$$\mathbf{g}_{GSE,in}(h_{GSE}(\mathbf{F}(\mathbf{x}))) \approx \mathbf{x}, \ \mathbf{g}_{GSE,out}(h_{GSE}(\mathbf{F}(\mathbf{x}))) \approx \mathbf{f}(\mathbf{x}), \qquad (12)$$

but the left part of the latter equation cannot be used for estimating the unknown function $\mathbf{f}(\mathbf{x})$ since it depends on the function $h_{GSE}(\mathbf{F}(\mathbf{x}))$, which in its turn depends on the function $\mathbf{f}(\mathbf{x})$.

According to the MLR approach we construct the estimator $\varphi(\mathbf{x})$ for the function $h_{GSE}(\mathbf{F}(\mathbf{x}))$ as follows. We have two parameterizations of the manifold points $\mathbf{Z} = \mathbf{F}(\mathbf{x}) \in \mathbf{M}(\mathbf{f})$: the "natural" parameterization by the input $\mathbf{x} \in \mathbf{X}$ and the GSE-parameterization $t = h_{GSE}(\mathbf{Z})$, which are linked by the unknown one-to-one mapping $t = \varphi(\mathbf{x})$, whose values $\{\varphi(\mathbf{x}_i) = t_i = h_{GSE}(\mathbf{Z}_i)\}$ are known at the sample inputs $\{\mathbf{x}_i\}$. The relations (5) and (12) imply that $\mathbf{g}_{GSE,in}(\varphi(\mathbf{x})) \approx \mathbf{x}$ and $\mathbf{G}_{GSE,in}(\varphi(\mathbf{x})) \cdot \mathbf{J}_\varphi(\mathbf{x}) \approx \mathbf{I}_q$. Thus we get that $\mathbf{J}_\varphi(\mathbf{x}) \approx \mathbf{G}_{GSE,in}^{-1}(\varphi(\mathbf{x}))$; here $\mathbf{J}_\varphi(\mathbf{x})$ is the Jacobian matrix of the mapping $\varphi(\mathbf{x})$. Therefore, the known matrices $\{\mathbf{G}_{GSE,in}^{-1}(\varphi(\mathbf{x}_i)) = \mathbf{G}_{GSE,in}^{-1}(t_i)\}$ estimate the Jacobian matrices $\{\mathbf{J}_\varphi(\mathbf{x}_i)\}$ at the sample inputs $\{\mathbf{x}_i\}$.

Based on the known values $\{(\varphi(\mathbf{x}_i), \mathbf{J}_\varphi(\mathbf{x}_i))\}$, $\varphi(\mathbf{x})$ is estimated at the arbitrary point $\mathbf{x}$ by $\varphi_{MLR}(\mathbf{x}) = \frac{1}{K_q(\mathbf{x})} \sum_{j=1}^n K_q(\mathbf{x}, \mathbf{x}_j) \cdot \{t_j + \mathbf{G}_{GSE,in}^{-1} \cdot (\mathbf{x} - \mathbf{x}_j)\}$; here $K_q(\mathbf{x}, \mathbf{x}')$ is a stationary kernel in $\mathbb{R}^q$ (like $K_{p,\varepsilon,\rho}$, but defined in $\mathbb{R}^q$).

The relations (12) imply that $\mathbf{G}_{GSE,out}(\varphi(\mathbf{x})) \cdot \mathbf{J}_\varphi(\mathbf{x}) \approx \mathbf{J}_f(\mathbf{x})$ and we get

$$\mathbf{f}_{MLR}(\mathbf{x}) = \mathbf{g}_{GSE,out}(\varphi_{MLR}(\mathbf{x})), \qquad (13)$$

$$\mathbf{G}_{MLR}(\mathbf{x}) = \mathbf{G}_{GSE,out}(\varphi_{MLR}(\mathbf{x})) \cdot \mathbf{G}_{GSE,in}^{-1}(\varphi_{MLR}(\mathbf{x})) \qquad (14)$$

as the estimators for the unknown function $\mathbf{f}(\mathbf{x})$ and its Jacobian matrix $\mathbf{J}_f(\mathbf{x})$.

Note that the estimators (13), (14) require constructing the aligned bases (matrices $\mathbf{H}_{GSE}(\mathbf{Z})$), the embedding mapping $h_{GSE}(\mathbf{Z})$, the recovery mapping $\mathbf{g}_{GSE}(t)$ and the estimator $\mathbf{G}_{GSE}(t)$ for its Jacobian matrix, and the reparameterization mapping $\varphi_{MLR}(\mathbf{x})$. These GSE steps are computationally expensive, even if the incremental version of GSE is used [29].

3   Modified Manifold Learning Regression

The proposed modified version of the MLR method consists of the following parts: constructing both the PCA-approximations for the tangent spaces at the sample points (as in case of the GSE algorithm) and the preliminary estimation of $\mathbf{f}(\mathbf{x})$ for arbitrary inputs (Sect. 3.1), constructing both the PCA-approximations $\mathbf{L}_{PCA}(\mathbf{Z})$ at the OoS points $\mathbf{Z} = \mathbf{F}(\mathbf{x})$ and the estimators $\mathbf{G}_{MLR}(\mathbf{x})$ of the Jacobian matrix $\mathbf{J}_f(\mathbf{x})$ for arbitrary inputs (Sect. 3.2), constructing the non-stationary kernels based on the preliminary MLR estimators and their usage for construction of both the new adaptive PCA-approximations and the final estimators $(\mathbf{f}_{mMLR}(\mathbf{x}), \mathbf{G}_{mMLR}(\mathbf{x}))$.

3.1 Preliminary Estimation of Unknown Functions

We start from the standard PCA-approximations for the tangent spaces $L(\mathbf{Z})$ at the sample points.

Step 1. Given the training dataset $\mathbf{Z}_{(n)}$ (1), $p \times q$ matrices $Q_{PCA}(\mathbf{Z}_i)$ and linear spaces $L_{PCA}(\mathbf{Z}_i) = \mathrm{Span}(Q_{PCA}(\mathbf{Z}_i))$, $i = 1, 2, \ldots, n$, are constructed as in Sect. 2.2.

Let $\{\mathbf{H}_{GSE}(\mathbf{Z}_i) = Q_{PCA}(\mathbf{Z}_i) \cdot v(\mathbf{Z}_i)\}$ (10) be the GSE-matrices, computed after the estimation of the aligning matrices $\{v(\mathbf{Z}_i)\}$. It follows from (7) and (9)–(11) that

$$\mathbf{G}_{GSE,in}(h_{GSE}(\mathbf{Z})) = \mathbf{H}_{GSE,in}(\mathbf{Z}) = Q_{PCA,in}(\mathbf{Z}) \cdot v(\mathbf{Z}),$$
$$\mathbf{G}_{GSE,out}(h_{GSE}(\mathbf{Z})) = \mathbf{H}_{GSE,out}(\mathbf{Z}) = Q_{PCA,out}(\mathbf{Z}) \cdot v(\mathbf{Z}).$$

Thus the estimator $\mathbf{G}_{MLR}(\mathbf{x})$ (14) at the sample inputs $\{\mathbf{x}_i\}$ is equal to

$$\mathbf{G}_{MLR}(\mathbf{x}_i) = \mathbf{H}_{GSE,out}(\mathbf{Z}_i) \cdot \mathbf{H}_{GSE,in}^{-1}(\mathbf{Z}_i)$$
$$= Q_{PCA,out}(\mathbf{Z}_i) v(\mathbf{Z}_i) v^{-1}(\mathbf{Z}_i) Q_{PCA,in}(\mathbf{Z}_i) = Q_{PCA,out}(\mathbf{Z}_i) Q_{PCA,in}^{-1}(\mathbf{Z}_i) \quad (15)$$

and depends only on the PCA-matrices $\{Q_{PCA}(\mathbf{Z}_i)\}$, not on the matrices $v(\mathbf{Z}_i)$.

Step 2. Compute the estimators $\{\mathbf{G}_{MLR}(\mathbf{x}_i)\}$ (15) for $i = 1, 2, \ldots, n$.

After the Step 2 we obtain values $\mathbf{G}_{MLR}(\mathbf{x}_i)$ of the Jacobian matrix of $\mathbf{f}(\mathbf{x})$ at the sample inputs. Using the Taylor series expansion we get that $\mathbf{f}(\mathbf{x}) \approx \mathbf{f}(\mathbf{x}') + \mathbf{J}_f(\mathbf{x}') \cdot (\mathbf{x} - \mathbf{x}')$ for the neighboring input points $\mathbf{x}, \mathbf{x}' \in \mathbf{X}$. We construct the estimator $\mathbf{f}^*(\mathbf{x})$ for $\mathbf{f}(\mathbf{x})$ at the arbitrary point $\mathbf{x}$ as a solution to the regression problem with known Jacobian values at sample points [30] by minimizing the residual $\sum_{j=1}^{n} K_q(\mathbf{x}, \mathbf{x}_j) \cdot |\mathbf{y} - \mathbf{y}_j - \mathbf{G}_{MLR}(\mathbf{x}_j) \cdot (\mathbf{x} - \mathbf{x}_j)|^2$ over $\mathbf{y}$.

Step 3. Compute the estimator $\mathbf{f}^*(\mathbf{x})$ at the arbitrary input $\mathbf{x} \in \mathbf{X}$

$$\mathbf{f}^*(\mathbf{x}) = \frac{1}{K_q(\mathbf{x})} \sum_{j=1}^{n} K_q(\mathbf{x}, \mathbf{x}_j) \cdot \{\mathbf{y}_j + \mathbf{G}_{MLR}(\mathbf{x}_j) \cdot (\mathbf{x} - \mathbf{x}_j)\}$$

$$= \mathbf{f}_{sKNR}(\mathbf{x}) + \frac{1}{K_q(\mathbf{x})} \sum_{j=1}^{n} K_q(\mathbf{x}, \mathbf{x}_j) \cdot \mathbf{G}_{MLR}(\mathbf{x}_j) \cdot (\mathbf{x} - \mathbf{x}_j). \quad (16)$$

Here $\mathbf{f}_{sKNR}(\mathbf{x}) = \frac{1}{K_q(\mathbf{x})} \sum_{j=1}^{n} K_q(\mathbf{x}, \mathbf{x}_j) \cdot \mathbf{y}_j$ is the KNR-estimator (2) with a stationary kernel.

Note that the estimators $\mathbf{f}^*(\mathbf{x})$ (16) and $\{\mathbf{G}_{MLR}(\mathbf{x}_i)\}$ (15) coincide with the MLR-estimators (13) and (14) but they have significantly lower computational complexity.

3.2 Estimation of Jacobian Matrix at Arbitrary Point

The $p \times q$ matrix $Q_{PCA}(\mathbf{Z})$ and the tangent space $L_{PCA}(\mathbf{Z})$ at the OoS point $\mathbf{Z} = \mathbf{F}(\mathbf{x})$ are computed using the estimator $\mathbf{f}^*(\mathbf{x})$ (16). Thus we can define $\mathbf{F}_{MLR}(\mathbf{x}) = (\mathbf{x}, \mathbf{f}^*(\mathbf{x}))$ (4).

Step 4. Compute the $p \times q$ matrix $Q_{PCA}(\mathbf{Z}^*)$ at the point $\mathbf{Z}^* = \mathbf{F}_{MLR}(\mathbf{x})$, such that its columns are the eigenvectors of the matrix $\Sigma(\mathbf{Z}^*|K_p)$ (8) corresponding to the q largest eigenvalues.

The matrix $Q_{PCA}(\mathbf{F}_{MLR}(\mathbf{x}))$ estimates the matrix $Q_{PCA}(\mathbf{F}(\mathbf{x}))$ at the arbitrary input $\mathbf{x} \in \mathbf{X}$. Thus, the relation (14) results in the next step.

Step 5. Compute the preliminary estimator $\mathbf{G}_{MLR}(\mathbf{x})$ for $\mathbf{J}_f(\mathbf{x})$ at the arbitrary input $\mathbf{x} \in \mathbf{X}$

$$\mathbf{G}_{MLR}(\mathbf{x}) = Q_{PCA,out}(\mathbf{F}_{MLR}(\mathbf{x})) \cdot Q_{PCA,in}^{-1}(\mathbf{F}_{MLR}(\mathbf{x})). \tag{17}$$

Then based on (17) we compute the preliminary estimators

$$\mathbf{f}_{MLR}(\mathbf{x}) = \frac{1}{K_q(\mathbf{x})} \sum_{j=1}^{n} K_q(\mathbf{x}, \mathbf{x}_j) \cdot \{\mathbf{y}_j + \mathbf{G}_{MLR}(\mathbf{x}) \cdot (\mathbf{x} - \mathbf{x}_j)\}$$

$$= \mathbf{f}_{sKNR}(\mathbf{x}) + \mathbf{G}_{MLR}(\mathbf{x}) \cdot (\mathbf{x} - \overline{\mathbf{x}}_{sKNR}) \tag{18}$$

for $\mathbf{f}(\mathbf{x})$ at the arbitrary input $\mathbf{x} \in \mathbf{X}$; here $\overline{\mathbf{x}}_{sKNR} = \frac{1}{K_q(\mathbf{x})} \sum_{j=1}^{n} K_q(\mathbf{x}, \mathbf{x}_j) \cdot \mathbf{x}_j$.

3.3 Estimation of Unknown Function at Arbitrary Point

The estimators $\mathbf{f}_{MLR}(\mathbf{x})$ (18) and $\mathbf{G}_{MLR}(\mathbf{x})$ (17) use the stationary kernels $K_q(\mathbf{x}, \mathbf{x}')$ in (18) and $K_p(\mathbf{Z}, \mathbf{Z}')$ in $\Sigma(\mathbf{Z}^*|K_p)$ (7), respectively; here we introduce their non-stationary analogues.

Let $L = \text{Span}(Q)$ and $L' = \text{Span}(Q')$ be q-dimensional linear spaces in $\mathbb{R}^p$ whose orthonormal bases are the columns of the $p \times q$ orthogonal matrices Q and Q', respectively. Considering them as elements of the Grassmann manifold $\text{Grass}(p, q)$, let us denote by

$$d_{BC}(L, L') - \{1 - \text{Det}^2[Q^T \cdot Q']\}^{1/2} \quad \text{and} \quad K_{BC}(L, L') = \text{Det}^2[Q^T \cdot Q']$$

the Binet-Cauchy metric and the Binet-Cauchy kernel on the Grassmann manifold, respectively [37,38]. Note that these quantities do not depend on a choice of the orthonormal bases Q and Q'. Let us introduce another Grassmann kernel depending on the threshold τ as

$$K_{G,\tau}(L, L') = \text{I}\{d_{BC}(L, L') \leq \tau\} \cdot K_{BC}(L, L').$$

The final mMLR estimators are constructed by modification of the Steps 1–5 above using the introduced non-stationary kernels. For $\mathbf{Z}, \mathbf{Z}' \in \mathbf{Z}_{(n)}$, we introduce the non-stationary kernel

$$K_{p,MLR}(\mathbf{Z}, \mathbf{Z}') = K_{p,\varepsilon,\rho}(\mathbf{Z}, \mathbf{Z}') \cdot K_{G,\tau}(L_{PCA}(\mathbf{Z}), L_{PCA}(\mathbf{Z}')). \tag{19}$$

Step 6 (modified Step 1). The columns of the orthogonal $p \times q$ matrices $Q_{mPCA}(\mathbf{Z}_i)$ at sample points consist of the eigenvectors of the matrices $\Sigma(\mathbf{Z}_i|K_{p,MLR})$ (8) corresponding to its q largest eigenvalues, $i = 1, 2, \ldots, n$.

When calculating the covariance matrices $\Sigma(\mathbf{Z}_i|K_{p,MLR})$ we use the non-stationary kernels $K_{p,MLR}$ (19) at the sample points.

Step 7 (modified Step 2). Using (17) with the matrices $\{Q_{PCA}(\mathbf{Z}_i)\}$ replaced by the matrices $\{Q_{mPCA}(\mathbf{Z}_i)\}$ we compute the modified $m \times q$ matrices $\{\mathbf{G}_{mMLR}(\mathbf{x}_i)\}$.

Step 8 (modified Step 3). The value $\mathbf{f}^{**}(\mathbf{x})$ at the arbitrary input $\mathbf{x} \in \mathbf{X}$ is computed by (16) with the matrices $\{\mathbf{G}_{MLR}(\mathbf{x}_i)\}$ replaced by the matrices $\{\mathbf{G}_{mMLR}(\mathbf{x}_i)\}$.

Step 9 (modified Step 4). We compute the $p \times q$ matrix $Q_{mPCA}(\mathbf{Z})$ at the point $\mathbf{Z} = \mathbf{F}_{mMLR}(\mathbf{x}) = (\mathbf{x}, \mathbf{f}^{**}(\mathbf{x}))$ with arbitrary input $\mathbf{x} \in \mathbf{X}$. Columns of this matrix are the eigenvectors of the matrix $\Sigma(\mathbf{F}_{mMLR}(\mathbf{x})|K_{p,MLR})$ (8) corresponding to its q largest eigenvalues with the non-stationary kernel $K_{p,MLR}(\mathbf{Z},\mathbf{Z}')$ (19), $\mathbf{Z},\mathbf{Z}' \in \mathbf{Z}_{(n)}$.

Let us denote $\mathbf{L}_{mPCA}(\mathbf{F}_{mMLR}(\mathbf{x})) = \mathrm{Span}(Q_{mPCA}(\mathbf{F}_{mMLR}(\mathbf{x})))$. For the arbitrary inputs $\mathbf{x}, \mathbf{x}' \in \mathbf{X}$ we introduce the non-stationary kernel

$$K_{q,MLR}(\mathbf{x},\mathbf{x}') = K_{q,\varepsilon,\rho}(\mathbf{x},\mathbf{x}') \cdot K_{G,\tau}(\mathbf{L}_{mPCA}(\mathbf{F}_{mMLR}(\mathbf{x})), \mathbf{L}_{mPCA}(\mathbf{F}(\mathbf{x}))). \quad (20)$$

Step 10 (modified Step 5). We compute the final estimators $\mathbf{G}_{mMLR}(\mathbf{x})$ for $\mathbf{J}_f(\mathbf{x})$ at the arbitrary input $\mathbf{x} \in \mathbf{X}$ by the formula (17), where $Q_{PCA}(\mathbf{F}_{MLR}(\mathbf{x}))$ is replaced by $Q_{mPCA}(\mathbf{F}_{mMLR}(\mathbf{x}))$.

After that, we compute the final estimators $\mathbf{f}_{mMLR}(\mathbf{x})$ for $\mathbf{f}(\mathbf{x})$ at the arbitrary input $\mathbf{x} \in \mathbf{X}$ by the formula (18) in which $\mathbf{G}_{MLR}(\mathbf{x})$ is replaced by $\mathbf{G}_{mMLR}(\mathbf{x})$, the KNR-estimators $\mathbf{f}_{sKNR}(\mathbf{x})$ and $\overline{\mathbf{x}}_{sKNR}$ with the stationary kernel K_q are replaced by the KNR-estimators $\mathbf{f}_{nsKNR}(\mathbf{x})$ and $\overline{\mathbf{x}}_{nsKNR}$ with the non-stationary kernel $K_{q,MLR}$ (20), respectively.

4   Conclusion

The initially proposed Manifold Learning Regression (MLR) method was based on the GSE-solution to the Tangent Bundle Manifold Learning problem, which is very computationally expensive. The paper proposes a modified version of the MLR method, which does not require to use the most of GSE/MLR steps (such as constructing the aligned bases at the estimated tangent spaces, the embedding and the recovery mappings, the reparameterization mapping, etc.). As a result the modified estimator has significantly smaller computational complexity while preserving its accuracy.

References

1. Seber, G., Wild, C.: Nonlinear Regression. Wiley, Hoboken (2003)
2. Vapnik, V.N.: Statistical Learning Theory. Wiley-Interscience, Hoboken (1998)
3. Loader, C.: Local Regression and Likelihood. Springer, New York (1999). https://doi.org/10.1007/b98858

4. Belyaev, M., Burnaev, E., Kapushev, E., et al.: GTApprox: surrogate modeling for industrial design. Adv. Eng. Softw. **102**, 29–39 (2016)
5. Rasmussen, C.E., Williams, C.K.I.: Gaussian Processes for Machine Learning. Adaptive Computation and Machine Learning. The MIT Press, Cambridge (2005)
6. Calandra, R., Peters, J., Rasmussen, C., Deisenroth, M.: Manifold Gaussian processes for regression. CoRR abs/1402.5876v4 (2014)
7. Wasserman, L.: All of Nonparametric Statistics. Springer, Berlin (2007). https://doi.org/10.1007/0-387-30623-4
8. Bishop, C.M.: Pattern Recognition and Machine Learning. Information Science and Statistics. Springer, New York (2006)
9. Burnaev, E., Vovk, V.: Efficiency of conformalized ridge regression. CoRR arXiv:abs/1404.2083 (2014)
10. Bernstein, A., Kuleshov, A., Yanovich, Y.: Manifold learning in regression tasks. In: Gammerman, A., Vovk, V., Papadopoulos, H. (eds.) SLDS 2015. LNCS (LNAI), vol. 9047, pp. 414–423. Springer, Cham (2015). https://doi.org/10.1007/978-3-319-17091-6_36
11. Bernstein, A.V., Kuleshov, A.P., Yanovich, Y.: Statistical learning via manifold learning. In: 14th IEEE International Conference on Machine Learning and Applications, ICMLA 2015, Miami, FL, USA, 9–11 December 2015, pp. 64–69 (2015)
12. Burnaev, E.V., Panov, M.E., Zaytsev, A.A.: Regression on the basis of nonstationary Gaussian processes with Bayesian regularization. J. Commun. Technol. Electron. **61**(6), 661–671 (2016)
13. Burnaev, E., Nazarov, I.: Conformalized kernel ridge regression. In: 15th IEEE International Conference on Machine Learning and Applications, ICMLA 2016, Anaheim, CA, USA, 18–20 December 2016, pp. 45–52 (2016)
14. Burnaev, E., Panov, M.: Adaptive design of experiments based on Gaussian processes. In: Gammerman, A., Vovk, V., Papadopoulos, H. (eds.) SLDS 2015. LNCS (LNAI), vol. 9047, pp. 116–125. Springer, Cham (2015). https://doi.org/10.1007/978-3-319-17091-6_7
15. Belyaev, M., Burnaev, E., Kapushev, Y.: Gaussian process regression for structured data sets. In: Gammerman, A., Vovk, V., Papadopoulos, H. (eds.) SLDS 2015. LNCS (LNAI), vol. 9047, pp. 106–115. Springer, Cham (2015). https://doi.org/10.1007/978-3-319-17091-6_6
16. Yang, Y., Dunson, D.: Bayesian manifold regression. CoRR abs/1305.0167v2 (2014)
17. Genton, M.G.: Classes of kernels for machine learning: a statistical perspective. JMLR **2**, 299–312 (2001)
18. Burnaev, E., Zaytsev, A., Spokoiny, V.: The Bernstein-von Mises theorem for regression based on Gaussian processes. Russ. Math. Surv. **68**(5), 954–956 (2013)
19. Zaitsev, A.A., Burnaev, E.V., Spokoiny, V.G.: Properties of the posterior distribution of a regression model based on Gaussian random fields. Autom. Remote Control **74**(10), 1645–1655 (2013)
20. Krige, D.: A statistical approach to some basic mine valuation problems on the witwatersrand. J. Chem. Metall. Mining Eng. Soc. South Affrica **52**(6), 119–139 (1951)
21. Sacks, J., Welch, W., Mitchell, T., Wynn, H.: Design and analysis of computer experiments. Stat. Sci. **4**(4), 409–435 (1989)
22. Simpson, T.W., et al.: Metamodels for computer-based engineering design: survey and recommendations. Eng. Comput. **7**(2), 129–150 (2001)
23. Wang, G., Gary, S.S.: Review of metamodeling techniques in support of engineering design optimization. J. Mech. Des. **129**(3), 370–381 (2007)

24. Forrester, A., Sobester, A., Keane, A.: Engineering Design via Surrogate Modelling. A Practical Guide. Wiley, New York (2008)
25. Xiong, Y., Chen, W., Apley, D., Ding, X.: A non-stationary covariance-based kriging method for metamodelling in engineering design. Int. J. Numerical Methods Eng. **71**(6), 733–756 (2006)
26. Toal, D.J., Keane, A.J.: Non-stationary kriging for design optimization. Eng. Optim. **44**(6), 741–765 (2012)
27. Sampson, P.D., Guttorp, P.: Nonparametric estimation of nonstationary spatial covariance structure. J. Am. Stat. Assoc. **87**(417), 108–119 (1992)
28. Bernstein, A., Kuleshov, A.: Low-dimensional data representation in data analysis. In: El Gayar, N., Schwenker, F., Suen, C. (eds.) ANNPR 2014. LNCS (LNAI), vol. 8774, pp. 47–58. Springer, Cham (2014). https://doi.org/10.1007/978-3-319-11656-3_5
29. Kuleshov, A., Bernstein, A.: Incremental construction of low-dimensional data representations. In: Schwenker, F., Abbas, H.M., El Gayar, N., Trentin, E. (eds.) ANNPR 2016. LNCS (LNAI), vol. 9896, pp. 55–67. Springer, Cham (2016). https://doi.org/10.1007/978-3-319-46182-3_5
30. Kuleshov, A.P., Bernstein, A.: Nonlinear multi-output regression on unknown input manifold. Ann. Math. Artif. Intell. **81**(1–2), 209–240 (2017)
31. Bernstein, A.V., Kuleshov, A.P.: Tangent bundle manifold learning via grassmann & stiefel eigenmaps. CoRR abs/1212.6031 (2012)
32. Bernstein, A., Kuleshov, A.: Manifold learning: generalization ability and tangent proximity. Int. J. Softw. Inf. **7**(3), 359–390 (2013)
33. Jost, J.: Riemannian Geometry and Geometric Analysis. Springer, Heidelberg (2002). https://doi.org/10.1007/978-3-642-21298-7
34. Zhang, Z., Zha, H.: Principal manifolds and nonlinear dimension reduction via local tangent space alignment. SIAM J. Sci. Comput. **26**(1), 313–338 (2005)
35. Belkin, M., Niyogi, P.: Laplacian eigenmaps for dimensionality reduction and data representation. Neural Comput. **15**, 1373–1396 (2003)
36. Tyagi, H., Vural, E., Frossard, P.: Tangent space estimation for smooth embeddings of riemannian manifold. CoRR abs/1208.1065v2 (2013)
37. Hamm, J., Lee, D.D.: Grassmann discriminant analysis: a unifying view on subspace-based learning. In: Koller, D., Schuurmans, D., Bengio, T., Bottou, L. (eds.) The 25th NIPS Conference, Advances in Neural Information Processing Systems 21, pp. 376–383. MIT Press, Cambridge (2009)
38. Wolf, L., Shashua, A.: Learning over sets using kernel principal angles. J. Mach. Learn. Res. **4**, 913–931 (2003)

F-Measure Curves for Visualizing Classifier Performance with Imbalanced Data

Roghayeh Soleymani[1], Eric Granger[1], and Giorgio Fumera[2(✉)]

[1] Laboratoire d'imagerie, de vision et d'intelligence artificielle,
École de technologie supérieure, Université du Québec, Montreal, Canada
rSoleymani@livia.etsmtl.ca, Eric.Granger@etsmtl.ca
[2] Pattern Recognition and Applications Lab,
Department of Electrical and Electronic Engineering, University of Cagliari,
Cagliari, Italy
fumera@diee.unica.it

Abstract. Training classifiers using imbalanced data is a challenging problem in many real-world recognition applications due in part to the bias in performance that occur for: (1) classifiers that are often optimized and compared using unsuitable performance measurements for imbalance problems; (2) classifiers that are trained and tested on a fixed imbalance level of data, which may differ from operational scenarios; (3) cases where the preference of correct classification of classes is application dependent. Specialized performance evaluation metrics and tools are needed for problems that involve class imbalance, including scalar metrics that assume a given operating condition (skew level and relative preference of classes), and global evaluation curves or metrics that consider a range of operating conditions. We propose a global evaluation space for the scalar F-measure metric that is analogous to the cost curves for expected cost. In this space, a classifier is represented as a curve that shows its performance over all of its decision thresholds and a range of imbalance levels for the desired preference of true positive rate to precision. Experiments with synthetic data show the benefits of evaluating and comparing classifiers under different operating conditions in the proposed F-measure space over ROC, precision-recall, and cost spaces.

Keywords: Class imbalance · Performance visualization tools
F-measure

1 Introduction

Evaluating performance is a critical step in classifier design and comparison. Classification accuracy is the most widely used performance metric, also used as the objective function of many state-of-the-art learning algorithms (e.g., support

© Springer Nature Switzerland AG 2018
L. Pancioni et al. (Eds.): ANNPR 2018, LNAI 11081, pp. 165–177, 2018.
https://doi.org/10.1007/978-3-319-99978-4_13

vector machines). However, when data from different classes are imbalanced, it favours the correct classification of the majority classes at the expense of high misclassification rates for the minority ones. This is an issue in many detection problems where samples of the class of interest ("positive" or "target" class) are heavily outnumbered by those of other ("negative" or "non-target") classes. The widely used ROC curve (which plots the true positive rate vs the false positive rate for two-class classification problems), is not suitable for imbalanced data either, since it is independent of the level of imbalance. The alternative Precision-Recall (PR) curve is more suitable than ROC space, since precision is sensitive to imbalance; however, the performance of a given classifier under different imbalance levels corresponds to different PR curves, which makes it difficult to evaluate and compare classifiers.

Alternatively, scalar performance metrics like the expected cost (EC) and the F-measure (widely used in information retrieval) are typically employed when data is imbalanced. Since they seek different trade-offs between positive and negative samples, the choice between them is application-dependent. EC allows to indirectly address class imbalance by assigning different misclassification costs to positive and negative samples. Two graphical techniques have recently been proposed to easily visualize and compare classifier performance in terms of EC under all possible operating conditions: cost curves (CC) [3] and Brier curves (BC) [5]. The F-measure, recently analyzed by many researchers [2,12–14] is defined as the weighted harmonic mean of precision and recall, and thus evaluates classifier performance using a weight that controls the relative importance of recall (i.e., the true positive rate) and precision, which is sensitive to class imbalance. However, no performance visualization tool analogous to CC or BC exists for the F-measure. One may use the PR space to this aim, but the isometrics of the F-measure in PR space are hyperbolic [7,9], which does not allow to easily evaluate classifiers under diverse operating conditions.

This paper introduces F-measure curves, a global visualization tool for the F-measure analogous to CC. It consists in plotting the F-measure of a given classifier versus two parameters – the level of imbalance and the preference between recall and precision – and allows to visualize and compare classifier performance in class imbalance problems for different decision thresholds, under different operating conditions. In this space, a crisp classifier corresponds to a curve that shows its F-measure over all possible imbalance levels, for a desired level of preference between recall and precision. A soft classifier corresponds to the upper envelope of such curves for all possible decision thresholds. This space allows to compare classifiers more easily than in the PR space for a given operating condition, analogously to CC or BC vs the ROC space. For a given preference level between precision and recall, one classifier may outperform another over all skew levels, or only for a specific range, which can be determined both analytically and empirically in the proposed space, as with the CC space. To clarify the benefits of the proposed space, experiments are performed on synthetic data.

2   Performance Metrics and Visualization Tools

In many real-world applications, the distribution of data is imbalanced [10]; correctly recognizing positive samples is the main requirement, while avoiding excessive misclassification of negative samples can also be important. If application requirements are given by misclassification costs, misclassification of positive samples usually exhibits a higher cost, which "indirectly" addresses class imbalance. Otherwise, assigning different "fictitious" costs to misclassifications of positive and negative samples can be an indirect means to achieve the same goal. Several performance metrics have been proposed so far for applications involving imbalanced classes [1,6,8,11,15]. This section provides a review of these metrics in terms of their sensitivity to imbalance, focusing on global spaces that consider different operating conditions and preference weights.

Scalar Performance Metrics. We focus on two-class problems, although some metrics can also be applied in multi-class cases. Let $P(+)$ and $P(-)$ be the prior probability of the positive and negative class, and $\lambda = {P(-)}/{P(+)}$ the class skew. From a given data set with n_+ positive and n_- negative samples, $P(+)$ can be estimated as ${n_+}/{(n_+ + n_-)}$, and similarly for $P(-)$, whereas λ can be estimated as ${n_-}/{n_+}$. As in [3], we focus on evaluating classifier performance as a function of the prior of the positive class when the classifier is deployed, which can be different than in the training and testing sets; accordingly, from now on we will use $P(+)$ (and $P(-)$) to denote the class prior during classifier deployment (use). Since this value is unknown during classifier design, we will evaluate classifier performance across all possible $P(+)$ values.

   Classifier performance on a given data set can be summarized by its confusion matrix, in terms of the true positives (TP), false positives (FP), true negatives (TN) and false negatives (FN) counts. Let N_+ and N_- the number of samples classified as positive and negative, respectively. The corresponding rates are defined as $TPR = TP/n_+$, $FNR = FN/n_+$, $TNR = TN/n_-$ and $FPR = FP/n_-$.

   Several scalar metrics can be defined from the above rates. The widely used error rate, defined as $(FP + FN)/(n_+ + n_-)$, is biased towards the correct classification of the negative (majority) class, which is not suitable to imbalanced data. When costs can be associated to classification outcomes (either correct or incorrect), the expected cost (EC) is used; denoting as C_{FN} and C_{FP} the misclassification costs of positive and negative samples (usually the cost of correct classifications is zero), EC is defined as:

$$EC = \text{FNR} \cdot P(+) \cdot C_{FN} + \text{FPR} \cdot P(-) \cdot C_{FP} \tag{1}$$

When data is imbalanced, usually $C_{FN} > C_{FP}$, which can also avoid the bias of the error probability toward the negative class. Accordingly, by setting suitable fictitious costs, EC can also be used to deal with class imbalance even if misclassification costs are not precisely known or difficult to define. However, as C_{FN}/C_{FP} increases, minimizing EC increases TPR at the expense of increasing FPR, which may be undesirable.

In information retrieval applications the complementary metrics Precision (Pr) and Recall (Re) are often used, instead: Re corresponds to TPR, whereas Pr is defined as $TP/(TP + FP)$ or TP/N_+. Pr depends on both TP and FP, and drops severely when correct classification of positive class is attained at the expense of a high fraction of misclassified negative samples, as can be seen by rewriting Pr as:

$$Pr = \frac{\frac{TP}{n_+}}{\frac{TP}{n_+} + \frac{FP}{n_+} \times \frac{n_-}{n_-}} = \frac{TPR}{TPR + \lambda FPR}. \tag{2}$$

This is useful to reveal the effect of class imbalance, compared to EC.

Pr and Re can be combined into the F-measure scalar metric [16], defined as their weighted harmonic mean:

$$F_\alpha = \frac{1}{\alpha \frac{1}{Pr} + (1 - \alpha)\frac{1}{Re}}, \tag{3}$$

where $0 < \alpha < 1$. By rewriting α as $(1+\beta^2)^{-1}$, $\beta \in [0, +\infty)$, F_α can be rewritten as:

$$F_\beta = \frac{(1 + \beta^2)Pr \cdot Re}{\beta^2 Pr + Re} = \frac{(1 + \beta^2)TP}{(1 + \beta^2)TP + FP + \beta^2 FN}. \tag{4}$$

When $\alpha \to 0$, $F_\alpha \to$ Re, and when $\alpha \to 1$, $F_\alpha \to$ Pr. Note that the sensitivity of the F-measure to the positive and negative classes can be adjusted by tuning α. This measure can be preferable to EC for imbalanced data, since it weighs the relative importance of TPR (i.e., Re) and Pr, rather than TPR and FPR.

Other metrics have been used, or specifically proposed, for class imbalance problems, although they are currently less used than EC and the F-measure [6,8].

Global Evaluation Curves. In many applications it is desirable for the classifier to perform well over a wide range of operating conditions, i.e., the misclassification costs or the relative importance between Pr and Re, and the class priors. Global curves depict the trade-offs between different evaluation metrics under different operating conditions, without reducing them to an incomplete scalar measure.

The ROC curve is widely used for two-class classifiers: it plots TPR vs FPR as a function of the decision threshold. A classifier with a specific threshold corresponds to a point in ROC space; a potentially optimal classifier lies on the ROC convex hull (ROCCH) of the available points, regardless of operating conditions. The best thresholds correspond to the upper-left point, corresponding to the higher TPR and the lower FPR (see Fig. 4(a)). A drawback of the ROC space is that it does not reflect the impact of imbalance, since TPR and FPR do not depend on class priors [4]. The performance of a classifier for a given skew level can be indirectly estimated in terms of EC, since in ROC space, each operating condition corresponds to a set of isoperformance lines with identical slope. An optimal classifier for a given operating condition is found by intersecting the ROCCH with the upper-left isoperformance line.

When Pr and Re are used, their trade-off across different decision thresholds can be evaluated by the precision-recall (PR) curve, which plots Pr vs Re. The PR curve is sensitive to class imbalance, given its dependence on Pr. However, different operating conditions (skew levels) lead to different PR curves, which makes classifier comparison difficult. Moreover, differently from ROC space, the convex hull of a set of points in PR space has no clear meaning [7]. If the F-measure is used, its isometrics can be analytically obtained in PR space, analogously to EC isometrics in ROC space; however they are hyperbolic [7,9], which makes it difficult to visualize classifier performance over a range of decision thresholds, skew levels, and preference of Pr to Re. In the case of EC this problem has been addressed by the CC visualization tool, described below, and by its BC extension. Inspired by CC, we propose in Sect. 3 an analogous visualization tool for the F-measure.

Expected Costs Visualization Tools. CCs [3] are used to visualize EC over a range of misclassification costs and skew levels. More precisely, CCs visualize the normalised EC (NEC), which is defined as EC divided by the maximum possible value of EC; the latter value turns out to be $P(+)C_{\mathrm{FN}} + P(-)C_{\mathrm{FP}}$, and NEC can be written as:

$$NEC = (\mathrm{FNR} - \mathrm{FPR})PC(+) + \mathrm{FPR} \in [0,1], \tag{5}$$

where $PC(+)$ is the "probability times cost" normalization term, which is defined as:

$$PC(+) = \frac{P(+) \cdot C_{\mathrm{FN}}}{P(+) \cdot C_{\mathrm{FN}} + P(-)C_{\mathrm{FP}}} \in [0,1]. \tag{6}$$

CCs are obtained by depicting NEC versus $PC(+)$ on a $[0,1] \times [0,1]$ plot, which is named "cost space". Note that $NEC = \mathrm{FPR}$, if $PC(+) = 0$, and $NEC = \mathrm{FNR} = 1 - \mathrm{TPR}$, if $PC(+) = 1$. The always positive and always negative classifiers correspond to two lines connecting points $(1,0)$ to $(0,1)$, and $(0,0)$ to $(1,1)$, respectively, in the cost space. The operating range of a classifier is the set of operating points for which it dominates both these lines [3]. By defining:

$$m = \frac{C_{\mathrm{FP}}}{C_{\mathrm{FP}} + C_{\mathrm{FN}}}, \text{ where } 0 < m \le 1 \tag{7}$$

m can be seens as weighing the importance of both classes, and Eq. (6) can be rewritten as:

$$PC(+) = \frac{(1/m - 1) \cdot P(+)}{(1/m - 2) \cdot P(+) + 1} \tag{8}$$

The CCs of two classifiers C_i and C_j may cross: in this case each classifier outperforms the other for a certain range of operating points.

Interestingly, there is a point-line duality between CC and ROC space: a point in ROC space is a line in cost space, and vice versa. The lower envelope of cost lines corresponds to the ROCCH in ROC space. In cost space quantitatively evaluating classifier performance for given operating conditions does not require geometric constructions as in ROC space, but only a quick visual inspection [3]. This helps users to easily compare classifiers to the trivial classifiers, to select between them, or to measure their difference in performance [3].

BCs [5] are a variant of CCs – they visualize classifier performance assuming that the classifier scores are estimates of the posterior class probabilities, without requiring optimal decision threshold for a given operating condition.

No performance visualization tools analogous to CCs or BCs exist for the F-measure: defining and investigating such a space is the subject of the next section.

3   The F-Measure Space

We propose a visualization tool analogous to CC for evaluating and comparing the F-measure of one or more classifiers under different operating conditions, i.e., class priors and α. To this aim we rewrite the F-measure from Eq. (3) to make the dependence on $P(+)$ and α explicit:

$$F_\alpha = \frac{\text{TPR}}{\alpha(\text{TPR} + \lambda \cdot \text{FPR}) + (1 - \alpha)} \tag{9}$$

$$= \frac{^1/_\alpha \text{TPR}}{^1/_\alpha + ^1/_{P(+)}\text{FPR} + \text{TPR} - \text{FPR} - 1} \tag{10}$$

In contrast to the EC of Eqs. (1) and (10) indicates that F_α cannot be written as a function of a single parameter. However, since our main focus is performance evaluation under class imbalance, we consider the F-measure as a function of $P(+)$ only, for a fixed α value. Accordingly, we define the F-measure curve of a classifier as the plot of F_α as a function of $P(+)$, for a given α.

F-Measure Curve of a Classifier. For a crisp classifier defined by given values of TPR and FPR, the F-measure curve is obtained by simply plotting F_α as a function of $P(+)$, for a given α, using Eq. (10). Equation (10) implies that, when $P(+) = 0$, $F_\alpha = 0$, and when $P(+) = 1$, $F_\alpha = \text{TPR}/(\alpha(\text{TPR} - 1) + 1)$. It is easy to see that, when TPR > FPR (which is always the case for a non-trivial classifier), F_α is an increasing and concave function of $P(+)$. For different values of α one gets a family of curves. For $\alpha = 0$ we have $F_\alpha = \text{TPR}$, and for $\alpha = 1$ we have $F_\alpha = \text{Pr}$. Thus, for any fixed $\alpha \in (0, 1)$, each curve starts at $F_\alpha = 0$ for $P(+) = 0$, and ends in $F_\alpha = \text{Pr}$ for $P(+) = 1$. By computing $dF_\alpha/d\alpha$ from Eq. (10), one also obtains that all curves (including the one for $\alpha = 0$) cross when $P(+) = FPR/(FPR - TPR + 1)$. Figure 1 shows an example for a classifier with TPR $= 0.8$ and FPR $= 0.15$, and for five α values. CCs are also shown for comparison.

Consider now changing the decision threshold for a given soft classifier and a given α value. Whereas a point in ROC space corresponds to a line in cost space, it corresponds to a (non-linear) curve in F-measure space. As the decision

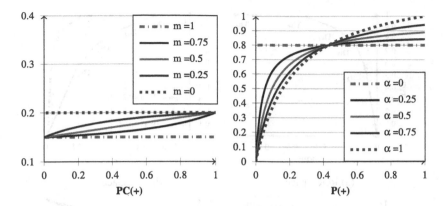

Fig. 1. Cost curves (left) and F-measure curves (right) for a given classifier with TPR = 0.8 and FPR = 0.15, for different values of m and α. Note that for all values of $P(+)$: (1) for $m = 0$, $EC = 1 - TPR$, (2) for $m = 1$, $EC = FPR$, (3) for $\alpha = 0$, $F_\alpha = TPR$.

threshold changes, one obtains a curve in ROC space, a family of lines in cost space, and a family of curves in F-measure space. More precisely, as the decision threshold increases (assuming that higher classifier scores correspond to a higher probability of the positive class), the ROC curve starts at $TPR = 0$ and $FPR = 0$, and proceeds towards $TPR = 1$ and $FPR = 1$. For a given value of α, the corresponding F-measure curves move away from the Y axis and get closer to the diagonal line connecting the lower-left point $P(+) = 0, F_\alpha = 0$ to the upper-right point $P(+) = 1, F_\alpha = 1$. An example is shown in Fig. 2. For any given operating condition (i.e., value of $P(+)$), only one decision threshold provides the highest F_α. Accordingly, the upper envelope of the curves that correspond to the available pairs of (TPR, FPR) values shows the best performance of the classifier with the most suitable decision threshold for each operating condition.

Comparing Classifiers in the F-Measure Space. Consider two classifiers with given values of (TPR_i, FPR_i) and (TPR_j, FPR_j), and a fixed value of α. From Eq. (10) one obtains that, if $\text{FPR}_j < \text{FPR}_i$ and $\text{TPR}_j < \text{TPR}_i$, or when $\text{FPR}_j > \text{FPR}_i$ and $\text{TPR}_j > \text{TPR}_i$, then the F-measure curves cross in a *single* point characterized by:

$$P_{i,j}^*(+) = \frac{\text{FPR}_i \cdot \text{TPR}_j - \text{FPR}_j \cdot \text{TPR}_i}{(1 - 1/\alpha)(\text{TPR}_j - \text{TPR}_i) + \text{FPR}_i \cdot \text{TPR}_j - \text{FPR}_j \cdot \text{TPR}_i}. \quad (11)$$

It is also easy to analytically determine which of the classifiers outperform the other for lower or higher $P(+)$ values than $P_{i,j}^*(+)$. If the above conditions do not hold, one of the classifiers dominates the other for all values of $P(+) > 0$; the detailed conditions under which $F_\alpha^j > F_\alpha^i$ or $F_\alpha^j < F_\alpha^i$ are not reported here for the sake of simplicity, but can be easily obtained as well. Examples of the two cases above are shown in Fig. 3.

In general, given any set of crisp classifiers, the best one for any given $P(+)$ value can be analytically determined in terms of the corresponding TPR and

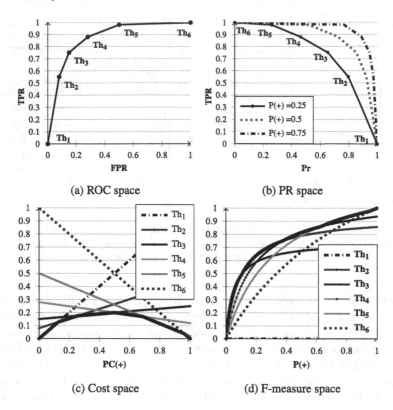

(a) ROC space

(b) PR space

(c) Cost space

(d) F-measure space

Fig. 2. A soft classifier in ROC space (ROCCH), inverted PR space (for three values of $P(+)$), cost space ($m = 0.5$) and F-measure space ($\alpha = 0.5$), for six threshold values $\text{Th}_1 > \text{Th}_2 > \ldots > \text{Th}_6$ corresponding to $\text{TPR}_1 = 0, 0.55, 0.75, 0.88, 0.98, 1$, and $\text{FPR}_1 = 0, 0.08, 0.15, 0.28, 0.5, 1$. The upper envelope of the cost and F-measure curves is shown as a thick, blue line. (Color figure online)

FPR values, and can be easily identified by the corresponding F-measure curve. Similarly, the overall performance of two or more soft classifiers can be easily compared by visually comparing the upper envelopes of their F-curves.

An example of the comparison of two soft classifiers, with six different threshold values, is shown in Fig. 4, where C_1 is the same as in Fig. 2. In ROC space, the ROCCH of C_1 and C_2 cross on a single point around $FPR = 0.3$. The lower envelopes of the corresponding CCs cross around $PC(+) = 0.7$, and thus C_1 and C_2 perform the same for approximately $0.6 < PC(+) < 0.7$, whereas C_1 outperforms C_2 for $PC(+) < 0.6$. When the F-measure is used, comparing C_1 and C_2 for different skew levels in PR space is more difficult, instead, as shown by the corresponding (inverted) PR curves. This task is much easier in the F-measure space; in this example it can be seen that the upper envelopes of the F-measure curves of C_1 and C_2 cross: C_2 outperforms C_1 for $P(+) < 0.4$, they perform the same for $0.4 < P(+) < 0.6$, and C_1 outperforms C_2 for $P(+) > 0.6$.

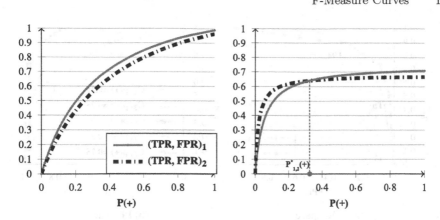

Fig. 3. F-measure curves of two classifiers, for $\alpha = 0.5$. Left: $(TPR_1, FPR_1) = (0.98, 0.5)$, $(TPR_2, FPR_2) = (0.93, 0.6)$: C_1 dominates C_2. Right: $(TPR_1, FPR_1) = (0.55, 0.08)$, $(TPR_2, FPR_2) = (0.5, 0.03)$: the two curves cross at the $P_{1,2}^*(+)$ value of Eq. (11) shown in red. (Color figure online)

These example shows that comparing the F-measure of two (or more) classifiers over all skew levels in F-measure space is as easy as comparing their EC in cost space.

Selecting the Best Decision Threshold or the Best Classifier. ROC curves can be used to set parameters like the optimal decision threshold, or to select the best classifier, for a given operating condition. To this aim, when the EC is used as the performance measure, the ROCCH of the classifier(s) is found and the optimal classifier (or parameter value) is selected by intersecting the upper-left EC iso-performance line corresponding to the given operating condition with the ROCCH. This process is easier in cost space, where the operating condition is shown on the X axis. Analogously, when the F-measure is used, this process is easier in the F-measure space than in PR space. For this purpose, the classifier(s) can be evaluated during design on a validation set (or on different validation sets with different imbalance levels, if the imbalance level during operation is unknown); then, during operation, the imbalance level of the data is estimated and the classification system is adapted based on its performance in cost or F-measure space.

4   Synthetic Examples

We give an example of classifier performance evaluation and comparison in F-measure space and, for reference, in ROC, PR, and cost spaces. In particular, we show how the effect of class imbalance can be observed using these global visualization tools. To this aim we generate a non-linear, 2D synthetic data set: the negative class is uniformly distributed, and surrounds the normally distributed positive class with mean $\mu_+ = (0.5, 0.5)$ and standard deviation $\sigma_+ = 0.33$. The

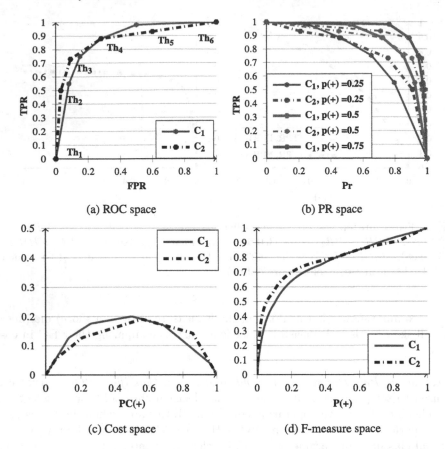

(a) ROC space

(b) PR space

(c) Cost space

(d) F-measure space

Fig. 4. Comparison between two soft classifiers (C_1 is the same as in Fig. 2) with six threshold values in ROC space, inverted PR space, cost space ($m = 0.5$) and F-measure space ($\alpha = 0.5$). Note that in cost and F-measure spaces the lower and upper envelopes of the curves corresponding to the six threshold values are shown, respectively.

class overlap is controlled by the minimum distance $\delta = 0.15$ of negative samples to μ_+. We consider three classifiers: Naive Bayes (C_1), 5-NN (C_2), and RBF-SVM (C_3). We draw 2000 samples from each class ($M^- = M^+ = 2000$), and use half of them for balanced training. To visualize classifier performance under different operating conditions, we consider different imbalance levels for testing (which simulates the classifier deployment phase). To this aim, we draw from the remaining 2000 samples different testing data subsets of fixed size equal to 1000. The number of testing samples from both classes is chosen as follows: for $P(+) < 0.5$, $M_+ = 500$, $M_- = \lambda M_+$, where $\lambda \in \{0.1, \ldots, 0.9\}$ with a step of 0.05; for $P(+) > 0.5$, $M_- = 500$, $M_+ = \lambda M_-$, with λ chosen in the same way; for $P(+) = 0.5$, $M_+ = M_- = 500$.

The performance of the three crisp classifiers, using a decision threshold of 0.5, is first compared in F-measure and cost spaces in Figs. 5a and b, for $0.1 < P(+) < 0.9$, $\alpha = 0.1, 0.5, 0.9$, and $m = 0.1, 0.5, 0.9$. It can be seen that

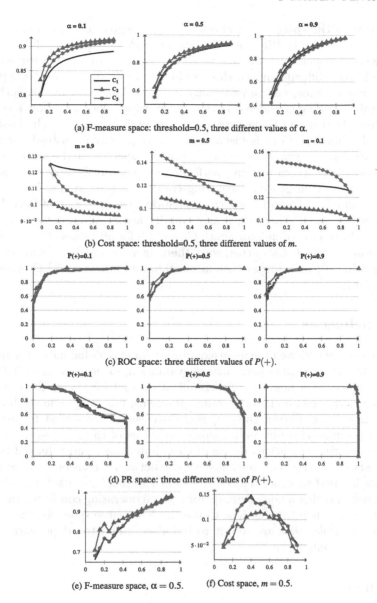

Fig. 5. Performance comparison among Naive Bayes (C_1), 5-NN (C_2) and RBF-SVM (C_3) in different spaces.

some of the corresponding curves cross, depending on α and m: in this case each classifier outperforms the other for a different range of values of $PC(+)$ or $P(+)$; these ranges can be easily determined analytically. The performance of the same, soft classifiers across different decision thresholds is then compared in ROC and PR spaces for three values of $P(+) = 0.1, 0.5, 0.9$ (Figs. 5c and d), and, for all possible values of $P(+)$, in cost and F-measure spaces (Figs. 5e and f).

As expected, ROC space is not affected by the degree of class imbalance, i.e., by changes in $P(+)$. In PR space each value of $P(+)$ leads to a different curve for a given classifier, instead, but visual comparison of the corresponding F-measure is very difficult: indeed this would require to draw also the hyperbolic iso-performance lines, and anyway only a small, finite number of both $P(+)$ and F_α values can be considered in this space, which does not allow a complete comparison. In cost and F-measure spaces the performance of each classifier for all possible values of $P(+)$ is visualized by a single curve, instead, for a given value of m (in cost space) or α (in F-measure space). In these spaces visual comparison of the corresponding performance measure is very easy, and can be carried out for all possible operating conditions (i.e., values $P(+)$). In this example, from Figs. 5e and f one can conclude that, in terms of both EC and F-measure, C_1 and C_3 perform nearly equally across all operating conditions. Moreover, classifier C_2 dominates both C_1 and C_3 for all values of $P(+)$; however the amount by which C_2 outperforms them is very small in terms of the F-measure, when $P(+)$ is higher than about 0.6, and in terms of EC, when $P(+)$ is around 0.7.

5   Conclusions

In this paper, we reviewed the main existing scalar and global measures and visualization tools for classifier performance evaluation, focusing on class imbalance. Then we proposed a new, specific visualization tool for the scalar F-measure, which is widely used for class imbalance problems, filling a gap in the literature.

Similarly to cost curves, the proposed F-measure curves allow to easily evaluate and compare classifier performance, in terms of the F-measure, across all possible operating conditions (levels of class imbalance) and values of the decision threshold, for a given preference weight between precision and recall. This space can be used to select the best decision threshold for a soft classifier, and the best soft classifier among a group, for a given operating condition. In ongoing research, we are investigating how to use the F-measure space for the design of classifier ensembles that are robust to imbalance, and to adapt learning algorithms to class imbalance.

References

1. Davis, J., Goadrich, M.: The relationship between precision-recall and ROC curves. In: ICML, pp. 233–240 (2006)
2. Dembczynski, K.J., Waegeman, W., Cheng, W., Hüllermeier, E.: An exact algorithm for F-measure maximization. In: NIPS, pp. 1404–1412 (2011)
3. Drummond, C., Holte, R.C.: Cost curves: an improved method for visualizing classifier performance. Mach. Learn. 65(1), 95–130 (2006)
4. Fawcett, T.: An introduction to ROC analysis. Pattern. Recognit. Lett. 27(8), 861–874 (2006)
5. Ferri, C., Hernández-orallo, J., Flach, P.A.: Brier curves: a new cost-based visualisation of classifier performance. In: ICML, pp. 585–592 (2011)

6. Ferri, C., Hernández-Orallo, J., Modroiu, R.: An experimental comparison of performance measures for classification. Pattern. Recognit. Lett. **30**(1), 27–38 (2009)
7. Flach, P., Kull, M.: Precision-recall-gain curves: PR analysis done right. In: NIPS, pp. 838–846 (2015)
8. García, V., Mollineda, R., Sánchez, J.: Theoretical analysis of a performance measure for imbalanced data. In: ICPR, pp. 617–620 (2010)
9. Hanczar, B., Nadif, M.: Precision-recall space to correct external indices for biclustering. In: ICML, pp. 136–144 (2013)
10. Krawczyk, B.: Learning from imbalanced data: open challenges and future directions. Prog. in AI **5**(4), 221–232 (2016)
11. Landgrebe, T.C., Paclik, P., Duin, R.P.: Precision-recall operating characteristic (P-ROC) curves in imprecise environments. In: ICPR, vol. 4, pp. 123–127 (2006)
12. Lipton, Z.C., Elkan, C., Naryanaswamy, B.: Optimal Thresholding of classifiers to maximize F1 measure. In: Calders, T., Esposito, F., Hüllermeier, E., Meo, R. (eds.) ECML PKDD 2014. LNCS (LNAI), vol. 8725, pp. 225–239. Springer, Heidelberg (2014). https://doi.org/10.1007/978-3-662-44851-9_15
13. Parambath, S.P., Usunier, N., Grandvalet, Y.: Optimizing F-measures by cost-sensitive classification. In: NIPS, pp. 2123–2131 (2014)
14. Pillai, I., Fumera, G., Roli, F.: Designing multi-label classifiers that maximize F measures: state of the art. Pattern. Recognit. **61**, 394–404 (2017)
15. Prati, R.C., Batista, G.E., Monard, M.C.: A survey on graphical methods for classification predictive performance evaluation. IEEE Trans. KDE **23**(11), 1601–1618 (2011)
16. Van Rijsbergen, C.: Information retrieval: theory and practice. In: Proceedings of the Joint IBM/University of Newcastle upon Tyne Seminar on Data Base Systems, pp. 1–14 (1979)

Maximum-Likelihood Estimation of Neural Mixture Densities: Model, Algorithm, and Preliminary Experimental Evaluation

Edmondo Trentin[✉]

Dipartimento di Ingegneria dell'Informazione e Scienze Matematiche,
Università degli Studi di Siena, Siena, Italy
trentin@dii.unisi.it

Abstract. Unsupervised estimation of probability density functions by means of parametric mixture densities (e.g., Gaussian mixture models) may improve significantly over plain, single-density estimators in terms of modeling capabilities. Moreover, mixture densities (and even mixtures of mixture densities) may be exploited for the statistical description of phenomena whose data distributions implicitly depend on the distinct outcomes of a number of non-observable, latent states of nature. In spite of some recent advances in density estimation via neural networks, no proper mixtures of neural component densities have been investigated so far. The paper proposes a first algorithm for estimating Neural Mixture Densities based on the usual maximum-likelihood criterion, satisfying numerically a combination of hard and soft constraints aimed at ensuring a proper probabilistic interpretation of the resulting model. Preliminary results are presented and their statistical significance is assessed, corroborating the soundness of the approach with respect to established statistical techniques.

Keywords: Density estimation · Mixture density
Unsupervised learning · Constrained learning · Mixture of experts

1 Introduction

Density estimation is fundamental to a number of (apparently unrelated) tasks. Firs and foremost, it is at the core of the search for a statistical description of populations represented in terms of a sample of data distributed according to an unknown probability density function (pdf) [10]. Then, it is involved (possibly only implicitly) in the estimation of the probabilistic quantities that are necessary in order to apply Bayes decision rule for pattern classification, in particular the class-conditional probabilities [10]. Other tasks include data compression and model selection [11], coding [3], etc. Even the estimation of regression models may rely implicitly on density estimation, since it can be described as the

© Springer Nature Switzerland AG 2018
L. Pancioni et al. (Eds.): ANNPR 2018, LNAI 11081, pp. 178–189, 2018.
https://doi.org/10.1007/978-3-319-99978-4_14

estimation of a model $p(\mathbf{y}|\mathbf{x})$ that captures the statistical relationship between an independent random vector $\mathbf{x}$ and the corresponding output vector $\mathbf{y}$ [4]. As pointed out by Vapnik [28], density estimation is an intrinsically difficult problem, and it is still open nowadays. This latter fact is mostly due to the shortcomings of established statistical approaches, either parametric or non-parametric (the reader is referred to [25] for a list of the major drawbacks of the statistical techniques), and by the technical difficulties that arise from attempting to use artificial neural networks (ANNs) or machine learning for pdf estimation. Such difficulties stem from: (1) the unsupervised nature of the learning task, (2) the numerical instability problems entailed by pdfs, whose codomains may span the interval $[0, +\infty)$ in the general case, and (3) the requirement of mathematical plausibility of the estimated model, i.e. the respect of the axioms of probability. Furthermore, the use of maximum-likelihood (ML) training in ANNs tends to result in the so-called "divergence problem", observed first in the realm of hybrid ANN/hidden Markov models [20]. It consists in the progressive divergence of the value of the ANN connection weights as ML training proceeds, resulting in an unbounded growth of the integral of the pseudo-pdf computed by the ANN. The problem does not affect radial basis functions (RBF) networks whose hidden-to-output weights were constrained to be positive and to sum to one, as in the RBF/echo state machine for sequences proposed in [26], or in the RBF/graph neural network presented in [6] for the estimation of generalized random graphs. Unfortunately, the use of RBFs in the latter contexts is justified by its allowing for a proper algorithmic hybridization with models devised specifically for sequence/structure processing, but using RBFs as a stand-alone paradigm for density estimation is of neglectable practical interest, since they end up realizing plain Gaussian mixture models (GMM) estimated via ML.

In spite of these difficulties, several approaches to pdf estimation via ANNs are found in the literature [23]. First of all, a ML technique is presented in [13] where the "integral equals 1" requirement is satisfied numerically dividing the output of a multilayer Perceptron (MLP) by the numerical integral of the function the MLP computes. No algorithms for computing the numerical integral over high-dimensional spaces are handed out in [13]. Nonetheless, this approach is related to the technique presented in this paper, insofar that ML will be exploited herein. Differently from [13], a multi-dimensional ad-hoc numeric integration method will be used in the following, jointly with hard constraints, over a mixture of ANNs. Other approaches found in the literature translated the estimation of univariate pdfs to the (theoretically equivalent) estimation of the corresponding cumulative distribution functions (cdf) [12, 27]. Regular backpropagation (BP) is applied, relying on the empirical cdf of the data for generating synthetic target outputs. After training the MLP model $\phi(\cdot)$ of the cdf, the pdf can be recovered by applying differentiation to $\phi(\cdot)$. The idea is sound, since the requirements that $\phi(\cdot)$ has to satisfy to be interpretable as a proper cdf (namely, that it ranges between 0 and 1, and that it is monotonically non-decreasing) appear to be more easily met than the corresponding constraints on pdf models (that is, the unit integral). Unfortunately, there are drawbacks to the cdf-based approaches (see [25]). In particular, a good approximation of the cdf does not necessarily translate into a similarly good estimate of its derivative. In fact, a small squared error between $\phi(\cdot)$ and the target cdf does not mean that $\phi(\cdot)$ is free from steep fluctuations that imply huge, rapidly changing values of its derivative. Negative values of $\frac{\partial \phi(x)}{\partial x}$ may occasionally occur, since a linear combination of logistics is not necessarily monotonically increasing. Besides, cdf-based algorithms naturally apply to univariate cases, whilst extension to multivariate pdfs is far less realistic. The idea of generating empirical target outputs was applied to non-parametric ANN-based pdf estimation in [22,24]. The former resorts to the k_n-Nearest Neighbor (k_n-NN) technique [10] for generating unbiased pdf estimates that are used to label the training set for a MLP. Like in the k_n-NN, the resulting model is not a proper pdf (the axioms of probability are not satisfied in the general case). On the other way around, the algorithm presented in [22] uses a modified criterion function to be minimized via gradient descent for pdf estimation via MLP. The criterion involves two terms: a loss between the MLP output and a synthetically-generated non-parametric estimate of the corresponding input pattern, and a loss between the integral of the function computed by the MLP and its target (i.e., unity) value. Numerical integration methods are used to compute the integral at hand and its derivatives w.r.t. the MLP parameters within the gradient-descent via backpropagation. The ideas behind such integration methods are exploited in this paper, as well.

A generalization of plain pdf estimation models stems from the adoption of mixture densities, where the unknown pdf is rather modeled in terms of a combination of any number of component densities [10]. GMMs are the most popular instance of mixture densities [5]. Traditionally, mixture densities were intended mostly as real-life extensions of the single-pdf parametric model, e.g. along the following line: one Gaussian may not be capable to explain the whole data distribution but K Gaussian pdfs might as well be, as long as K is large enough. Nevertheless, there is much more than this to the very notion of mixture density. In fact, different components are specialized to explain distinct latent phenomena (e.g., stochastic processes) that underlie the overall data generation process, each such phenomenon having different likelihood of occurrence w.r.t. others at diverse regions of the feature space. This suites particularly those situations where the statistical population under analysis is composed of several sub-populations, each having different distribution. Examples of practical relevance include (among many others) the statistical study of heterogeneity in meta-analysis [7], where samples drawn from disjoint populations (e.g., adults and children, male and female subjects, etc.) are collectively collected and have to be analyzed as a whole; the modeling of unsupervised or partially-supervised [17] data samples in statistical pattern recognition [10], where each sub-population corresponds to a class or category; the distribution of financial returns on the stock market depending on latent phenomena such as a political crisis or a war [8]; the assessment of projectile accuracy in the military science of ballistics when shots at the same target come from multiple locations and/or from different munition types [18], etc. In general, the sub-populations in a mixture are unlikely to be individually distributed according to simple (e.g., Gaussian) pdfs, therefore parametric models (e.g., GMMs) are seldom a good fit to these scenarios. In fact, let $\xi_1, \ldots, \xi_K$ be K disjoint states of nature (the outcomes of a discrete, latent random variable Ξ, each outcome corresponding to a specific sub-population), and let $p(\mathbf{x}|\xi_i)$ be the pdf that explains the distribution of the random observations $\mathbf{x}$ given the i-th state of the latent variable, for $i = 1, \ldots, K$. At the whole population level the data will the distributed according to the mixture $p(\mathbf{x}) = \sum_{i=1}^{K} P(\xi_i)p(\mathbf{x}|\xi_i)$. Attempts to apply a GMM to model $p(\mathbf{x})$ will not necessarily result in a one-to-one relationship between the Gaussian components in the GMM and the state-specific generative models $p(\mathbf{x}|\xi_i)$. In general, at the very least, more than one Gaussian component will be needed to model $p(\mathbf{x}|\xi_i)$. Although mixtures of mixture models offer increased modeling capabilities over plain mixture models to this end, they turned out to be unpopular due to the difficulties of estimation of their parameters [2].

Given the aforementioned relevance and difficulties of estimating pdfs in general and mixture models in particular, and in the light of the above-named shortcomings of the established approaches, the paper contributes (for the first time) a plausible solution in the form of a mixture model built on ANNs. The model, presented in Sect. 2 and called neural mixture model (NMM), is a convex combination of component densities estimated by component-specific MLPs. The NMM is intrinsically non-parametric, since no prior assumptions on the form of the underlying component densities is made [10]. In fact, due to the "universality" of MLPs [9], the model may approximate any (bounded and continuous) multimodal multivariate pdf to any degree of precision[1]. Besides, due to the learning and generalization capabilities of ANNs, the NMM can actually learn a smooth and general form for the mixture at hand, overcoming the drawbacks of the traditional non-parametric techniques, as well. A ML training algorithm is devised, satisfying (at least numerically) a combination of hard and soft constraints required in order to guarantee a proper probabilistic interpretation of the estimated model. The resulting machine can also be seen as a novel, special case of mixture of experts [29] having a specific task, a ML-based unsupervised training algorithm, and a particular probabilistic strategy for assigning credit to its individual experts. A preliminary experimental evaluation is reported in Sect. 3, while Sect. 4 draws some pro tempore conclusions.

2   Model and Estimation Algorithm

Let us consider an unlabeled training set $\mathcal{T} = \{\mathbf{x}_1, \ldots, \mathbf{x}_n\}$ of n independent random vectors (i.e., patterns) in a d-dimensional feature space, say $\mathbb{R}^d$. The patterns are assumed to be identically distributed according to an unknown pdf $p(\mathbf{x})$. In order to estimate $p(\mathbf{x})$ from $\mathcal{T}$ we introduce a neural mixture model $\tilde{p}(\mathbf{x}|W)$ defined as

---

[1] According to the meaning of "approximation" and under the conditions required in order for (e.g.) Cybenko's theorem to hold true [9].

$$\tilde{p}(\mathbf{x}|W) = \sum_{i=1}^{K} c_i \tilde{p}_i(\mathbf{x}|W_i) \tag{1}$$

where W denotes the overall set of parameters in the NMM (that is $c_1, \ldots, c_K, W_1, \ldots, W_k$). The mixing coefficients c_i are such that $c_i \in [0,1]$ for $i = 1, \ldots, K$ and $\sum_{i=1}^{K} c_i = 1$, and the generic i-th component density $\tilde{p}_i(\mathbf{x}|W_i)$ is defined, in turn, as

$$\tilde{p}_i(\mathbf{x}|W_i) = \frac{\varphi_i(\mathbf{x}, W_i)}{\int \varphi_i(\mathbf{x}, W_i)d\mathbf{x}} \tag{2}$$

where $\varphi_i(\mathbf{x}, W_i)$ represents the function computed by a component-specific MLP having adaptive parameters W_i. We say that this MLP realizes the i-th neural component of the NMM. A constraint on $\int \varphi_i(\mathbf{x}, W_i)d\mathbf{x}$ will be imposed shortly to enure satisfaction of the axioms of probability. Clearly, each MLP in the NMM has d input units and a single output unit, and it is expected to have one or more hidden layers. Without loss of generality for all the present intents and purposes, we assume that the patterns of interest are confined within a compact $S \subset \mathbb{R}^d$ (in practice, any data normalization technique may be applied in order to guarantee that this assumption holds true) such that, in turn, S can be seen as the definition domain of $\varphi_i(\mathbf{x}, W_i)$ for all $i = 1, \ldots, K$. As a consequence, numerical integration techniques can be used to compute $\int \varphi_i(\mathbf{x}, W_i)d\mathbf{x}$ and the other integrals required shortly. In so doing, Eq. (2) reduces to $\tilde{p}_i(\mathbf{x}|W_i) = \frac{\varphi_i(\mathbf{x}, W_i)}{\int_S \varphi_i(\mathbf{x}, W_i)d\mathbf{x}}$.

Some precautions are to be taken in regard to the nature of the activation function $f_i(.)$ used in the output layer of the i-th MLP. In fact, $f_i(.)$ shall be capable of spanning a codomain that fits the general definition of pdf, that is (in principle) any range in $[0, +\infty)$. Although this may be granted in several different ways, herein we opt for a logistic sigmoid with component-specific adaptive amplitude $\lambda_i \in \mathbb{R}^+$, namely $f_i(a_i) = \lambda_i/(1 + \exp(-a_i))$ as described in [19], where a_i represents the current activation value for the output unit of the i-th neural component. Consequently, each MLP in the NMM can stretch its output over any required component-specific interval $[0, \lambda_i)$, which is not bounded a priori but is rather learned (along with the other parameters in W_i) so as to fit the nature of the specific component density at hand. Other general advantages entailed by the use of adaptive amplitudes are pointed out in [19].

The training algorithm is expected to revolve around a proper learning rule for the mixture parameters W given the unlabeled sample $\mathcal{T}$, such that eventually $\tilde{p}(\mathbf{x}|W)$ results in a proper estimate of $p(\mathbf{x})$. This requires pursuing two purposes: (1) exploiting the information encapsulated in $\mathcal{T}$ to approximate the unknown pdf; (2) preventing the MLPs in the NMM from developing spurious solutions, by enforcing the constraints $\int_S \varphi_i(\mathbf{x}, W_i)d\mathbf{x} = 1$ for all $i = 1, \ldots, K$. To this end, a constrained stochastic gradient-ascent algorithm is devised that aims at the maximization of the point-wise likelihood $\tilde{p}(\mathbf{x}_j|W)$ of the NMM given the current training pattern $\mathbf{x}_j$, to be applied iteratively for $j = 1, \ldots, n$. This is achieved by means of an on-line, differentiable criterion function $C(.)$ defined as

$$C(W, \mathbf{x}_j) = \tilde{p}(\mathbf{x}_j|W) - \rho \sum_{i=1}^{K} \frac{1}{2} \left(1 - \int_S \varphi_i(\mathbf{x}, W_i) d\mathbf{x} \right)^2 \tag{3}$$

that has to be maximized with respect to the NMM parameters W under the (hard) constraints that $c_i \in [0,1]$ for $i = 1, \ldots, K$ and $\sum_{i=1}^{K} c_i = 1$. The second term in the criterion, instead, is a "soft" constraint that enforces a unit integral of $\tilde{p}_i(\mathbf{x}, W_i)$ over S for all $i = 1, \ldots, K$, as sought, resulting in $\int_S \tilde{p}(\mathbf{x}|W) d\mathbf{x} \simeq 1$. The hyper-parameter $\rho \in \mathbb{R}^+$ controls the importance of the constraints, and it is used in practical applications to tackle numerical issues. The gradient-ascent learning rule Δw for a generic parameter w in the NMM is then defined as $\Delta w = \eta \frac{\partial C(.)}{\partial w}$, where $\eta \in \mathbb{R}^+$ is the learning rate. Different calculations are needed, according to the fact that w is either: (i) a mixing coefficient, say $w = c_k$; or (ii) a parameter (connection weight, bias, or adaptive amplitude) within any of the neural component densities. In case (i), we first introduce K unconstrained latent variables $\gamma_1, \ldots, \gamma_K$, and we let

$$c_k = \frac{\varsigma(\gamma_k)}{\sum_{i=1}^{K} \varsigma(\gamma_i)} \tag{4}$$

for $k = 1, \ldots, K$, where $\varsigma(x) = 1/(1 + e^{-x})$. Each γ_k is then treated as the unknown parameter to be actually estimated instead of the corresponding c_k. In so doing, higher-likelihood mixing coefficients that satisfy the required constraints are implicitly obtained from application of the learning rule. The latter takes the following form:

$$\begin{aligned}
\Delta \gamma_k &= \eta \frac{\partial C(.)}{\partial \gamma_k} \tag{5} \\
&= \eta \frac{\partial \tilde{p}(\mathbf{x}_j|W)}{\partial \gamma_k} \\
&= \eta \frac{\partial \sum_{i=1}^{K} c_i \tilde{p}_i(\mathbf{x}_j|W_i)}{\partial \gamma_k} \\
&= \eta \sum_{i=1}^{K} \tilde{p}_i(\mathbf{x}_j|W_i) \frac{\partial}{\partial \gamma_k} \left(\frac{\varsigma(\gamma_i)}{\sum_{\ell=1}^{K} \varsigma(\gamma_\ell)} \right) \\
&= \eta \left\{ \tilde{p}_k(\mathbf{x}_j|W_k) \frac{\varsigma'(\gamma_k)}{\sum_{\ell=1}^{K} \varsigma(\gamma_\ell)} - \sum_{i=1}^{K} \tilde{p}_i(\mathbf{x}_j|W_i) \frac{\varsigma(\gamma_i)\varsigma'(\gamma_k)}{[\sum_\ell \varsigma(\gamma_\ell)]^2} \right\} \\
&= \eta \frac{\varsigma'(\gamma_k)}{\sum_\ell \varsigma(\gamma_\ell)} \left\{ \tilde{p}_k(\mathbf{x}_j|W_k) - \tilde{p}(\mathbf{x}_j|W) \right\}
\end{aligned}$$

Secondly, let us move to scenario (ii), that is where w is a parameter within one of the neural components. In this case, taking the partial derivative of $C(W, \mathbf{x}_j)$ with respect to w requires calculating the derivatives of the first and the second terms in the right-hand side of Eq. (3). In the following calculations we assume that w belongs to the (generic) k-th neural component. For the first term we have:

$$\frac{\partial \tilde{p}(\mathbf{x_j}|W)}{\partial w} = \frac{\partial}{\partial w} \sum_{i=1}^{K} c_i \tilde{p}_i(\mathbf{x_j}|W_i) \tag{6}$$

$$= \frac{\partial}{\partial w}\{c_k \tilde{p}_k(\mathbf{x_j}|W_k)\}$$

$$= c_k \frac{\partial}{\partial w}\left\{\frac{\varphi_k(\mathbf{x}_j, W_k)}{\int_S \varphi_k(\mathbf{x}, W_k)d\mathbf{x}}\right\}$$

$$= c_k \left\{\frac{1}{\int_S \varphi_k(\mathbf{x}, W_k)d\mathbf{x}} \frac{\partial \varphi_k(\mathbf{x}_j, W_k)}{\partial w} - \frac{\tilde{p}_k(\mathbf{x}_j, W_k)}{\int_S \varphi_k(\mathbf{x}, W_k)d\mathbf{x}} \frac{\partial}{\partial w}\int_S \varphi_k(\mathbf{x}, W_k)d\mathbf{x}\right\}$$

$$= \frac{c_k}{\int_S \varphi_k(\mathbf{x}, W_k)d\mathbf{x}} \left\{\frac{\partial \varphi_k(\mathbf{x}_j, W_k)}{\partial w} - \frac{\varphi_k(\mathbf{x}_j, W_k)}{\int_S \varphi_k(\mathbf{x}, W_k)d\mathbf{x}} \int_S \frac{\partial \varphi_k(\mathbf{x}, W_k)}{\partial w}d\mathbf{x}\right\}$$

where Leibniz rule was exploited in the last step of the calculations. Note that Eq. (6) is a mathematical statement of the rationale behind the different impact that current training pattern $\mathbf{x}_j$ has on the learning process in distinct neural components of the NMM. First, the amount of parameter change Δw is proportional to the probabilistic "credit" c_k of the component at hand. Second (and foremost), the quantities within brackets in Eq. (6) depend on the value of the k-th MLP output over $\mathbf{x}_j$, and on its derivative. If, at any time during the training, $\varphi_k(.)$ does not change significantly in a neighborhood of $\mathbf{x}_j$ (e.g. if $\mathbf{x}_j$ lies in a high-likelihood plateau or, vice versa, in a close-to-zero plateau of $\varphi_k(.)$) then the contribution of the first quantity within brackets is neglectable. Moreover, if $\varphi_k(\mathbf{x}_j) \simeq 0$ then the second term within brackets turns out to be neglectable, as well. To the contrary, the contribution of $\mathbf{x}_j$ to the parameter adaptation of k-th component network will be paramount if $\varphi_k(.)$ returns high likelihood over $\mathbf{x}_j$ and significant variations in its surroundings.

At this point Leibniz rule is used again in the calculation of the derivative of the second term in the right-hand side of Eq. (3), which can be written as

$$\frac{\partial}{\partial w}\left\{\rho \sum_{i=1}^{K} \frac{1}{2}\left(1 - \int_S \varphi_i(\mathbf{x}, W_i)d\mathbf{x}\right)^2\right\} \tag{7}$$

$$= \frac{\partial}{\partial w}\left\{\frac{\rho}{2}\left(1 - \int_S \varphi_k(\mathbf{x}, W_k)d\mathbf{x}\right)^2\right\}$$

$$= -\rho\left(1 - \int_S \varphi_k(\mathbf{x}, W_k)d\mathbf{x}\right)\frac{\partial}{\partial w}\int_S \varphi_k(\mathbf{x}, W_k)d\mathbf{x}$$

$$= -\rho\left(1 - \int_S \varphi_k(\mathbf{x}, W_k)d\mathbf{x}\right)\int_S \frac{\partial \varphi_k(\mathbf{x}, W_k)}{\partial w}d\mathbf{x}.$$

Algorithms for the computation of $\frac{\partial \varphi_k(\mathbf{x}_j, W_k)}{\partial w}$, $\int_S \varphi_k(\mathbf{x}, W_k)d\mathbf{x}$, and $\int_S \frac{\partial}{\partial w}\varphi_k$ $(\mathbf{x}, W_k)d\mathbf{x}$ are needed in order to compute the right-hand side of Eqs. (6) and (7). The quantity $\frac{\partial \varphi_k(\mathbf{x}_j, W_k)}{\partial w}$ is the usual partial derivative of the output of a MLP with respect to one of its parameters, and it is computed via plain BP (or, as in [19] if $w = \lambda_k$). As for the integrals, deterministic numerical quadrature integration techniques (e.g., Simpson's method, trapezoidal rule, etc.) are viable only for $d = 1$, since they do not scale up computationally to higher dimensions ($d \geq 2$) of the feature space. This is all the more critical if we bear in mind that $\int_S \frac{\partial}{\partial w} \varphi_k(\mathbf{x}, W_k) d\mathbf{x}$ has to be computed iteratively and individually for each parameter of each neural component in the NMM. Besides, deterministic integration methods do not exploit the very nature of the function to be integrated. In fact, in the present context the integrand is expected to be closely related to the pdf (say, $p_k(\mathbf{x})$) that explains the distribution of the specific sub-sample of data drawn from the k-th component of the mixture. In fact, accounting for the pdf of the data should drive the integration algorithm towards integration points that cover "interesting" (i.e., having high component-specific likelihood) regions of the domain of the integrand. For these reasons, we use a component-oriented variant of the approach we proposed in [21], that can be seen as an instance of Markov chain Monte Carlo [1]. It is a non-deterministic, multi-dimensional integration technique which actually accounts for the component-specific probability distribution of the data. Let $\phi_k(\mathbf{x})$ denote the integrand of interest (either $\varphi_k(\mathbf{x}, W_k)$ or $\frac{\partial \varphi_k(\mathbf{x}, W_k)}{\partial w}$). An approximation of the integral of $\phi_k(\mathbf{x})$ over S is obtained via Monte Carlo with importance sampling [16] as $\int_S \phi_k(\mathbf{x}) d\mathbf{x} \simeq \frac{V(S)}{m} \sum_{\ell=1}^{m} \phi_k(\dot{\mathbf{x}}_\ell)$ where m properly sampled integration points $\dot{\mathbf{x}}_1, \ldots, \dot{\mathbf{x}}_m$ are used. Sampling of the ℓ-th integration point $\dot{\mathbf{x}}_\ell$ (for $\ell = 1, \ldots, m$) is attained by drawing it at random from the mixture density $p_u^{(k)}(\mathbf{x})$ defined as

$$p_u^{(k)}(\mathbf{x}) = \alpha(t)u(\mathbf{x}) + (1 - \alpha(t))\tilde{p}_k(\mathbf{x}|W_k) \qquad (8)$$

where $u(\mathbf{x})$ is the uniform distribution over S, and $\alpha : \mathbb{N} \to (0, 1)$ is a decaying function of the number t of the NMM training epochs (a training epoch is a completed re-iteration of Eqs. (6) and (7) for all the parameters of the NMM over all the observations in $\mathcal{T}$) for $t = 1, \ldots, T$, such that $\alpha(1) \sim 1.0$ and $\alpha(T) \simeq 0.0$. As in [21] we let $\alpha(t) = 1/(1 + e^{\frac{t/T - 1/2}{\theta}})$, where θ is a hyperparameter. Eq. (8) is such that the importance sampling mechanism it implies accounts for the (estimated) component density $\tilde{p}_k(\mathbf{x}|W_k)$ of the k-th latent subsample of the data, therefore respecting the natural distribution of such sub-population and focusing integration on the relevant (i.e., having high component-specific likelihood) integration points. On the other hand, since the estimates of this component densities are not reliable during the early stage of the NMM training process, Eq. (8) prescribes a (safer) sampling from a uniform distribution at the beginning, like in the plain Monte Carlo algorithm. As the robustness of the NMM estimates increases, i.e. as t increases, sampling from $\tilde{p}_k(\mathbf{x}|W_k)$ replaces progressively the sampling from $u(\mathbf{x})$, ending up in entirely non-uniform importance sampling. The form of $\alpha(t)$ is such that the algorithm basically sticks with the uniform sampling for quite some time before beginning crediting the estimated pdfs, but it ends up relying mostly on the neural component densities throughout the advanced stage of training. Since $\varphi_k(\mathbf{x}, W_k)$ is non-negative by construction, for $t \to T$ the sampling occurs substantially from $|\varphi_k(\mathbf{x}, W_k)| / \int_S |\varphi_k(\mathbf{x}, W_k)| d\mathbf{x}$, that is a sufficient condition for granting that the variance of the estimated integral vanishes and the corresponding error goes to zero [15].

Sampling from $p_u^{(k)}(\mathbf{x})$ requires an effective method for sampling from the output of the k-th MLP in the NMM. A specialization of Markov chain Monte Carlo, namely the Metropolis–Hastings (M-H) algorithm [14], is used herein. M-H is robust to the fact that, during training, $\varphi_k(\mathbf{x}, W_k)$ may not be properly normalized but it is proportional by construction to the corresponding pdf estimate (which is normalized properly, instead, by definition) [14]. Due to its efficiency and ease of sampling, a multivariate logistic pdf with radial basis, having location $\mathbf{x}$ and scale σ, is used as the *proposal* pdf $q(\mathbf{x}'|\mathbf{x})$ required by M-H to generate a new candidate sample $\mathbf{x}' = (x_1', \ldots, x_d')$ from the current sample $\mathbf{x} = (x_1, \ldots, x_d)$. Formally, such a proposal pdf is defined as $q(\mathbf{x}'|\mathbf{x}, \sigma) = \prod_{i=1}^{d} \frac{1}{\sigma} e^{(x_i'-x_i)/\sigma}(1 + e^{(x_i'-x_i)/\sigma})^{-2}$ which can be sampled readily by means of the inverse transform sampling technique. Any approach to model selection can be applied to fix empirically the values for the present hyperparameters (that is, the scale σ of the proposal pdf and the burn-in period for M-H).

3   Preliminary Experimental Evaluation

This Section presents a preliminary evaluation of the NMM behavior on random samples generated synthetically from a multimodal pdf $p(\cdot)$ having known form. The random samples used in the experiments were drawn from mixtures $p(x)$ of c Fisher-Tippett pdfs, that is $p(x) = \sum_{i=1}^{c} \frac{P_i}{\beta_i} \exp\left(-\frac{x-\mu_i}{\beta_i}\right) \exp\left\{-\exp\left(-\frac{x-\mu_i}{\beta_i}\right)\right\}$. The mixing parameters $P_1, \ldots, P_c$ were drawn at random (any time a new dataset had to be generated) from the uniform distribution over [0.1] and normalized such that $\sum_{i=1}^{c} P_i = 1$. The component densities of the Fisher-Tippett mixture are identified by their locations μ_i and scales β_i, respectively, for $i = 1, \ldots, c$. The locations were drawn at random from the uniform distribution over $(0, 10)$, while the scales were randomly and uniformly distributed over $(0.01, 0.9)$. The estimation tasks we faced involved 1200 patterns randomly drawn from $p(x)$, and a variable number c of component densities, namely $c = 5, 10, 15$ and 20, respectively. Each c-specific dataset was split into a training set ($n = 800$ patterns) and a validation set (the remaining 400 patterns). The integrated squared error (ISE) between $p(x)$ and its estimate $\tilde{p}(x)$, that is $\int (p(x) - \tilde{p}(x))^2 dx$, is used as the criterion of assessment of the performance of the estimation techniques. The ISE is computed numerically via Simpson's method.

In the present experiments we used NMMs involving MLPs with a single hidden layer of 9 units. Logistic sigmoid activation functions were used, having layer-wise [19] trainable amplitudes λ. All the parameters in the NMM were initialized at random over small intervals centered at zero, except for the amplitudes λ that were initially set to 1 and the mixing coefficients that were set to $c_i = 1/K$ for $i = 1, \ldots, K$. As in [22] we let $\theta = 0.07$ in the definition of the function $\alpha(t)$, and the number of integration points was set to $m = 400$, sampled at the beginning of each training epoch using a scale $\sigma = 9$ for the logistic proposal pdf in M-H. The burn-in period of the Markov chain in M-H was stretched over the first 500 states. The other hyperparameters of the NMM were fixed via random-search based on the cross-validated likelihood criterion exploiting the validation set. Input data were not normalized. Results are reported in Table 1. NMMs having different values of K (that is $K = 4$, 8 and 12) were evaluated, and compared with the most prominent statistical techniques, parametric and non-parametric, namely: 8-, 16- and 32-GMM initialized via k-means and refined via iterative maximum-likelihood re-estimation [10], k_n-NN with unbiased $k_n = 1\sqrt{n}$ [10], and Parzen Window (PW) with unbiased bandwidth $h_n = 1/\sqrt{n}$ of the Gaussian kernels [10].

Table 1. Estimation of the Fisher-Tippett mixture $p(x)$ (with $n = 800$) in terms of integrated squared error as a function of the number c of Fisher-Tippett component densities. Best results are shown in boldface. (*Legend*: k-GMM and k-NMM denote the GMM and the NMM with k components, respectively).

c	5	10	15	20	*Avg. ± std. dev.*
8-GMM	9.60×10^{-3}	1.12×10^{-2}	4.57×10^{-2}	7.99×10^{-2}	$(3.66 \pm 2.89) \times 10^{-2}$
16-GMM	6.33×10^{-3}	9.29×10^{-3}	3.78×10^{-2}	4.24×10^{-2}	$(2.40 \pm 1.63) \times 10^{-2}$
32-GMM	7.15×10^{-3}	9.82×10^{-3}	2.41×10^{-2}	3.03×10^{-2}	$(1.78 \pm 0.97) \times 10^{-2}$
k_n-NN	6.54×10^{-3}	8.70×10^{-3}	2.03×10^{-2}	2.36×10^{-2}	$(1.48 \pm 0.73) \times 10^{-2}$
PW	6.02×10^{-3}	8.94×10^{-3}	2.14×10^{-2}	1.98×10^{-2}	$(1.40 \pm 0.67) \times 10^{-2}$
4-NMM	6.41×10^{-3}	7.06×10^{-3}	1.09×10^{-2}	1.40×10^{-2}	$(9.59 \pm 3.07) \times 10^{-3}$
8-NMM	$\mathbf{5.89 \times 10^{-3}}$	$\mathbf{6.02 \times 10^{-3}}$	8.11×10^{-3}	1.01×10^{-2}	$(\mathbf{7.53 \pm 1.73}) \times 10^{-3}$
12-NMM	6.38×10^{-3}	6.27×10^{-3}	$\mathbf{8.05 \times 10^{-3}}$	$\mathbf{9.64 \times 10^{-3}}$	$(7.59 \pm \mathbf{1.38}) \times 10^{-3}$

It is seen that all the models yield values of the resulting ISE that tend to increase as a function of c. Nevertheless, the NMM improves systematically and significantly over all the statistical approaches regardless of c. On average, both the 8-NMM and the 16-NMM offer a 46% relative ISE reduction over the PW (the PW turns out to be the closest competitor to NMM in the present setup). Welch's t-test returns a level of confidence >90% on the statistical significance of the gap between the results yielded by the 8-NMM (or, by the 12-NMM) and by the PW, respectively. Moreover, the NMM results in the stablest estimators overall, as shown by the values of the standard deviations (last column of the table). This is evidence of the fact that the degree of fitness of the NMMs to the true pdf is less sensitive to the complexity of the underlying Fisher-Tippett mixture (that is, to c) than that yielded by the traditional statistical models. As expected, ISE differences are observed among the k-NMMs depending on the different values of k. In the present setup, differences between the 8-NMM and the 12-NMM turn out to be mild (especially if compared to the overall gaps between NMMs and the established statistical techniques), and they depend on the complexity of the underlying pdf to be estimated, at least to some extent, as expected.

4   Pro Tempore Conclusions and On-Going Research

Density estimation remains an open, non-trivial task: in fact, significant improvement over established approaches may still be expected of novel techniques, capable of increased robustness of the resulting pdf estimates. In spite of the relative simplicity of the data used in the previous Section, the empirical evidence reported therein confirms that the traditional statistical techniques may yield inaccurate estimates, whereas the NMM may result in a model of the data that is closer to the true pdf underlying the unlabeled samples at hand. For the time being, we are in the early stages of investigating the behavior of the technique over multivariate random vectors. Model selection (in particular, selection of proper ANN architectures) is under further investigation, as well, by exploiting the implicit availability of a mathematically grounded assessment criterion, namely the likelihood of the model (e.g., of its architecture) given a validation set. Finally, another facet of the NMM that is currently under development lies in the initialization procedure: non-uniform initialization of the mixing coefficients may turn out to be helpful in breaking potential ties, and initializing the individual neural components via BP learning of a subset of the components of a pre-estimated reference mixture model (i.e., a GMM) may also fruitfully replace the bare random initialization of the MLPs parameters.

References

1. Andrieu, C., de Freitas, N., Doucet, A., Jordan, M.I.: An introduction to MCMC for machine learning. Mach. Learn. **50**(1–2), 5–43 (2003)
2. Aste, M., Boninsegna, M., Freno, A., Trentin, E.: Techniques for dealing with incomplete data: a tutorial and survey. Pattern Anal. Appl. **18**(1), 1–29 (2015)
3. Beirami, A.: Wireless network compression via memory-enabled overhearing helpers. IEEE Trans. Wirel. Commun. **15**(1), 176–190 (2016)
4. Bishop, C.M.: Neural Networks for Pattern Recognition. Oxford University Press, Oxford (1995)
5. Bishop, C.M.: Pattern Recognition and Machine Learning. Information Science and Statistics. Springer, Heidelberg (2006). https://doi.org/10.1007/978-1-4615-7566-5
6. Bongini, M., Rigutini, L., Trentin, E.: Recursive neural networks for density estimation over generalized random graphs. IEEE Trans. Neural Netw. Learn. Syst. (2018). https://doi.org/10.1109/TNNLS.2018.2803523
7. Borenstein, M., Hedges, L.V., Higgins, J.P.T., Rothstein, H.R.: Introduction to MetaAnalysis. Wiley-Blackwell, New York (2009)
8. Cuthbertson, K., Nitzsche, D.: Quantitative Financial Economics: Stocks, Bonds and Foreign Exchange, 2nd edn. Wiley, New York (2004)
9. Cybenko, G.: Approximation by superposition of sigmoidal functions. Math. Control Signal Syst. **2**(4), 303–314 (1989)
10. Duda, R.O., Hart, P.E., Stork, D.G.: Pattern Classification, 2nd edn. Wiley-Interscience, New York (2000)
11. Liang, F., Barron, A.: Exact minimax strategies for predictive density estimation, data compression, and model selection. IEEE Trans. Inf. Theory **50**(11), 2708–2726 (2004)

12. Magdon-Ismail, M., Atiya, A.: Density estimation and random variate generation using multilayer networks. IEEE Trans. Neural Netw. **13**(3), 497–520 (2002)
13. Modha, D.S., Fainman, Y.: A learning law for density estimation. IEEE Trans. Neural Netw. **5**(3), 519–23 (1994)
14. Newman, M.E.J., Barkema, G.T.: Monte Carlo Methods in Statistical Physics. Oxford University Press, Oxford (1999)
15. Ohl, T.: VEGAS revisited: adaptive Monte Carlo integration beyond factorization. Comput. Phys. Commun. **120**, 13–19 (1999)
16. Rubinstein, R.Y., Kroese, D.P.: Simulation and the Monte Carlo Method, 2nd edn. Wiley, Hoboken (2012)
17. Schwenker, F., Trentin, E.: Pattern classification and clustering: a review of partially supervised learning approaches. Pattern Recognit. Lett. **37**, 4–14 (2014)
18. Spall, J.C., Maryak, J.L.: A feasible bayesian estimator of quantiles for projectile accuracy from non-i.i.d. data. J. Am. Stat. Assoc. **87**(419), 676–681 (1992)
19. Trentin, E.: Networks with trainable amplitude of activation functions. Neural Netw. **14**(45), 471–493 (2001)
20. Trentin, E.: Maximum-likelihood normalization of features increases the robustness of neural-based spoken human-computer interaction. Pattern Recognit. Lett. **66**, 71–80 (2015)
21. Trentin, E.: Soft-constrained nonparametric density estimation with artificial neural networks. In: Schwenker, F., Abbas, H.M., El Gayar, N., Trentin, E. (eds.) ANNPR 2016. LNCS (LNAI), vol. 9896, pp. 68–79. Springer, Cham (2016). https://doi.org/10.1007/978-3-319-46182-3_6
22. Trentin, E.: Soft-constrained neural networks for nonparametric density estimation. Neural Process. Lett. (2017). https://doi.org/10.1007/s11063-017-9740-1
23. Trentin, E., Freno, A.: Probabilistic interpretation of neural networks for the classification of vectors, sequences and graphs. In: Bianchini, M., Maggini, M., Scarselli, F., Jain, L.C. (eds.) Innovations in Neural Information Paradigms and Applications. SCI, vol. 247, pp. 155–182. Springer, Heidelberg (2009). https://doi.org/10.1007/978-3-642-04003-0_7
24. Trentin, E., Freno, A.: Unsupervised nonparametric density estimation: a neural network approach. In: Proceedings of the International Joint Conference on Neural Networks, IJCNN 2009, pp. 3140–3147 (2009)
25. Trentin, E., Lusnig, L., Cavalli, F.: Parzen neural networks: fundamentals, properties, and an application to forensic anthropology. Neural Netw. **97**, 137–151 (2018)
26. Trentin, E., Scherer, S., Schwenker, F.: Emotion recognition from speech signals via a probabilistic echo-state network. Pattern Recognit. Lett. **66**, 4–12 (2015)
27. Vapnik, V.N., Mukherjee, S.: Support vector method for multivariate density estimation. In: Advances in Neural Information Processing Systems, pp. 659–665. MIT Press (2000)
28. Vapnik, V.: The Nature of Statistical Learning Theory. Springer, New York (1995). https://doi.org/10.1007/978-1-4757-2440-0
29. Yuksel, S.E., Wilson, J.N., Gader, P.D.: Twenty years of mixture of experts. IEEE Trans. Neural Netw. Learn. Syst. **23**, 1177–1193 (2012)

Generating Bounding Box Supervision for Semantic Segmentation with Deep Learning

Simone Bonechi, Paolo Andreini$^{(\boxtimes)}$, Monica Bianchini, and Franco Scarselli

Department of Information Engineering and Mathematics,
University of Siena, Siena, Italy
paolo.andreini@yahoo.it

Abstract. Most of the leading Convolutional Neural Network (CNN) models for semantic segmentation exploit a large number of pixel–level annotations. Such a human based labeling requires a considerable effort that complicates the creation of large–scale datasets. In this paper, we propose a deep learning approach that uses bounding box annotations to train a semantic segmentation network. Indeed, the bounding box supervision, even though less accurate, is a valuable alternative, effective in reducing the dataset collection costs. The proposed method is based on a two stage training procedure: first, a deep neural network is trained to distinguish the relevant object from the background inside a given bounding box; then, the output of the network is used to provide a weak supervision for a multi–class segmentation CNN. The performances of our approach have been assessed on the Pascal–VOC 2012 segmentation dataset, obtaining competitive results compared to a fully supervised setting.

Keywords: Deep learning · Semantic segmentation
Weak supervision · Bounding box

1 Introduction

Image semantic segmentation is one of the fundamental topic in computer vision. Its goal is to make dense predictions, inferring the label of every pixel within an image. In the last few years, the use of Convolutional Neural Networks (CNNs) has lead to an impressive progress in this field [1–3], yet based on the use of large datasets of fully annotated images. The human annotation procedure for semantic segmentation is particularly expensive, since it requires a pixel–level characterization of images. For this reason, the available datasets are normally orders of magnitude smaller than image classification datasets (f.i. ImageNet [4]). Such a limitation is important, since the performance of CNNs is largely affected by the amount of training examples. On the other hand, bounding box annotations are less accurate than per–pixel annotations, but they are cheaper

L. Pancioni et al. (Eds.): ANNPR 2018, LNAI 11081, pp. 190–200, 2018.
https://doi.org/10.1007/978-3-319-99978-4_15

and easier to be obtained. In this paper, we propose a simple method, called BBSDL – for Bounding Box Supervision with Deep Learning –, to train CNNs for semantic segmentation using only a bounding box supervision (or a mix of bounding box and pixel–level annotations). Figure 1 provides a general overview of our method, that can be sketched as follows.

- A background–foreground network (BF–Net) is trained on a relatively large dataset with a full pixel–level supervision. The aim of the BF–Net is to recognize the most relevant object inside a bounding box.
- A multi–class segmentation CNN is trained on a target dataset, in which the supervision is obtained exploiting the output of the BF–Net.

The rationale behind this approach is that realizing a background–foreground segmentation, constrained to a bounding box, is significantly simpler than obtaining a multi–class semantic segmentation on the whole image. Following this intuition, we consider a scenario in which only bounding box annotations are available on a target dataset. The pixel–level supervision, on such dataset, can be produced from the bounding boxes exploiting the BF–Net trained on a different dataset. In particular, multi–class annotations can be generated in many ways from the output of the BF–Net and, indeed, a set of different solutions were tested, in order to produce the best target for the multi–class segmentation network. The effectiveness of the proposed method has also been compared with other existing techniques [5,6].

The paper is organized as follows. In Sect. 2, we briefly review the state–of–the–art research in semantic segmentation and weakly supervised approaches. Section 3 presents the details of our method, whereas Sect. 4 describes the experimental setup and collects the obtained results. Finally, some conclusions and future perspectives are drawn in Sect. 5.

2   Related Works

Semantic segmentation describes the process of associating each pixel of an image with a class label. Over the past few years, impressive results in image semantic segmentation, so as in many other visual recognition tasks, have been obtained thanks to deep learning techniques [1–3]. Recent semantic segmentation algorithms often convert existing CNN architectures, designed for image classification, to fully convolutional networks. In this framework, semantic segmentation is generally formulated as a pixel–level labeling problem, which requires hand–made fully annotated images. Sadly, producing this kind of supervision is highly demanding and costly. In order to reduce the annotation efforts, some deep learning methods exploit weak supervision. In contrast to learning under strong supervision, these methods are able to learn from weaker annotations, such as image–level tags, partial labels, bounding boxes, etc. In particular, weak supervised learning has been addressed through Multiple Instance Learning (MIL) [7]. MIL deals with training data arranged in sets, called bags, with the supervision provided only at the set level, while single instances are not individually labeled

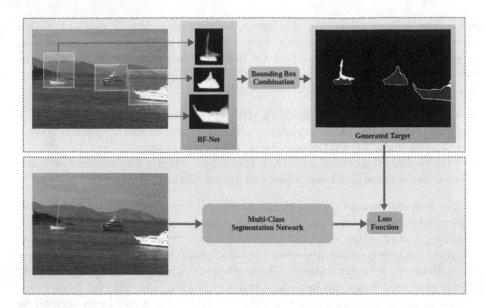

Fig. 1. The training scheme. On the top, weak segmentation annotations are generated from the BF–Net output. At the bottom, the multi–class network is trained based on the generated weak supervision.

[8]. For instance, in [9], a semantic texton forest approach—based on ensembles of decision trees that act directly on image pixels—, revisited in the MIL framework, is proposed for semantic segmentation. Instead, a MIL formulation of multi–class semantic segmentation, by a fully convolutional network, is presented in [10]. MIL extensions to classical segmentation approaches are also introduced in [11] and [12]. Finally, the recently proposed WILDCAT method [13] exploits only global image labels to train deep convolutional neural networks to perform image classification, point–wise object localization, and semantic segmentation.

On the other hand, following an approach which is something similar to our proposal, i.e. that of using bounding box labeling to aid semantic segmentation, in [5], the BoxSup method is proposed, where the core idea is that to iterate between automatically generating region proposals and training convolutional networks. Similarly, in [6], an Expectation–Maximization algorithm was used to iteratively update the training supervision. Nevertheless, while both the above described methods rely on an iterative procedure, our approach directly produces the segmentation supervision, exploiting a deep convolutional network.

3   The BBSDL Method

In the following, we delve into the details of the multi stage training algorithm proposed in this paper (see Fig. 1).

BF–Net Training. The first step in the proposed approach consists in training a deep neural network, capable of recognizing the most relevant object inside a bounding box, thus separating the background from the foreground, called BF–Net (top of Fig. 1). Our experiments are conducted using the Pyramid Scene Parsing architecture [3] (PSP, see Fig. 2).

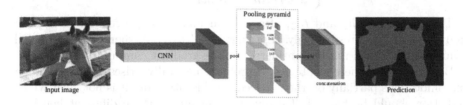

Fig. 2. Scheme of the pyramid scene parsing network, proposed by [3], used in this paper.

The PSP net is a deep fully convolutional neural network which re–purposes the ResNet [14], originally designed for image classification, to perform semantic segmentation. Differently from the original ResNet, a set of dilated convolutions [2] replaces standard convolutions to enlarge the receptive field of the neural network. To gather context information, the PSP exploits a pyramid of pooling with different kernel size. Both upsampling and concatenation produce the final feature representation, which is fed into a convolutional layer to get the desired per–pixel predictions. ResNets of different depths (i.e. with a different number of convolutional layers) were proposed in the original paper [14]. We chose to use the ResNet50 architecture, due to computational issues.

To train this network a dataset composed by image crops is required. Each crop should contain only a single relevant object, in which pixels are anno-tated either as foreground or background; the information about the object class is not needed and indeed it is not used during training. We employ the COCO dataset [15], which collects instance–level fully annotated images (i.e., in which objects of the same category are labeled separately). Such supervi-sion can be used to extract the bounding box that encloses each object and its background–foreground pixel–wise annotation. The images are then cropped, using the obtained bounding boxes. Moreover, in order to include more context information, each crop is enlarged by 5%, compared with the corresponding box dimensions. Image crops are finally used for training and validating the BF–Net.

Multiclass Dataset Generation. Once the BF–Net has been trained, the pixel–level supervision for the multi–class segmentation network training is gen-erated (bottom of Fig. 1). All the bounding box annotations in the target dataset need to be replaced with a multi–class pixel–level supervision. To this aim, the BF–Net is used to produce predictions over each bounding box. Different strate-gies can be employed in order to convert such predictions into the final seg-mentation supervision. In particular, if the naïve approach consists in directly replacing each bounding box with the pixel–level classification given by the BF–Net, a more refined strategy suggests to use the value of the BF–Net output probability $prob(x, y)$, at position (x, y), to obtain the label $l(x, y)$ for the same point:

$$l(x, y) = \begin{cases} background & \text{if} \quad prob(x, y) < th_1 \\ foreground & \text{if} \quad prob(x, y) > th_2 \\ uncertain & \text{otherwise} \end{cases} \tag{1}$$

The thresholds th_1 and th_2, after a trial–and–error procedure, have been fixed to 0.3 and 0.7, respectively. If $prob(x, y) \in (th_1, th_2)$, then (x, y) is labeled as uncertain and will not be considered for the gradient computation.

Based on both these strategies, a problem naturally arises when bounding box annotations partially overlap. Indeed, in this situation, it is not clear which prediction should be trusted. To solve the ambiguity, three different heuristic approaches were used in the experiments, which are sketched in the following.

- **Ignore Intersection** – Overlapping regions are labeled as "uncertain", so that the gradient will not be propagated in these regions.
- **Smallest Box** – Overlapping regions are considered to belong to the smallest bounding box, which is supposed to coincide with the foreground object.
- **Fixed Threshold** – Overlapping regions are considered to belong to the bounding box with the highest foreground probability prediction.

In Sect. 4.2, we review the experimental results obtained using the three different strategies.

Multiclass Segmentation Network Training. Once the pixel–level supervision is provided, the multi–class network can be trained. In all the experiments, the Pascal–VOC 2012 dataset [16] has been exploited for the PSP training and validation. Similarly to the BF–Net, we used the PSP50 as the multi–class segmentation network, with 21 probability output maps. The experimental details are reported in Sect. 4.3.

Implementation Details. Both the BF–Net and the multi–class segmentation network are implemented in TensorFlow. All the experiments follow the same training procedure that will be explained in the following. Actually, the training phase is composed of two different stages. First, the images are resized at a fixed resolution of 233×233, using padding to maintain the original aspect ratio; early stopping is implemented based on the validation set. Then, the training continues using random crops of 233×233 pixels to obtain a more accurate prediction. The Adam optimizer [17], with learning rate set to 10^{-6} and a mini–batch of 15 examples, has been used to train the network. The evaluation phase relies on a sliding window approach. The experimentation was carried out in a Debian environment, with a single NVIDIA GeForce GTX 1080 Ti GPU, with 128 GB of RAM. The average inference time for each image is about 1.6 s and depends on its size.

4   Experiments and Results

In Sect. 4.1, we describe the datasets used in our experiments, whereas the weak supervision generation is presented in Sect. 4.2. Finally the experimental results are discussed in Sect. 4.3.

4.1   The Datasets

COCO–2017. The COCO–2017 dataset [15], firstly released by Microsoft Corporation, collects 115000 training and 5000 validation instance–level fully annotated images. Also a test set of 41000 images is provided. The object categories are 80, plus the background. However in our experiments, the class supervision is not used. From the given annotations, 816590 and 34938 bounding boxes have been extracted, respectively, for training and evaluating the BF–Net.

Pascal–VOC 2012. The original Pascal–VOC 2012 segmentation dataset collects 1464 training and 1449 validation pixel–level fully annotated images. A test set of 1456 images is also provided, yet without a publicly available labeling. The object categories are 20, plus the background class and a "don't care" class, to account for uncertain regions. Finally, a set of 14212 additional images are provided, with only bounding box annotations. Following the procedure reported in [18], an augmented Pascal–VOC segmentation set was also devised, which provides full pixel–level annotations for 9118 out of the 14212 images originally weakly annotated, yielding a total of 10582 training images. The Pascal–VOC dataset is used for training and evaluating the multi–class segmentation network.

4.2   Weak Supervision Generation for Pascal–VOC 2012

The generation of weak supervisions for the Pascal–VOC dataset follows the procedure described in Sect. 3. First, the BF–Net is trained on the COCO dataset. All the bounding box annotations of the 10582 augmented Pascal–VOC images are then replaced with the multi–class pixel–level supervision obtained from the output of the BF–Net.

Table 1 compares the generated weak supervisions with the strong annotations provided by the Pascal–VOC dataset. Based on the reported results, the best performances are obtained using the "Fixed Threshold" approach, providing an improvement of more than 4% of the mean Intersection over Union (mean IoU)[1], compared to the other methods. It is also worth noting that, when the probability falls between the two thresholds, an uncertainty region is produced. This region, as depicted in Fig. 3, mostly coincides with the uncertainty class present in the Pascal–VOC annotations.

---

[1] The Mean Intersection over Union is a common measure used to evaluate the quality of a segmentation algorithm, and adopted by the Pascal–VOC competitions. The mean IoU is defined as the average of the ratios $|T \cap P|/|T \cup P|$ for all the images in the test set, where P is the set of pixels predicted as foreground, T is the set of pixels actually annotated as foreground, and $|\cdot|$ denotes the set cardinality operator.

Table 1. Comparison between the Pascal–VOC annotations and the annotations generated by BBSDL.

Supervision generation approach	Mean IoU
Ignore intersection	76.22%
Smallest box	78.01%
Fixed threshold	82.32%

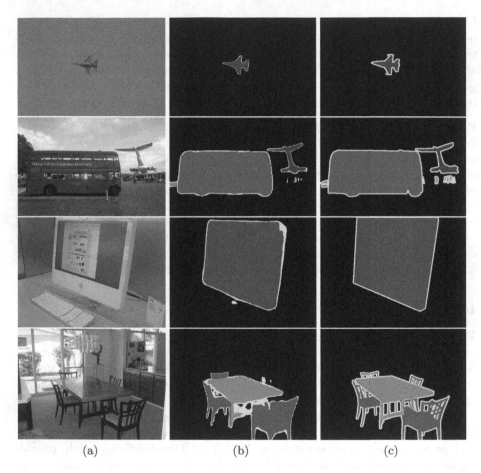

(a)   (b)   (c)

Fig. 3. Qualitative comparison between Ground–Truth segmentation and generated annotations. (a) Original image. (b) Generated annotations with a fixed threshold. (c) Ground–Truth segmentation.

Table 2 reports the results obtained by training the multi–class segmentation network on the Pascal–VOC validation set, confirming the Fixed Threshold approach as the most effective. For this reason, this setup will be used in all the following experiments.

Table 2. Results of the multi–class segmentation on the Pascal–VOC validation set, obtained with different strategies, for the pixel–level weak supervision generation.

Supervision generation approach	Mean IoU
Ignore intersection	60.93%
Smallest box	60.64%
Fixed threshold	65.28%

4.3 Experimental Results

In order to evaluate the proposed framework, we set up the following experiments, that simulate a different availability of pixel–level and bounding box annotations.

- **Mask supervised setting** – This is the baseline method, in which all the 10582 pixel–level annotations of the Pascal–VOC training set are used.
- **BoundingBox supervised setting** – The pixel–wise labeling provided by the Pascal–VOC dataset is totally disregarded. All the bounding boxes are replaced with the supervision provided by the BF–Net.
- **Semi supervised setting** – This simulate the situation in which a relatively reduced number of pixel–wise annotations is available, whereas it is possible to rely on a greater set of bounding box annotations. As in [5] and [6], we used 1464 strongly supervised pixel–level annotations, replacing the bounding boxes in the remaining 9118 images with the supervision provided by BBSDL.

Table 3 shows the results obtained by BBSDL on the Pascal–VOC 2012 validation set, with the three different experimental setups, compared with other state–of–the–art methods, namely BoxSup [5] and Box–EM [6]. A qualitative evaluation is reported in Fig. 4.

Training with strong annotations produces the best mean IoU on the validation set (70.41%). Instead, the mean IoU drops to 65.28%, using only weak bounding box annotations. On the other hand, the semi–supervised setup allows to obtain a mean IoU of 69.20%[2], which is just 1.21 point worse than the strongly supervised setup. As expected, the performance achieved by using only bounding box annotations is less than that obtainable with a strong supervision. However, the produced results show that BBSDL is viable to be used in practical applications, where strong annotations are not available or, in general, are too expensive to be produced. On the validation set, the difference in performance

---

[2] We report the results of the semi–supervised approach just for the sake of completeness, since the real purpose of this paper is to present a method that can work on a dataset where no strong supervision is available.

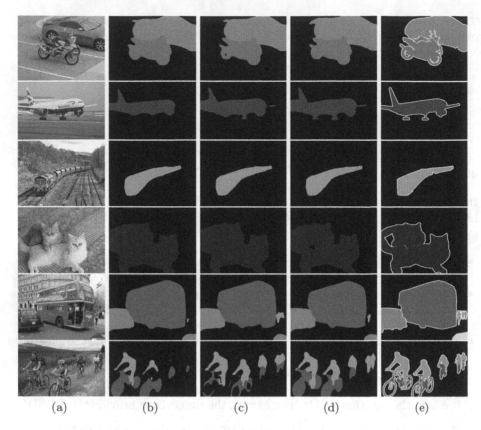

(a)   (b)   (c)   (d)   (e)

Fig. 4. Qualitative comparison of the results obtained with the three different supervision strategies. (a) Original image. Segmentation obtained by weak bounding box annotations (b), based on the semi–supervised setting (c), and by pixel–level annotations (d). (e) Ground–Truth segmentation.

of BBSDL compared to the strong–supervised (Mask) and the weakly supervised (BoundingBox) cases is 5.13%. This result outperforms that obtained by the Box–EM approach (with a difference of 7%), but it is worse with respect to BoxSup. However, BoxSup employs the MCG segmentation proposal mechanism [19], previously trained on the pixel–level annotations of the Pascal–VOC training set. In Table 4, the results on the Pascal–VOC test set are reported, which look similar to those obtained on the validation set. Unfortunately, the baseline results for BoxSup on the test set are not reported in [5], whereas Box–EM uses a different number of training images—differently from BBSDL and BoxSup, Box–EM also uses the validation images to train the model. For this reason, the comparative evaluation is possible only on the validation set.

Table 3. Comparative results on the Pascal–VOC 2012 validation set.

Method	Supervision	Num. of strong tag	Num. of weak tag	Mean IoU
BoxSup [5]	Mask	10582	-	63.80%
BoxSup [5]	BoundingBox	-	10582	62.00%
BoxSup [5]	Semi	1464	9118	63.50%
Bbox–EM [6]	Mask	10582	-	67.60%
Bbox–EM [6]	BoundingBox	-	10582	60.60%
Bbox–EM [6]	Semi	1464	9118	65.10%
BBSDL	Mask	10582	-	70.41%
BBSDL	BoundingBox	-	10582	65.28%
BBSDL	Semi	1464	9118	69.20%

Table 4. Comparative results on the Pascal–VOC 2012 test set.

Method	Supervision	Num. of strong tag	Num. of weak tag	Mean IoU
BoxSup [5]	Mask	10582	-	-
BoxSup [5]	BoundingBox	-	10582	64.4%
BoxSup [5]	Semi	1464	9118	66.2%
Bbox–EM [6]	Mask	12031	-	70.3%
Bbox–EM [6]	BoundingBox	-	12031	62.2%
Bbox–EM [6]	Semi	1464	10567	66.6%
BBSDL	Mask	10582	-	70.36%
BBSDL	BoundingBox	-	10582	66.24%
BBSDL	Semi	1464	9118	70.25%

5   Conclusions and Future Perspectives

This paper explores the use of bounding box annotations for the training of a state–of–the–art semantic segmentation network. The output of a background–foreground network, capable of recognizing the most relevant object inside a region, has been used to deduce pixel–wise annotations. A fixed threshold strategy has been employed in order to convert the background–foreground network output into the final segmentation supervision. Actually, the obtained weak supervision allowed to train a multi–class segmentation network, whose performances are competitive with respect to approaching the semantic segmentation problem in a strongly–supervised framework. In perspective, how to avoid the use of predefined thresholds for the multi–class dataset generation represents an important issue to deal with, in order to improve the BBSDL performances. Moreover, also training the BF–Net based on unsupervised data should be a matter of future work, capturing images from videos and exploiting the temporal information related to successive frames.

References

1. Long, J., Shelhamer, E., Darrell, T.: Fully convolutional networks for semantic segmentation. In: Proceedings of IEEE CVPR 2015, pp. 3431–3440 (2015)
2. Chen, L., Papandreou, G., Kokkinos, I., Murphy, K., Yuille, A.: Semantic image segmentation with deep convolutional nets and fully connected CRFs. In: Proceedings of ICLR 2015 (2015)
3. Zhao, H., Shi, J., Qi, X., Wang, X., Jia, J.: Pyramid scene parsing network. In: Proceedings of IEEE CVPR 2017, pp. 6230–6239 (2017)
4. Deng, J., Dong, W., Socher, R., Li, L.-J., Li, K., Fei-Fei, L.: ImageNet: a large-scale hierarchical image database. In: Proceedings of IEEE CVPR 2009, pp. 248–255 (2009)
5. Dai, J., He, K., Sun, J.: Boxsup: exploiting bounding boxes to supervise convolutional networks for semantic segmentation. In: Proceedings of IEEE ICCV 2015, pp. 1635–1643 (2015)
6. Papandreou, G., Chen, L.-C., Murphy, K., Yuille, A.L.: Weakly and semi-supervised learning of a DCNN for semantic image segmentation. In: Proceedings of IEEE ICCV 2015, pp. 1742–1750 (2015)
7. Dietterich, T.G., Lathrop, R.H., Lozano-Pérez, T.: Solving the multiple instance problem with axis-parallel rectangles. Artif. Intell. 89(1–2), 31–71 (1997)
8. Carbonneau, M.-A., Cheplygina, V., Granger, E., Gagnon, G.: Multiple instance learning: a survey of problem characteristics and applications. Pattern Recognit. 77, 329–353 (2018)
9. Vezhnevets, A., Buhmann, J.M.: Towards weakly supervised semantic segmentation by means of multiple instance and multitask learning. In: Proceedings of IEEE CVPR 2010, pp. 3249–3256 (2010)
10. Pathak, D., Shelhamer, E., Long, J., Darrell, T.: Fully convolutional multi-class multiple instance learning. In: Proceedings of ICLR 2015 (2015)
11. Pinheiro, P.O., Collobert, R.: From image-level to pixel-level labeling with convolutional networks. In: Proceedings of IEEE CVPR 2015, pp. 1713–1721 (2015)
12. Pathak, D., Krahenbuhl, P., Darrell, T.: Constrained convolutional neural networks for weakly supervised segmentation. In: Proceedings of IEEE CVPR 2015, pp. 1796–1804 (2015)
13. Durand, T., Mordan, T., Thome, N., Cord, M.: Wildcat: weakly supervised learning of deep convnets for image classification, pointwise localization and segmentation. In: Proceedings of IEEE CVPR 2017, pp. 5957–5966 (2017)
14. He, K., Zhang, X., Ren, S., Sun, J.: Deep residual learning for image recognition. In: Proceedings of IEEE CVPR 2016, pp. 770–778 (2016)
15. Lin, T.-Y., et al.: Microsoft COCO: common objects in context. In: Fleet, David, Pajdla, Tomas, Schiele, Bernt, Tuytelaars, Tinne (eds.) ECCV 2014. LNCS, vol. 8693, pp. 740–755. Springer, Cham (2014). https://doi.org/10.1007/978-3-319-10602-1_48
16. Everingham, M., Eslami, S.M.A., Van Gool, L., Williams, C.K.I., Winn, J., Zisserman, A.: The PASCAL visual object classes challenge: a retrospective. Int. J. Comput. Vis. 111(1), 98–136 (2015)
17. Kingma, D., Ba, J.: Adam: a method for stochastic optimization. In: Proceedings of ICLR 2015 (2015)
18. Hariharan, B., Arbeláez, P., Bourdev, L., Maji, S., Malik, J.: Semantic contours from inverse detectors. In: Proceedings of IEEE ICCV 2011, pp. 991–998 (2011)
19. Arbeláez, P., Pont-Tuset, J., Barron, J.T., Marques, F., Malik, J.: Multiscale combinatorial grouping. In: Proceedings of IEEE CVPR 2014, pp. 328–335 (2014)

Inductive–Transductive Learning with Graph Neural Networks

Alberto Rossi[1,2]([✉]), Matteo Tiezzi[1]([✉]), Giovanna Maria Dimitri[3],
Monica Bianchini[1], Marco Maggini[1], and Franco Scarselli[1]([✉])

[1] Department of Information Engineering and Mathematics,
University of Siena, Siena, Italy
{mtiezzi,monica,maggini,franco}@diism.unisi.it
[2] Department of Information Engineering, University of Florence, Florence, Italy
alberto.rossi@unifi.it
[3] Department of Computer Science, Computer Laboratory,
University of Cambridge, Cambridge, UK
gmd43@cam.ac.uk
http://sailab.diism.unisi.it

Abstract. Graphs are a natural choice to encode data in many real–world applications. In fact, a graph can describe a given pattern as a complex structure made up of parts (the nodes) and relationships between them (the edges). Despite their rich representational power, most of machine learning approaches cannot deal directly with inputs encoded by graphs. Indeed, Graph Neural Networks (GNNs) have been devised as an extension of recursive models, able to process general graphs, possibly undirected and cyclic. In particular, GNNs can be trained to approximate all the "practically useful" functions on the graph space, based on the classical inductive learning approach, realized within the supervised framework. However, the information encoded in the edges can actually be used in a more refined way, to switch from inductive to transductive learning. In this paper, we present an inductive–transductive learning scheme based on GNNs. The proposed approach is evaluated both on artificial and real–world datasets showing promising results. The recently released GNN software, based on the Tensorflow library, is made available for interested users.

Keywords: Graph Neural Networks · Transductive learning
Graph representations

1 Introduction

Graphs are a rich structured model that can be exploited to encode data from many different domains, which range from bioinformatics [1,2] to neuroscience [3], and social networks [4]. Despite the simplicity of the concepts at the basis of the definition of a graph, the possibility to encode complex data as a set of parts, i.e. the graph nodes, and a set of relationships between these parts, i.e.

© Springer Nature Switzerland AG 2018
L. Pancioni et al. (Eds.): ANNPR 2018, LNAI 11081, pp. 201–212, 2018.
https://doi.org/10.1007/978-3-319-99978-4_16

the graph edges, allows a good compromise between the need of a compact data representation and the preservation of most of the original input information. This model is quite natural in many of the aforementioned applications and the encoding is often straightforward. As an example, the Web can be naturally seen and represented as a graph, with nodes corresponding to web pages (and storing their content), and edges standing for the hyper-links between them [5].

Many classical machine learning approaches assume to deal with *flat* data, encoded, for instance, as real valued vectors. Hence, complex data, such as graphs, need to be transformed into these simpler encodings—typically with some sort of graph traversal—often loosing useful information [6]. In this way, the input to the machine learning tool is no more the original graph but instead a linearized representation, in which the topological and structural information is usually encoded in an unnatural way, that may hinder the learning process itself. Moreover, the natural variability within graphs requires artificial solutions. For instance, the mapping of a graph to a real valued vector can be implemented by concatenating the features stored in each node, following an order derived from the connection topology. However, this approach has many drawbacks. First, in order to have a fixed dimensionality for the vector, all the input graphs should have the same number of nodes or, at least, a maximum cardinality must be chosen for this set, filling the vector elements with padding values when the graph has a lower number of nodes. Second, the encoding of the topology by the position of the node inside the vector is not well defined for any category of graphs. Indeed, if for Directed Ordered Acyclic Graphs (DOAGs) such a topological order is uniquely defined, this does not hold for generic cyclic graphs, where the mapping between nodes and related elements occupying particular positions is arbitrary.

The Graph Neural Network Model (GNN), which was introduced in [6], is able to process graphs directly, without the need of a preprocessing step and without any limitation on the graph type. GNNs are supervised architectures, designed as an extension to Recursive Neural Networks [17–19] and Markov Random Chain Models. The original GNN model is based on the classical inductive learning scheme, where a training set is used to adapt a parametric model. Actually, inductive learning assumes the existence of some rules, that can be implemented by the model, allowing us to classify a pattern given its properties. In this framework, GNNs have been successfully used in different applications, from the classification of Web pages (in Spam or Non–Spam) to the prediction of chemical properties of drug molecules [7].

On the other hand, transductive learning adopts a more direct approach, by which a pattern is classified according to its relationships with the examples available in the training set. In this case, the training patterns are used directly in the classification procedure, without adapting a parametric model, and even without relying on the existence of classification rules and pattern features. In the standard inductive approach, GNNs exclusively employ the parameters learnt during the training procedure. Vice versa, in the transductive approach, the available targets are added to the node labels, and they are directly diffused through the graph in the classification phase.

In this paper, we present a mixed transductive–inductive GNN model that exhibits characteristics common to both the learning frameworks. This model is evaluated on synthetic (clique detection and subgraph matching) and real (traffic flow prediction and web–spam prediction) problems, involving structured inputs encoded with graphs. In particular, we exploit a new implementation of GNNs based on the TensorFlow platform [8].

The paper is organized as follows. In the next section, the GNN model and the related learning algorithms are briefly sketched. Then, in Sect. 3 the transductive approach for GNNs is described. Section 4 presents the experimental settings and reports the obtained results. Finally Sect. 5 draws some conclusions and delineates future perspectives.

2   The Graph Neural Network Model

A graph G is defined as a pair $G = (V, E)$, where V represents the finite set of *nodes* and $E \subseteq V \times V$ denotes the set of *edges*. An edge is identified by the unordered pair of nodes it connects, i.e. $e = (a, b)$, $e \in E$ and $a, b \in V$. In the case in which an asymmetric relationship must be encoded, the pair of nodes that define an edge must be considered as ordered, so as (a, b) and (b, a) represent different connections. In this case, it is preferable to use the term *arc*, while the corresponding graph will be referred as *directed*. The GNN model has been devised to deal with either directed or undirected graphs. Both edges and nodes can be enriched by attributes that are collected into a *label*. In the following we will assume that labels are vectors of predefined dimensionality (eventually different for nodes and edges) that encode features describing each individual node (f.i. average color, area, shape factors for nodes representing homogeneous regions in an image) and each edge (f.i. the distance between the barycenters of two adjacent regions and the length of the common boundary), respectively.

Graph Neural Networks are supervised neural network architectures, able to face classification and regression tasks, where inputs are encoded as graphs [6]. The computation is driven by the input graph topology, which guides the network unfolding. The computational scheme is based on a diffusion mechanism, by which the GNN updates the state vector at each node as a function of the node label, and of the informative contribution of its neighborhood (edge labels and states of the neighboring nodes), as defined by the input graph topology. The state is supposed to summarize the information relevant to the task to be learnt for each node and, given the diffusion process, it will finally take into account the whole information attached to the input graph. Afterwards, the state is used to compute the node output, f.i. the node class or a target property.

More formally, let $x_n \in \mathbb{R}^s$ and $o_n \in \mathbb{R}^m$ be the state and the output at node n, respectively, being f_w the state transition function that drives the diffusion process, while g_w represents the output function. Then, the computation locally performed at each node during the diffusion process can be described by the following equation:

$$x_n = \sum_{(n,v) \in E} f_w(l_n, l_{(n,v)}, x_v, l_v) \tag{1}$$

$$o_n = g_w(x_n, l_n) \tag{2}$$

where $l_n \in \mathbb{R}^q$ and $l_{(n,v)} \in \mathbb{R}^p$ are the labels attached to n and (n,v), respectively. As previously stated, the computation considers the neighborhood of n, defined by its edges $(n,v) \in E$. In particular for each neighbor node v, the state x_v and the label l_v are used in the computation (Fig. 1). The summation in Eq. 1 allows us to deal with any number of neighbors without the need of specifying a particular position for each of them.

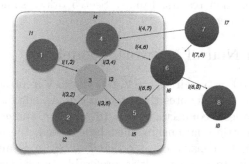

Fig. 1. The neighborhood of node 3. The state x_3 depends on the node neighborhood as $x_3 = f_w(l_3, l_{(3,1)}, x_1, l_1) + f_w(l_3, l_{(3,2)}, x_2, l_2) + f_w(l_3, l_{(3,4)}, x_4, l_4) + f_w(l_3, l_{(3,5)}, x_5, l_5)$.

Equation 1, replicated on all the nodes in the graph, defines a system of non-linear equations in the unknowns $x_n, n \in V$. The solution can be computed by the Jacobi iterative procedure as

$$x_n(t+1) = \sum_{(n,v) \in E} f_w(l_n, l_{(n,v)}, x_v(t), l_v) \tag{3}$$

that implements the diffusion process for the state computation. If the state transition function f_w is a contraction mapping, the Banach Theorem guarantees that the iterative procedure (Eq. 3) converges to a unique solution [6]. In practice, the required iterations can be limited to a maximum number.

Both f_w and g_w can be implemented by simple multilayer perceptrons (MLPs), with a unique hidden layer. The computation of Eq. 3 represents the unfolding of the so called *encoding network* (Fig. 2), where f_w and g_w are computed for each node. Basically, at each node in the graph, there is a replica of the MLP realizing f_w. Each unit stores the state at time t, i.e. $x_n(t)$. The set of states stored in all the nodes at time t are then used to compute the states at time $t+1$. The module g_w is also applied at each node for calculating the output, but only after the state computation has converged.

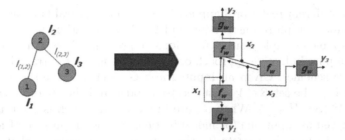

Fig. 2. Construction of the encoding network corresponding to a given input graph (from left to right). The processing units f_w and g_w are replicated for each node and connected following the graph topology.

During training, the network weights are adapted to reduce the error between the network outputs and the expected targets on the set of supervised nodes. The gradient computation is performed following the error Backpropagation scheme on the unfolding network (see [21] for more details).

3   Transductive Learning with GNNs

In the *inductive* learning approach, a model I_w is learnt by adjusting its weights w based on a set of labeled data, namely the training set [9]. Each example is processed independently of the others, but the overall statistics allow the learning algorithm to induce a general model to solve the task. Model prediction is based only on the features describing each different input object. Once the model is learnt, new unseen inputs can be processed one at a time to compute the model output (f.i. the predicted class of the pattern).

Instead, in the *transductive* framework, the algorithm is designed to exploit both labeled and unlabeled examples, taking advantage from relationships between different samples, such as, for instance, some kind of spatial regularization in the feature space (e.g. manifold regularization). The relationships among data can be exploited either in the learning or in the prediction phase, or in both of them. Basically, the prediction on the unlabeled data is obtained by propagating the information available for the "near" examples, through the given relationships between them. For instance, if n is an example at test time, then the targets available in its neighborhood may be exploited, together with the local features of n, as inputs to compute the transduced output [9]. This approach is especially useful and natural when only a small set of labeled data, that comes from an unknown stochastic process, is available. Indeed, a small sample cannot be statistically relevant for inducing a general predictive rule [10] based only on local features.

Most of the transductive approaches, available in literature, are based on graphs (see e.g. [11,12]). In recent years, these methods have been widely applied and implemented in many domains, thanks to their capability of being adaptable to different real–world applications, such as natural language processing, surveillance, graph reconstruction and ECG classification [9].

Being capable of implementing functions on graphs, GNNs can be employed either with a pure inductive approach or with a mixture of the transductive and the inductive schemes. Given an input graph $G = (V, E)$, the set of nodes V can be split into the set $S \subset V$ of supervised nodes and the set of unsupervised nodes $U \subset V$ ($S \cap U = \emptyset$). When a pure inductive approach is used, the GNN network is given as input one (or more) instance of the graph to be learnt (e.g. the Web graph) and the targets for the supervised nodes in S, that are used only to learn the GNN parameters. The trained GNN can then be exploited to process both the original graph(s) in the learning set, to compute the output predictions for the unlabeled nodes in U, and to process unseen graphs without supervised nodes. For example, when considering a Web Spam problem in which the input is a Web graph, the class of a given page is computed by the GNN considering only the node features (e.g. the page contents) and its context in the whole graph. The labels available at the nodes in S are not considered in the computation. Basically it is assumed that the learning process was able to embed the classification rules in the trained model.

However, it should be noted that also in this case the GNN exploits the topology of the relationships among the nodes through the diffusion process used to compute the states, as defined by Eq. 3. Both the nodes in S and U are involved in this computation, but no information on the targets of the nodes in S is exploited. In this sense, we cannot consider this scheme as a proper transduction, since at test time only the node features and its context in the data manifold affect the result of the computation.

In inductive–transductive GNNs, we assume to enrich the node features with the target label such that it is explicitly exploited in the diffusion process, yielding a direct transductive contribution. The way in which targets are diffused and contribute to the final outputs is learnt from examples. We assume that the learning set contains partially supervised graphs. For each graph, we split the set of supervised nodes S into two disjoint subsets: the set of nodes used to compute the loss L and the set of transductive nodes T. For the nodes in T the available target is concatenated to the input feature vectors, whereas for the nodes in L and in U a special `null` target is used (f.i. a vector of zeros). This setting corresponds to a transduction case in which only the targets on the nodes in T are available. Given a graph in the learning set, different training examples can be generated by different splits of S into L and T. The splits can be randomly generated. The nodes in L are used to define the training loss. This way the GNN learns how to exploit the features of the nodes in V, the topology of the relationships in E and the transductive targets in T to approximate the output targets for the nodes in L. During the test phase, the set of supervised nodes S is not split and all the targets are added as features for the corresponding nodes. As before, the features for the nodes in U are obtained by concatenating the original features with the `null` label, so that the trained model computes the outputs on the nodes in U exploiting also the learnt transductive process.

Notice that, during training, it is important that L and T have no intersection, otherwise the GNN would easily learn to produce the correct output at a node by propagating directly the added target feature. The criterion used to generate the sets L and U should try to approximate the actual distribution of these two sets of nodes in the test phase.

4   Experimental Evaluation

In this section, we describe the overall methodology applied to evaluate the proposed GNN transductive–inductive scheme. In particular, we describe the datasets, synthetic and real, the setup for the experiments, and finally we present the results.

4.1   Datasets

The evaluation of the inductive–transductive approach for GNNs has been performed on two synthetic datasets. The first one for subgraph matching, the other one for clique detection. Moreover, we tested the model on real–world benchmarks, i.e. the WEBSPAM–UK2006 dataset[1] and the traffic–flow graph of England[2].

Subgraph Localization. Given a graph G, the subgraph matching problem consists in finding a subgraph S, of a given dimension, in G. In a more formal way, the task is that of learning a function τ, such that $\tau_S(G, n) = 1, n \in V$, when the node n belongs to the given subgraph S, and $\tau_S(G, n) = -1$, otherwise [15]. The problem of finding a given subgraph is common in many practical problems and corresponds, for instance, to finding a particular small molecule inside a greater compound [16]. An example of a subgraph structure is shown in Fig. 3. Our dataset is composed of 700 different graphs, each one having 30 nodes. Instead, the considered subgraphs contain 15 nodes.

Clique Localization. A clique is a complete graph [4], i.e. a graph in which each node is connected with all the others. In a network, overlapping cliques (i.e. cliques that share some nodes) are admitted. In a social network for example, cliques could represent friendship ties. In bioinformatics and computational biology, cliques could be used for identifying similarities between different molecules or for understanding protein–protein interactions [13]. Clique localization is a particular instance of the subgraph matching problem [14]. A clique example is shown in Fig. 4. In the experiments, we consider a dataset composed by 700 different graphs having 15 nodes each, where the dimension of the maximal clique is 7 nodes.

WEBSPAM–UK2006—The dataset has been collected by a web crawl based on the .uk domain [5]. The nodes of the network represent 11402 hosts, and more than 730775 edges (links) are present. Many sets of features are available,

---

[1] http://webspam.lip6.fr/wiki/pmwiki.php?n=Main.PhaseII.
[2] https://github.com/weijianzhang/EvolvingGraphDatasets/tree/master/traffic.

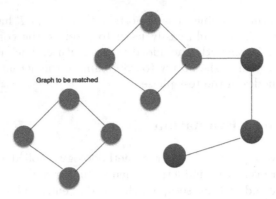

Fig. 3. An example of a subgraph matching problem, where the graph with the blue nodes is matched against the bigger graph. This task corresponds to finding the isomorphic function that maps the blue graph into the bigger one. (Color figure online)

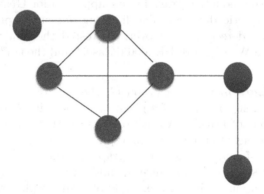

Fig. 4. An example of a graph containing a clique. The blue nodes represent a fully connected subgraph of dimension 4, whereas the red nodes do not belong to the clique. (Color figure online)

grouped into three categories: basic, link–based, and content–based. We consider only the first two categories, exploiting simple properties of the hosts, such as the number of pages and the length of their name, while in the link–based set we find also information on their in–degree, out–degree, PageRank, edge reciprocity, assortativity coefficient, TrustRank, Truncated PageRank, estimation of supporters, etc.

Traffic–Flow Prediction—This task consists in the prediction of the traffic–flow over all motorways and 'A' roads, managed by the Highways Agency in England. The problem is formulated as an edge–regression problem, since the roads are encoded as the arcs of the graph and the nodes represent the crossroads. In this case, nodes are not labeled, whereas a set of features (a label) is attached to each edge. In particular, such features represent the journey times and speeds, estimated using a combination of sources, including Automatic Number Plate Recognition (ANPR) cameras, in–vehicle Global Positioning Systems (GPS) and inductive loops installed on the road surface. Journey times are derived from real vehicle observations and computed using adjacent time periods or the same time period on different days. The data are collected every 15 min, based on a snapshot of the traffic at that time. The problem is that of predicting the traffic flow across a certain road. We focus on a single time–stamp, obtaining 1002 nodes and 2499 edges, representing the roads.

4.2 Experimental Setup

The main goal of this paper consists in providing a comparison between the transductive–inductive and the purely inductive learning frameworks. Hence, the reported results are not to be intended as the state of the art.

In the experiments, the available datasets were split into a training, a validation, and a test set, and different conditions were defined by varying the percentage of labeled nodes: these nodes are assigned to the set T and are not exploited in the performance evaluation. In fact, it is assumed that their output is given and they are only exploited for the transduction, thanks to the diffusion mechanism that characterizes the GNN model.

We evaluated all the models with five different percentages of labeled nodes: 0, 10, 20, 30, 50. In every task, we exploited a state function implemented by a feedforward neural network with two hyperbolic tangent layers, composed by 15 and 5 neurons, respectively. Consequently, the dimension of the state is 5. For the tasks of clique searching, subgraph and WebSpam detection, the output function consists of a single softmax layer. For the flow–traffic detection we employed a linear layer.

The learning procedure was based on a simple Gradient Descent Optimizer with learning rate of 10^{-3}, except for the WebSpam task, for which we used the Adam optimizer with the same learning rate, in order to speed–up the learning procedure. We set the threshold for the convergence of the state to 10^{-3}. This cut–off is used to stop the state update loop when the difference of the state vectors in two subsequent iterations is below this value.

Moreover, in the WebSpam problem, we used the softmax output of a simple MLP as node label, inspired by the work in [20]. This feedforward network is composed by two hyperbolic tangent layers and a softmax output layer. Their dimensions are 100, 20, and 2, respectively. We adopted the cross–entropy as the loss function for the classification problems, whereas we used the mean squared error function for the traffic–flow task, which is a regression problem.

In the comparisons, we considered also the learning time, since for some tasks (f.i. subgraph and clique detection) the differences in classification performances are not so evident when giving no time constraints. Hence, we set an appropriate maximum number of epochs for each problem. For the subgraph detection problem the limit was set to 20000 epochs, whereas we used 3000 as the number of epochs to train the GNN for the WebSpam and clique detection problems, and 5000 epochs for the traffic–flow prediction benchmark.

A full–batch learning has been used in all the tasks, meaning that we adapt the weights once for each epoch (i.e. after processing all the examples in the learning set). In the case of WebSpam, the learning set consists of the whole Web graph that must clearly be processed as a unique batch. In the case of synthetic datasets, we simply considered all the graphs as belonging to a bigger disconnected graph.

4.3 Results

Figure 5 shows the trend of the accuracy on the validation set for all the tasks during the learning process.

Table 1 reports the results for the addressed tasks, when varying the percentage of labeled nodes exploited in the transductive phase. The first column, corresponding to the value 0%, represents the purely inductive case.

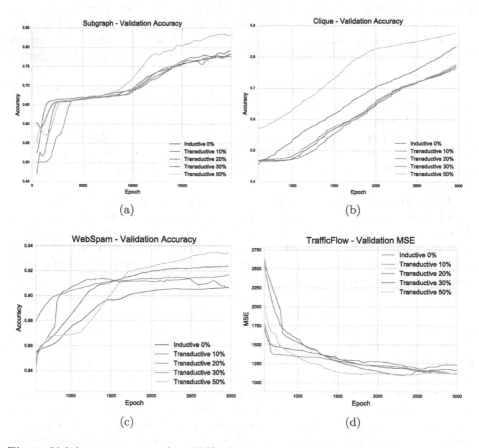

Fig. 5. Validation accuracy (or MSE) obtained varying the exploited percentage of labeled nodes, in the four different tasks.

Table 1. Mean accuracy over five runs (mean squared error for the traffic–flow benchmark), together with the standard deviation.

% of labels	0%		10%		20%		30%		50%	
Score	Mean	Std	Mean	Std	Mean	Std	Mean	Std	Mean	Std
SubGraph	77.26	0.44	77.58	1.10	77.94	1.14	78.87	0.53	**83.12**	1.50
Clique	82.16	2.31	83.55	3.62	84.9	2.95	84.1	2.95	83.56	1.86
WebSpam	91.46	0.49	91.54	0.48	91.94	0.60	92.23	0.46	**92.49**	0.65
Traffic (MSE)	1123	232	1159	152	968	114	865	178	**811**	105

Transductive learning demonstrated its effectiveness on all the benchmarks. For some simple problems, like subgraph and clique detection, it is anyway difficult to obtain evident differences in absolute performance, for all the percentages of labeled nodes exploited in the transduction.

5 Conclusions

In this paper, we presented a transductive learning framework based on GNNs applied to graphs. We showed how this paradigm may improve the performances with an experimental evaluation both on synthetic and real–world problems, belonging to different domains.

Given the increasing amount of available structured data, it would be interesting to test these techniques in other application domains, ranging from mobile communications to the biomedical field, or to the large graphs provided by the online social networks, like Facebook and Twitter. It would be also of interest to investigate the properties of the diffusion process and the influence of a subset of labeled nodes over the neighbors, in order to have a deeper understanding of the GNN model.

References

1. Grindrod, P., Kibble, M.: Review of uses of network and graph theory concepts within proteomics. Expert Rev. Proteomics **1**(2), 229–238 (2004)
2. Barabasi, A.-L., Oltvai, Z.N.: Network biology: understanding the cell's functional organization. Nature Rev. Genet. **5**, 101–113 (2004)
3. Sporns, O.: Graph theory methods for the analysis of neural connectivity patterns. In: Kötter, R. (ed.) Neuroscience Databases, pp. 171–185. Springer, Boston (2003). https://doi.org/10.1007/978-1-4615-1079-6_12
4. Newman, M.E.J.: Networks: An Introduction. Oxford University Press, Oxford (2010)
5. Belahcen, A., Bianchini, M., Scarselli, F.: Web spam detection using transductive–inductive graph neural networks. In: Bassis, S., Esposito, A., Morabito, F.C. (eds.) Advances in Neural Networks: Computational and Theoretical Issues. SIST, vol. 37, pp. 83–91. Springer, Cham (2015). https://doi.org/10.1007/978-3-319-18164-6_9

6. Scarselli, F., Gori, M., Tsoi, A.-C., Hagenbuchner, M., Monfardini, G.: The graph neural network model. IEEE Trans. Neural Netw. **20**(1), 61–80 (2009)
7. Bianchini, M., Dimitri, G.M., Maggini, M., Scarselli, F.: Deep neural networks for structured data. In: Pedrycz, W., Chen, S.-M. (eds.) Computational Intelligence for Pattern Recognition. SCI, vol. 777, pp. 29–51. Springer, Cham (2018). https://doi.org/10.1007/978-3-319-89629-8_2
8. Abadi, M., et al.: TensorFlow: a system for large-scale machine learning. In: Proceedings of OSDI 2016, pp. 265–283 (2016)
9. Bianchini, M., Belahcen, A., Scarselli, F.: A comparative study of inductive and transductive learning with feedforward neural networks. In: Adorni, G., Cagnoni, S., Gori, M., Maratea, M. (eds.) AI*IA 2016. LNCS (LNAI), vol. 10037, pp. 283–293. Springer, Cham (2016). https://doi.org/10.1007/978-3-319-49130-1_21
10. Vapnik, V.N.: The Nature of Statistical Learning Theory. Statistics for Engineering and Information Science. Springer, New York (2013). https://doi.org/10.1007/978-1-4757-3264-1
11. Arnold, A., Nallapati, R., Cohen, W.W.: A comparative study of methods for transductive transfer learning. In: Proceedings of IEEE ICDMW 2007, pp. 77–82 (2007)
12. Zhou, D., Burges, C.J.C.: Spectral clustering and transductive learning with multiple views. In: Proceedings of ICML 2007, pp. 1159–1166. ACM (2007)
13. Bu, D., et al.: Topological structure analysis of the protein–protein interaction network in budding yeast. Nucl. Acids Res. **31**(9), 2443–2450 (2003)
14. Wood, D.R.: An algorithm for finding a maximum clique in a graph. Oper. Res. Lett. **21**(5), 211–217 (1997)
15. Bandinelli, N., Bianchini, M., Scarselli, F.: Learning long-term dependencies using layered graph neural networks. In: Proceedings of IEEE IJCNN 2010, pp. 1–8 (2010)
16. Hu, H., Yan, X., Huang, Y., Han, J., Zhou, X.J.: Mining coherent dense subgraphs across massive biological networks for functional discovery. Bioinformatics **21**(Suppl. 1), i213–i221 (2005)
17. Di Massa, V., Monfardini, G., Sarti, L., Scarselli, F., Maggini, M., Gori, M.: A comparison between recursive neural networks and graph neural networks. In: Proceedings of IEEE IJCNN 2006, pp. 778–785 (2006)
18. Bianchini, M., Maggini, M., Sarti, L.: Recursive neural networks and their applications to image processing. In: Hawkes, P.W. (ed.) Advances in Imaging and Electron Physics, vol. 140, pp. 1–60. Elsevier/Academic Press (2006)
19. Bianchini, M., Maggini, M., Sarti, L., Scarselli, F.: Recursive neural networks for processing graphs with labelled edges: theory and applications. Neural Netw. **18**(8), 1040–1050 (2005)
20. Scarselli, F., Tsoi, A.-C., Hagenbuchner, M., Di Noi, L.: Solving graph data issues using a layered architecture approach with applications to web spam detection. Neural Netw. **48**, 78–90 (2013)
21. Bianchini, M., Maggini, M.: Supervised neural network models for processing graphs. In: Bianchini, M., Maggini, M., Jain, L. (eds.) Handbook on Neural Information Processing, pp. 67–96. Springer, Heidelberg (2013). https://doi.org/10.1007/978-3-642-36657-4_3

Bounded Rational Decision-Making with Adaptive Neural Network Priors

Heinke Hihn$^{(\boxtimes)}$, Sebastian Gottwald, and Daniel A. Braun

Faculty of Engineering, Computer Science, and Psychology,
Institute for Neural Information Processing, Ulm University,
Ulm, Germany
{heinke.hihn,sebastian.gottwald,daniel.braun}@uni-ulm.de

Abstract. Bounded rationality investigates utility-optimizing decision-makers with limited information-processing power. In particular, information theoretic bounded rationality models formalize resource constraints abstractly in terms of relative Shannon information, namely the Kullback-Leibler Divergence between the agents' prior and posterior policy. Between prior and posterior lies an anytime deliberation process that can be instantiated by sample-based evaluations of the utility function through Markov Chain Monte Carlo (MCMC) optimization. The most simple model assumes a fixed prior and can relate abstract information-theoretic processing costs to the number of sample evaluations. However, more advanced models would also address the question of learning, that is how the prior is adapted over time such that generated prior proposals become more efficient. In this work we investigate generative neural networks as priors that are optimized concurrently with anytime sample-based decision-making processes such as MCMC. We evaluate this approach on toy examples.

Keywords: Bounded rationality · Variational Autoencoder
Adaptive priors · Markov Chain Monte Carlo

1 Introduction

Intelligent agents are usually faced with the task of optimizing some utility function $\mathbf{U}$ that is a priori unknown and can only be evaluated sample-wise. We do not restrict ourselves on the form of this function, thus in principle it could be a classification or regression loss, a reward function in a reinforcement learning environment or any other utility function. The framework of information-theoretic bounded rationality [16,17] and related information-theoretic models [3,14,20,21,23] provide a formal framework to model agents that behave in a computationally restricted manner by modeling resource constraints through information-theoretic constraints. Such limitations also lead to the emergence of hierarchies and abstractions [5], which can be exploited to reduce computational and search effort. Recently, the main principles have been successfully applied

© The Author(s) 2018
L. Pancioni et al. (Eds.): ANNPR 2018, LNAI 11081, pp. 213–225, 2018.
https://doi.org/10.1007/978-3-319-99978-4_17

to spiking and artificial neural networks, in particular feedforward-neural network learning problems, where the information-theoretic constraint was mainly employed as some kind of regularization [7,11,12,18]. In this work we introduce bounded rational decision-making with adaptive generative neural network priors. We investigate the interaction between anytime sample-based decision-making processes and concurrent improvement of prior policies through learning, where the prior policies are parameterized as Variational Autoencoders [10]—a recently proposed generative neural network model.

The paper is structured as follows. In Sect. 2 we discuss the basic concepts of information-theoretic bounded rationality, sampled-based interpretations of bounded rationality in the context of Markov Chain Monte Carlo (MCMC), and the basic concepts of Variational Autoencoders. In Sect. 3 we present the proposed decision-making model by combining sample-based decision-making with concurrent learning of priors parameterized by Variational Autoencoders. In Sect. 4 we evaluate the model with toy examples. In Sect. 5 we discuss our results.

2 Methods

2.1 Bounded Rational Decision Making

The foundational concept in decision-making theory is Maximum Expected Utility [22], whereby an agent is modeled as choosing actions such that it maximizes its expected utility

$$\max_{p(a|w)} \sum_w \rho(w) \sum_a p(a|w) \mathbf{U}(w,a), \tag{1}$$

where a is an action from the action space A and w is a world state from the world state space W, and $\mathbf{U}(w,a)$ is a utility function. We assume that the world states are distributed according to a known and fixed distribution $\rho(w)$ and that the world sates w are finite and discrete. In the case of a single world state or world state distribution $\rho(w) = \delta(w - w_0)$, the decision-making problem simplifies into a single function optimization problem $a^* = \arg\max_a \mathbf{U}(a)$. In many cases, solving such optimization problems may require an exhaustive search, where simple enumeration is extremely expensive.

A bounded rational decision maker tackles the above decision-making problem by settling on a good enough solution. Finding a bounded optimal policy requires to maximize the utility function while simultaneously remaining within some given constraints. The resulting policy is a conditional probability distribution $p(a|w)$, which essentially consists of choosing an action a given a particular world state w. The constraints of limited information-processing resources can be formalized by setting an upper bound on the D_{KL} (say B bits) that the decision-maker is maximally allowed to spend to transform its prior strategy into a posterior strategy through deliberation. This results in the following constrained optimization problem [5]:

$$\max_{p(a|w)} \sum_w \rho(w) \sum_a p(a|w) \mathbf{U}(w,a), \text{ s.t. } D_{KL}(p(a|w)\|p(a)) \le B. \tag{2}$$

This constrained optimization problem can be formulated as an unconstrained problem [16]:

$$\max_{p(a|w)} \left(\sum_w \rho(w) \sum_a p(a|w)\mathbf{U}(w,a) - \frac{1}{\beta} D_{KL}(p(a|w)\|p(a)) \right), \qquad (3)$$

where the inverse temperature $\beta \in \mathbb{R}^+$ is a Lagrange multiplier that influences the trade off between expected utility gain and information cost. For $\beta \to \infty$ the agent behaves perfectly rational and for $\beta \to 0$ the agent can only act according to the prior policy. The optimal prior policy in this case is given by the marginal $p(a) = \sum_{w \in W} \rho(w)p(a|w)$ [5], in which case the Kullback-Leibler divergence becomes equal to the mutual information, i.e. $D_{KL}(p(a|w)\|p(a)) = I(W; A)$. The solution to the optimization problem (3) can be found by iterating the following set of self-consistent equations [5]:

$$\begin{cases} p(a|w) = \frac{1}{Z(w)} p(a) \exp(\beta_1 \mathbf{U}(w,a)) \\ p(a) = \sum_w \rho(w)p(a|w), \end{cases}$$

where $Z(w) = \sum_a p(a) \exp(\beta_1 \mathbf{U}(w,a))$ is normalization factor. Computing such a normalization factor is usually computationally expensive as it involves summing over spaces with high cardinality. We avoid this by Monte Carlo approximation.

2.2 MCMC as Sample-Based Bounded Rational Decision-Making

Monte Carlo methods are mostly used to solve two related kinds of problems. One is to generate samples x from a given distribution $q(x)$ and the other is to estimate the expectation of a function. For example, if $g(x)$ is a function for which we need to compute the expectation $\Phi = \mathbb{E}_{q(x)}[g(x)]$ we can draw N samples $\{x_i\}_{i=1}^N$ to obtain the estimate $\hat{\Phi} = \frac{1}{N} \sum_{i=1}^N g(x_i)$ [15]. Samples can be drawn by employing Markov Chains to simulate stochastic processes. A Markov Chain can be defined by an initial probability $p^0(x)$ and a transition probability $\mathbf{T}(x', x)$, which gives the probability of transitioning from state x to x'. The probability of being in state x' at the $(t+1)$-th iteration is given by:

$$p^{t+1}(x') = \sum_x \mathbf{T}(x', x)p^t(x). \qquad (4)$$

Such a chain can be used to generate sample proposals from a desired target distribution $q(x)$, if the following prerequisites are met [15]. Firstly, the chain must be ergodic, i.e. the chain must converge to $q(x)$ independent of the initial distribution $p^0(x)$. Secondly, the desired distribution must be an invariant distribution of the chain. A distribution $q(x)$ is an invariant of $\mathbf{T}(x', x)$ if its probability vector is an eigenvector of the transition probability matrix. A sufficient, but not necessary condition to fulfill this requirement is detailed balance, i.e. the probability of going from state x to x' is the same as going from x' to x: $q(x)\mathbf{T}(x', x) = q(x')\mathbf{T}(x, x')$.

An MCMC chain can be viewed as a bounded rational decision-making process for a single context w in the sense that it performs an anytime optimization of a utility function $\mathbf{U}(a)$ with some precision γ and that it is initialized with a prior $p(a)$. The target distribution has to be chosen as $q(a) \propto e^{\gamma \mathbf{U}(a)}$ in this case. A decision is made with the last sample when the chain is stopped. The resource corresponds then to the number of steps the chain has taken to evaluate the function $\mathbf{U}(a)$. To find the transition probabilities $\mathbf{T}(x', x)$ of the chain, we assume detailed balance and a Metropolis-Hastings scheme $\mathbf{T}(x', x) = g(x'|x)A(x'|x)$ such that

$$\frac{\mathbf{T}(x', x)}{\mathbf{T}(x, x')} = \frac{g(x'|x)A(x'|x)}{g(x|x')A(x|x')} = e^{\gamma(\mathbf{U}(x') - \mathbf{U}(x))} \tag{5}$$

with a proposal distribution $g(x'|x)$ and an acceptance probability $A(x'|x)$. One common choice that satisfies Eq. (5) is

$$A(x'|x) = \min\left\{1, \frac{g(x'|x)}{g(x|x')}e^{\gamma(\mathbf{U}(x') - \mathbf{U}(x))}\right\}, \tag{6}$$

which can be further simplified when using a symmetric proposal distribution with $g(x'|x) = g(x|x')$, resulting in $A(x'|x) = \min\left\{1, e^{\gamma(\mathbf{U}(x') - \mathbf{U}(x))}\right\}$.

Note that the decision of the chain will in general follow a non-equilibrium distribution, but that we can use the bounded rational optimum as a normative baseline to quantify how efficiently resources are used by analyzing how closely the bounded rational equilibrium is approximated.

2.3 Representing Prior Strategies with Variational Autoencoders

While an anytime optimization process such as MCMC can be regarded as a transformation from prior to posterior, the question remains how to choose the prior. While the prior may be assumed to be fixed, it would be far more efficient if the prior itself were subjected to an optimization process that minimizes the overall information-processing costs. Since in the case of multiple world states w the optimal prior is given by the marginal $p(a) = \sum_w \rho(w)p(a|w)$, we can use the outputs a of the anytime decision-making process to train a generative model of the prior $p(a)$. If the generative model was chosen from a parametric family such as a Gaussian distribution, then training would consist in updating mean and variance of the Gaussian. Choosing such a parametric family imposes restrictions on the shape of the prior, in particular in the continuous domain. Therefore, we investigate non-parametric generative models of the prior, in particular neural network models such as Variational Autoencoders (VAEs).

VAEs were introduced by [10] as generative models that use a similar architecture as deterministic autoencoder networks. Their functioning is best understood as variational Bayesian inference in a latent variable model $p(x|z, \theta)$ with prior $p(z)$, where x is observable data, and z is the latent variable that explains the data, but that cannot be observed directly. The aim is to find a parameter $\hat{\theta}_{ML}$ that maximizes the likelihood of the data $p(x|\theta) = \int p(x|z, \theta)p(z)dz$. Samples from $p(x|\theta)$ can then be generated by first sampling z and then sampling an x from $p(x|z, \theta)$. As the maximum likelihood optimization may prove difficult due to the integral, we may express the likelihood in a different form by assuming a distribution $q(z|x, \eta)$ such that

$$\log p(x|\theta, \eta) = \int q(z|x, \eta) \log \frac{p(x|z, \theta)p(z)}{q(z|x, \eta)} \, dz + \underbrace{\int q(z|x, \eta) \log \frac{q(z|x, \eta)}{p(z|x, \theta)} \, dz}_{=D_{KL}(q||p) \geq 0}$$

$$\geq \int q(z|x, \eta) \log \frac{p(x|z, \theta)p(z)}{q(z|x, \eta)} \, dz =: F(\theta, \eta). \tag{7}$$

Assuming that the distribution $q(z|x, \eta)$ is expressive enough to approximate the true posterior $p(z|x, \theta)$ reasonably well, we can neglect the D_{KL} between the two distributions, and directly optimize the lower bound $F(\theta, \eta)$ through gradient descent. In VAEs $q(z|x, \eta)$ is called the encoder that translates from x to z and $p(x|z, \theta)$ is called the decoder that translates from z to x. Both distributions and the prior $p(z)$ are assumed to be Gaussian

$$p(x|z, \theta) = \mathcal{N}\left(x|\mu_\theta(z), \sigma^2 \mathbb{I}\right)$$
$$q(z|x, \eta) = \mathcal{N}\left(z|\mu_\eta(x), \Sigma_\eta(x)\right)$$
$$p(z) = \mathcal{N}(z|0, \mathbb{I}),$$

where $\mu_\theta(z)$, $\mu_\eta(x)$ and $\Sigma_\eta(x)$ are non-linear functions implemented by feed-forward neural networks and where it is ensured that $\sigma^2 \searrow 0$ and that $\Sigma_\eta(x)$ is a covariance matrix.

Note that the optimization of the autoencoder itself can also be viewed as a bounded rational choice

$$\max_{\theta, \eta} \left(\mathbb{E}_{q(z|x, \eta)} \left[\log p(x|z, \theta)\right] - D_{KL}\left(q(z|x, \eta)||p(z)\right) \right), \tag{8}$$

where the expected likelihood is maximized while the encoder distribution $q(z|x, \eta)$ is kept close to the prior $p(z)$.

3   Modeling Bounded Rationality with Adaptive Neural Network Priors

In this section we combine MCMC anytime decision-processes with adaptive autoencoder priors. In the case of a single world state, the combination is straightforward in that each decision selected by the MCMC process is fed as an observable input to an autoencoder. The updated autoencoder is then used as an improved prior to initialize the next MCMC decision. In case of multiple world states, there are two straightforward scenarios. In the first scenario there are as many priors as world states and each of them is updated independently. For each world state we obtain exactly the same solution as in the single world

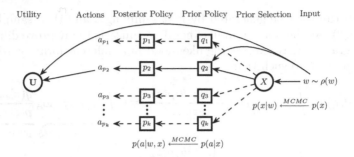

Fig. 1. For each incoming world state w our model samples a prior indexed by $x_i \sim p(x|w)$. Each prior $p(a|x)$ is represented by a VAE. To arrive at the posterior policy $p(a|w,x)$, an anytime MCMC optimization is seeded with $a_0 \sim p(a|x)$ to generate a sample from $p(a|w,x)$. The prior selection policy is also implemented by an MCMC chain and selects agents that have achieved high utility on a particular w.

state case. In the second scenario there is only a single prior over actions for all world states. In this case the autoencoder is trained with the decisions by all MCMC chains such that the autoencoder should converge to the optimal rate distortion prior. A third, more interesting scenario occurs when we allow multiple priors, but less than world states—compare Fig. 1. This is especially plausible when dealing with continuous world states, but also in the case of large discrete spaces.

3.1 Decision Making with Multiple Priors

Decision-making with multiple priors can be regarded as a multi-agent decision-making problem where several bounded rational decision-makers are combined into a single decision-making process [5]. In our case the most suitable arrangement of decision-makers is a two-step process where first each world state is assigned probabilistically to a prior which is then used in the second step to initialize an MCMC chain—compare Fig. 1. The output of that chain is then used to train the autoencoder corresponding to the selected prior. As each prior may be responsible for multiple world states, each prior will learn an abstraction that is specialized for this subspace of world states. This two-stage decision-process can be formalized as a bounded rational optimization problem

$$\max_{p(a|w,x),p(x|w)} \left(\mathbb{E}_{p(a|w,x)}[\mathbf{U}(w,a)] - \frac{1}{\beta_1}I(W;X) - \frac{1}{\beta_2}I(W;A|X) \right), \quad (9)$$

where $p(x|w)$ is selecting the responsible prior $p(a|x)$ indexed by x for world state w. The resource parameter for the first selection stage is given by β_1 and by β_2 for the second decision made by the MCMC process. The solution of optimization (9) is given by the following set of equations:

$$\left\{ \begin{array}{l} p(x|w) = \frac{1}{Z(w)} p(x) \exp(\beta_1 \Delta \, \mathrm{F}_{\mathrm{par}}(w,x)) \\ p(x) = \sum_w \rho(w) p(x|w) \\ p(a|w,x) = \frac{1}{Z(w,x)} p(a|x) \exp(\beta_2 \mathbf{U}(w,a)) \\ p(a|x) = \sum_w p(w|x) p(a|w,x) \\ \Delta \, \mathrm{F}_{\mathrm{par}}(w,x) = \mathbb{E}_{p(a|w,x)}[\mathbf{U}(w,a)] - \frac{1}{\beta_2} \mathrm{D}_{\mathrm{KL}}(p(a|w,x)\|p(a|x)), \end{array} \right. \tag{10}$$

where $Z(w)$ and $Z(w,x)$ are the normalization factors and $\Delta \, \mathrm{F}_{\mathrm{par}}(w,x)$ is the free energy of the action selection stage. The marginal distribution $p(a|x)$ encapsulates an action selection policy consisting of the priors $p(a|w,x)$ weighted by the responsibilities given by the Bayesian posterior $p(w|x)$. Note that the Bayesian posterior is not determined by a given likelihood model, but is the result of the optimization process (9).

3.2 Model Architecture

Equation (10) describe abstractly how a two-step decision process with bounded rational decision-makers should be optimally partitioned. In this section we propose a sample-based model of a bounded rational decision process that approximately corresponds to Eq. (10) such that the performance of the decision process can be compared against its normative baseline. To translate Eq. (10) into a stochastic process we proceed in three steps. First, we implement the priors $p(a|x)$ as Variational Autoencoders. Second, we formulate an MCMC chain that is initialized with a sample from the prior and generates a decision $a \sim p(a|x,w)$. Third, we design an MCMC chain that functions as a selector between the different priors.

Autoencoder Priors. Each prior $p(a|x)$ in Eq. (10) is represented by a VAE that learns to generate action samples that mimic the samples given by the MCMC chains—compare Fig. 2. The functions $\mu_\theta(z)$, $\mu_\eta(a)$ and $\Sigma_\eta(a)$ are implemented as feed-forward neural networks with one hidden layer. The units in the hidden layer were all chosen with sigmoid activation function, the output units in the case of the μ-functions were also chosen as sigmoids and for the Σ-function as ReLU. During training the weights η and θ are adapted to optimize the expected log-likelihood of the action samples that are given by the decisions made by the MCMC chains for all world states that have been assigned to the prior $p(a|x)$. Due to the Gaussian shape of the decoder distribution, optimizing the log-likelihood corresponds to minimizing quadratic loss of the reconstruction error. After training, the network can generate sample actions itself by feeding the decoder network with samples from $\mathcal{N}(z|0,\mathbb{I})$.

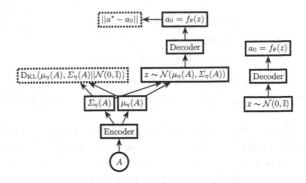

Fig. 2. The encoder translates the observed action into a latent variable z, whereas the decoder translates the latent variable z into a proposed action a. During training the weights η and θ are adapted to optimize the expected log-likelihood of the observed action samples. After training, the network can generate actions by feeding the decoder network with samples from $\mathcal{N}(z|0, \mathbb{I})$.

MCMC Decision-Making. To implement the bounded rational decision-maker $p(a|w, x)$ we obtain an action sample $a \sim p(a|x)$ from the autoencoder prior to initialize an MCMC chain that optimizes the target utility $\mathbf{U}(w, a)$ for the given world state. We run the MCMC chain for $N_{\max}$ steps. In each step we generate a proposal from a Gaussian distribution with $g(a'|a) = \mathcal{N}(a'|a, \sigma^2)$ and accept with probability

$$A(a'|a) = \min\left\{1, \exp(\gamma(\mathbf{U}(w, a') - \mathbf{U}(w, a)))\right\}. \qquad (11)$$

Over the course of $N_{\max}$ time steps, the precision γ is adjusted following an annealing schedule conditioned on the maximum number of steps $N_{\max}$. We use an inverse Boltzmann annealing schedule, i.e. $\gamma^{(k)} = \gamma^0 + \alpha \log(1 + k)$, where α is a tuning parameter. The rationale behind this is that we assume the sampling process to be coarse grained in the beginning and is getting finer during the search.

Prior Selection. To implement the bounded rational prior selection $p(x|w)$ through an MCMC process, we first sample an x from the prior $p(x)$ and start an MCMC chain that (approximately) optimizes $\Delta \mathrm{F}_{\mathrm{par}}(w, x)$ for a given world state w sampled from $\rho(w)$. The prior $p(x)$ is represented by a multinomial and updated by the frequencies of the selected prior indices x. The number of steps in the prior selection MCMC chain was kept constant at a value of $N_{\max}^{\mathrm{sel}}$ and similarly the precision γ^{sel} was annealed over the course of $N_{\max}^{\mathrm{sel}}$ time steps. The target $\Delta \mathrm{F}_{\mathrm{par}}(w, x)$ comprises a trade-off between expected utility and information resources. However, it cannot be directly evaluated and would require the computation of $\mathrm{D}_{\mathrm{KL}}(p(a|x, w) \| p(a|x))$. Here we use number of steps in the downstream MCMC process as a resource measure. As the number of

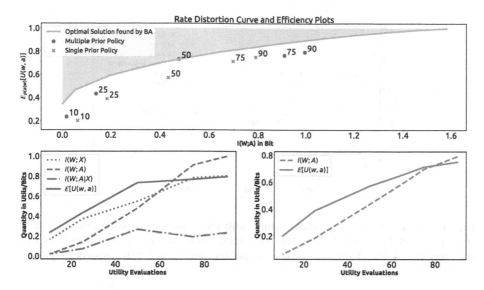

Fig. 3. Top: The line is given by the Rate Distortion Curve that forms a theoretical efficiency frontier, characterized by the ratio between mutual information and expected utility. Crosses represent single-prior agents and dots multi-prior systems. The labels indicate how many steps were assigned to the second MCMC chain of a total of 100 steps. Bottom: Information processing and expected utility is increasing in the number of utility evaluations, as we expected.

downstream steps was constant, the model selector's choice only depended on the average utility achieved by each decision-maker, which results in the acceptance rule

$$A(x'|x) = \min\left\{1, \exp(\gamma^{\mathrm{sel}}(\mathbb{E}_{p(a|w,x)}[\mathbf{U}(w,a)] - \mathbb{E}_{p(a|w,x')}[\mathbf{U}(w,a)]))\right\}.$$

As the priors are discrete choices the proposal distribution $q(x_{\mathrm{p}}|x_{\mathrm{p}})$ samples globally with $p(x) = \frac{1}{|X|}$ for all x.

4 Empirical Results

To demonstrate our approach we evaluate two scenarios. First, a simple agent, which is equipped with a single prior policy $p_{\eta}(a)$, as introduced in Sect. 2. In case of a single agent there is no need for a prior selection stage. Second, we evaluated a multi-prior decision-making system and compared the results to the single prior agent. For the mutli-prior agent, we split a fixed number of MCMC steps between the prior selection and the action selection. The task we designed consists of six world states where each world state has a Gaussian utility function in the interval $[0,1]$ with a unique optimum. In both settings, we equipped the Variational Autoencoders with one hidden layer consisting of 16 units with ReLU activations. We implemented the experiments using Keras [2]. We show the results in Fig. 3.

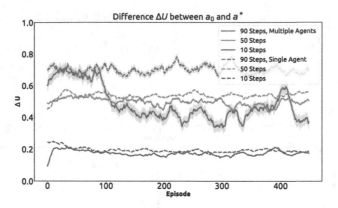

Fig. 4. Our results indicate that having multiple priors is more beneficial, if more steps are available in total. Note that the stochasticity of our method decreases with the number of allowed steps, as shown by the uncertainty band (transparent regions).

Our results indicate that using MCMC evaluation steps as a surrogate for information processing costs can be interpreted as bounded rational decision-making. In Fig. 3 we show the efficiency of several agents with different processing constraints. To compare our results to the theoretical baseline, we discretized the action space into 100 equidistant slices and solved the problem using the algorithm proposed in [5] to implement Eq. (10). Furthermore our results indicate that the multi-prior system generally outperforms the single-prior system in terms of utility.

To illustrate the differences in efficiency between the single prior agent and the multi-prior agents, we plotted in Fig. 4 utility gained through the second MCMC optimization. For multi-prior agents this is caused by specialized priors which provide initializations to the MCMC chains that are close to the optimal action. In this particular case, ΔU does not become zero because we allow only three priors to cover six world states, thus leading to abstraction, i.e. specializing on actions that fit well for the assigned world states. In single-prior agents, the prior is adapting to all world states, thus providing, on average, an initial action that is suboptimal for the requested world state.

5   Discussion

In this study we implemented bounded rational decision makers with adaptive priors. We achieved this with Variational Autoencoder priors. The bounded rational decision-making process was implemented by MCMC optimization to find the optimal posterior strategy, thus giving a computationally simple way of generating samples. As the number of steps in the optimization process was constrained, we could quantify the information processing capabilities of the resulting decision-makers using relative Shannon entropy. Our analysis may have interesting implications, as it provides a normative framework for this kind of combined optimization of adaptive priors and decision-making processes. Prior to our work there have been several attempts to apply the framework of information-theoretic bounded rationality to machine learning tasks [7,11,12, 18]. The novelty of our approach is that we design adaptive priors for both the single-step case and the multi-agent case and we demonstrate how to transform information-theoretic constraints into computational constraints in the form of MCMC steps.

Recently, the combination of Monte Carlo optimization and neural networks has gained increasing popularity. These approaches include both using MCMC processes to find optimal weights in ANNs [1,4] and using ANNs as parametrized proposal distributions in MCMC processes [8,13]. While our approach is more similar to the latter, the important difference is that in such adaptive MCMC approaches there is only a single MCMC chain with a single (adaptive) proposal to optimize a single task, whereas in our case there are multiple adaptive priors to initialize multiple chains with otherwise fixed proposal, which can be used to learn multiple tasks simultaneously. In that sense our work is more related to mixture-of-experts methods and divide-and-conquer paradigms [6,9,24], where we employ a selection policy rather than a blending policy, as we design our model specifically to encourage specialization. In mixture-of-experts models, there are multiple decision-makers that correspond to multiple priors in our case, but experts are typically not modeled as anytime optimization processes. The possibly most popular combination of neural network learning with Monte Carlo methods was achieved by AlphaGo [19], which beat the leading Go champion by optimizing the strategies provided by value networks and policy networks with Monte Carlo Tree Search, leading to a major breakthrough in reinforcement learning. An important difference here is that the neural network is used to directly approximate the posterior and MCMC is used to improve performance by concentrating on the most promising moves during learning, whereas in our case ANNs are used to represent the prior. Moreover, in our work we assumed the utility function (i.e. the value network) to be given. For future work it would be interesting to investigate how to incorporate learning the utility function into our model to investigate more complex scenarios such as in reinforcement learning.

Acknowledgement. This work was supported by the European Research Council Starting Grant *BRISC*, ERC-STG-2015, Project ID 678082.

References

1. Andrieu, C., De Freitas, N., Doucet, A.: Reversible jump MCMC simulated annealing for neural networks. In: Proceedings of the Sixteenth Conference on Uncertainty in Artificial Intelligence, pp. 11–18. Morgan Kaufmann Publishers Inc. (2000)
2. Chollet, F., et al.: Keras (2015). https://keras.io
3. Vul, E., Goodman, N., Griffiths, T.L., Tenenbaum, J.B.: One and done? Optimal decisions from very few samples. Cogn. Sci. **38**(4), 599–637 (2014)

4. Freitas, J., Niranjan, M., Gee, A.H., Doucet, A.: Sequential Monte Carlo methods to train neural network models. Neural Comput. **12**(4), 955–993 (2000)
5. Genewein, T., Leibfried, F., Grau-Moya, J., Braun, D.A.: Bounded rationality, abstraction, and hierarchical decision-making: an information-theoretic optimality principle. Front. Robot. AI **2**, 27 (2015)
6. Ghosh, D., Singh, A., Rajeswaran, A., Kumar, V., Levine, S.: Divide-and-conquer reinforcement learning. arXiv preprint arXiv:1711.09874 (2017)
7. Grau-Moya, J., Leibfried, F., Genewein, T., Braun, D.A.: Planning with information-processing constraints and model uncertainty in Markov decision processes. In: Frasconi, P., Landwehr, N., Manco, G., Vreeken, J. (eds.) ECML PKDD 2016. LNCS (LNAI), vol. 9852, pp. 475–491. Springer, Cham (2016). https://doi.org/10.1007/978-3-319-46227-1_30
8. Gu, S., Ghahramani, Z., Turner, R.E.: Neural adaptive sequential Monte Carlo. In: Advances in Neural Information Processing Systems, pp. 2629–2637 (2015)
9. Haruno, M., Wolpert, D.M., Kawato, M.: Mosaic model for sensorimotor learning and control. Neural Comput. **13**(10), 2201–2220 (2001)
10. Kingma, D.P., Welling, M.: Auto-encoding variational bayes. arXiv preprint arXiv:1312.6114 (2013)
11. Leibfried, F., Braun, D.A.: A reward-maximizing spiking neuron as a bounded rational decision maker. Neural Comput. **27**(8), 1686–1720 (2015)
12. Leibfried, F., Grau-Moya, J., Ammar, H.B.: An information-theoretic optimality principle for deep reinforcement learning. arXiv preprint arXiv:1708.01867 (2017)
13. Levy, D., Hoffman, M.D., Sohl-Dickstein, J.: Generalizing Hamiltonian Monte Carlo with neural networks. In: International Conference on Learning Representations (2018)
14. Lewis, R.L., Howes, A., Singh, S.: Computational rationality: linking mechanism and behavior through bounded utility maximization. Top. Cogn. Sci. **6**(2), 279–311 (2014)
15. MacKay, D.J.C.: Introduction to Monte Carlo methods. In: Jordan, M.I. (ed.) Learning in Graphical Models. ASID, vol. 89, pp. 175–204. Springer, Dordrecht (1998). https://doi.org/10.1007/978-94-011-5014-9_7
16. Ortega, P.A., Braun, D.A.: Thermodynamics as a theory of decision-making with information-processing costs. Proc. R. Soc. Lond. A: Math. Phys. Eng. Sci. **469**(2153) (2013)
17. Ortega, P.A., Braun, D.A., Dyer, J., Kim, K.E., Tishby, N.: Information-theoretic bounded rationality. arXiv preprint arXiv:1512.06789 (2015)
18. Peng, Z., Genewein, T., Leibfried, F., Braun, D.A.: An information-theoretic online update principle for perception-action coupling. In: 2017 IEEE/RSJ International Conference on Intelligent Robots and Systems (IROS), pp. 789–796. IEEE (2017)
19. Silver, D., et al.: Mastering the game of go with deep neural networks and tree search. Nature **529**(7587), 484–489 (2016)
20. Tishby, N., Polani, D.: Information theory of decisions and actions. In: Cutsuridis, V., Hussain, A., Taylor, J. (eds.) Perception-Action Cycle: Models, Architectures, and Hardware. SSCNS, pp. 601–636. Springer, New York (2011). https://doi.org/10.1007/978-1-4419-1452-1_19
21. Todorov, E.: Efficient computation of optimal actions. Proc. Natl. Acad. Sci. **106**(28), 11478–11483 (2009)
22. Von Neumann, J., Morgenstern, O.: Theory of Games and Economic Behavior, Commemorative edn. Princeton University Press, Princeton (2007)

23. Wolpert, D.H.: Information theory - the bridge connecting bounded rational game theory and statistical physics. In: Braha, D., Minai, A., Bar-Yam, Y. (eds.) Complex Engineered Systems: Science Meets Technology. UCS, pp. 262–290. Springer, Heidelberg (2006). https://doi.org/10.1007/3-540-32834-3_12
24. Yuksel, S.E., Wilson, J.N., Gader, P.D.: Twenty years of mixture of experts. IEEE Trans. Neural Netw. Learn. Syst. **23**(8), 1177–1193 (2012)

Feature Selection with Rényi Min-Entropy

Catuscia Palamidessi[1] and Marco Romanelli[1,2(✉)]

[1] Inria, École Polytechnique and University of Paris Saclay, Paris, France
`marco.romanelli@inria.fr`
[2] Università di Siena, Siena, Italy

Abstract. We consider the problem of feature selection, and we propose a new information-theoretic algorithm for ordering the features according to their relevance for classification. The novelty of our proposal consists in adopting Rényi min-entropy instead of the commonly used Shannon entropy. In particular, we adopt a notion of conditional min-entropy that has been recently proposed in the field of security and privacy, and which is strictly related to the Bayes error. We evaluate our method on two classifiers and three datasets, and we show that it compares favorably with the corresponding one based on Shannon entropy.

1 Introduction

The identification of the "best" features for classification is a problem of increasing importance in machine learning. The size of available datasets is becoming larger and larger, both in terms of samples and in terms of features of the samples, and keeping the dimensionality of the data under control is necessary for avoiding an explosion of the training complexity and for the accuracy of the classification. Several authors have considered this problem, including [5, 6, 8, 14, 15, 17–19, 24, 28].

The known methods for reducing the dimensionality can be divided in two categories: those which transform the feature space by reshaping the original features into new ones (*feature extraction*), and those which select a subset of the features (*feature selection*). The second category can in turn be divided in three groups: the *wrapper*, the *embedded*, and the *filter* methods. The last group has the advantage of being classifier-independent, more robust with respect to the risk of overfitting, and more amenable to a principled approach. In particular, several proposals for feature selection have successfully applied concepts and techniques from information theory [4–6, 13, 22, 28, 29]. The idea is that the smaller is the conditional (aka residual) entropy of the classes given a certain set of features, the more likely the classification of a sample is to be correct. Finding a good set of features corresponds therefore to identifying a set of features, as small as possible, for which such conditional entropy is below a certain threshold.

In this paper, we focus on the filter approach and we propose a new information-theoretical method for feature selection. The novelty consists in the

L. Pancioni et al. (Eds.): ANNPR 2018, LNAI 11081, pp. 226–239, 2018.
https://doi.org/10.1007/978-3-319-99978-4_18

use of Rényi min-entropy H_∞ rather than Shannon entropy. As far as we know, all the previous proposals are based on Shannon entropy, with the notable exception of [12] who considered the Rényi entropies but, however, used a different notion of min-entropy, and reported experimental results only on other orders of Rényi entropies.

For feature selection we need the conditional version of entropy. Rényi did not define it, but there have been various proposals, in particular by Arimoto [3], Sibson [25], Csiszár [11], and Cachin [7]. In particular, [7] defined the conditional min-entropy of X given Y along the lines of conditional Shannon entropy, namely as the expected value of the entropy of X for each given value of Y. Such definition, however, violates the *monotonicity property*. Namely, knowing the value of Y could increase the entropy of X rather than diminishing it.

Recently, some advances in the fields of security and privacy have revived the interest for the Rényi min-entropy. The reason is that it models a basic notion of attacker: the (*one-try*) *eavesdropper*. Such attacker tries to infer a secret (e.g., a key, a password, etc.) from the observable behavior of the system trying to minimize the probability of error. Note the similarity with the classification problem, where we choose a class on the basis of the observed features, trying to minimize the probability of mis-classification.

Driven by the motivation of providing an information-theoretic interpretation of the eavesdropper operational behavior, [26] proposed a definition of conditional min-entropy $H_\infty(X|Y)$ which is consistent with the rest of the theory, models all the expected properties of an eavesdropper, and corresponds closely to the Bayes risk of guessing the wrong secret. (The formal definition of $H_\infty(X|Y)$ will be given in Sect. 2.) This definition can be shown equivalent to the one of Arimoto [3]. It is then natural to investigate whether this notion can be useful also for feature selection.

We could state the problem of feature selection as finding a minimum-size subset S of the whole set of features F such that the min-entropy $H_\infty(C|S)$ of the classification C given S is below a given threshold. Because of the correspondence with the Bayes risk, this would mean that the set S is optimal (i.e., minimal) among the subsets for which the Bayes classifier achieves the desired level of accuracy. However, is that the construction of such an optimal S would be NP-hard. This is not due to the kind of entropy that we choose, but simply to the fact that it is a combinatorial problem. In [16] it was shown that the problem of feature selection can be modeled as search problem on a decision tree, and it was argued that finding the optimal subtree which is able to cover F is an NP-hard problem. The same intractability was claimed in [14] with respect to wrappers and embedded methods, on the basis of the proof of [2].

We then adopt a greedy strategy to approximate the minimal subset of features: following [6,28], we construct a sequence of subsets $S^0, S^1, \ldots, S^t, \ldots,$ where $S^0 = \emptyset$ and at each subsequent step S^{t+1} is obtained from S^t by adding the next feature in order of relevance for the classification, taking into account the ones already selected. In other words, we select the feature f such that $H_\infty(C|S^t \cup \{f\})$ is minimal, and we define S^{t+1} as $S^t \cup \{f\}$. The construction of this series should be interleaved with a test on the accuracy of the intended classifier(s): when we obtain an S^T that achieves the desired level of accuracy, we can stop. The difference with respect to [6,28] is that the relevance is measured by Rényi min-entropy rather than Shannon entropy.

Note that, because of the relation between the conditional min-entropy and the Bayes risk, our method is *locally optimal*. Namely, for any other possible feature f' (including the one that would be selected using Shannon entropy), the set S^{t+1} is at least as good as $S^t \cup \{f'\}$ in terms of accuracy of the Bayes classifier (the ideal classifier giving the best accuracy). This does not necessarily mean that the set S^T is the smallest one: since we are not making an exhaustive search on all possible subsets of F, and we add the features one by one, we may not find the "shortest path" to achieve sufficient accuracy. The same applies to the analogous algorithms based on Shannon entropy. Hence we have no guarantee that our method is better than that of [6,28], nor vice versa. In the experiments we have performed, however, our method outperforms almost always the one based on Shannon entropy (cfr. Sect. 4).

2   Preliminaries

In this section we briefly review some basic notions from probability and information theory. We refer to [10] for more details.

Let X, Y be discrete random variables with respectively n and m possible values: $\mathcal{X} = \{x_1, x_2, \ldots, x_n\}$ and $\mathcal{Y} = \{y_1, y_2, \ldots, y_m\}$. Let $p_X(\cdot)$ and $p_Y(\cdot)$ indicate the probability distribution associated to X and Y respectively, and let $p_{Y,X}(\cdot, \cdot)$ and $p_{Y|X}(\cdot|\cdot)$ indicate the joint and the conditional probability distributions, respectively. Namely, $p_{Y,X}(x, y)$ represents the probability that $X = x$ and $Y = y$, while $p_{Y|X}(y|x)$ represents the probability that $Y = y$ given that $X = x$. For simplicity, when clear from the context, we will omit the subscript, and write for instance $p(x)$ instead of $p_X(x)$.

Conditional and joint probabilities are related by the chain rule $p(x, y) = p(x) \, p(y|x)$, from which (by the commutativity of $p(x, y)$) we can derive the Bayes theorem: $p(x|y) = {}^{p(y|x) \, p(x)}/_{p(y)}$.

The Rényi entropies [23] are a family of functions representing the uncertainty associated to a random variable. Each Rényi entropy is characterized by a non-negative real number α (order), with $\alpha \neq 1$, and is defined as $H_\alpha(X) \overset{\text{def}}{=} \frac{1}{1-\alpha} \log(\sum_i p(x_i)^\alpha)$. If $p(\cdot)$ is uniform then all the Rényi entropies are equal to $\log |X|$. Otherwise they are weakly decreasing in α. Shannon and min-entropy are particular cases:

$$\alpha \to 1 \qquad H_1(X) = -\sum_x p(x) \log p(x) \qquad \text{Shannon entropy}$$
$$\alpha \to \infty \qquad H_\infty(X) = -\log \max_x p(x) \qquad \text{min-entropy}$$

Let $H_1(X, Y)$ represent the joint entropy X and Y. Shannon *conditional entropy* of X given Y is the average residual entropy of X once Y is known, and it is defined as $H_1(Y|X) \overset{\text{def}}{=} \sum_{xy} p(x, y) \log p(x|y) = H_1(X, Y) - H_1(Y)$. Shannon *mutual information* of X and Y represents the correlation of information between X and Y, and it is defined as $I_1(X;Y) \overset{\text{def}}{=} H_1(X) - H_1(X|Y) = H_1(X) + H_1(Y) - H_1(X,Y)$. It is possible to show that $I_1(X;Y) \geq 0$, with $I_1(X;Y) = 0$ iff X and Y are independent, and that $I_1(X;Y) = I_1(Y;X)$.

As for Rényi conditional min-entropy, we use the version of [26]: $H_\infty(X|Y) \overset{\text{def}}{=} -\log \sum_y \max_x (p(y|x)p(x))$. This definition closely corresponds to the Bayes risk, i.e., the expected error when we try to guess X once we know Y, formally defined as $\mathcal{B}(X|Y) \overset{\text{def}}{=} 1 - \sum_y p(y) \max_x p(x|y)$. The "Rényi mutual information" is defined as: $I_\infty(X;Y) \overset{\text{def}}{=} H_\infty(X) - H_\infty(X|Y)$. It is possible to show that $I_\infty(X;Y) \geq 0$, and that $I_\infty(X;Y) = 0$ if X and Y are independent (the reverse is not necessarily true). Contrary to Shannon mutual information, I_∞ is not symmetric. The conditional mutual information is defined as $I_\infty(X;Y|Z) \overset{\text{def}}{=} H_\infty(X|Z) - H_\infty(X|Y,Z)$, and analogously for Shannon conditional mutual information.

3 Our Proposed Algorithm

Let F be the set of features at our disposal, and let C be the set of classes. Our algorithm is based on forward feature selection and dependency maximization: it constructs a monotonically increasing sequence $\{S^t\}_{t\geq0}$ of subsets of F, and, at each step, the subset S^{t+1} is obtained from S^t by adding the next feature in order of importance (i.e., the informative contribution to classification), taking into account the information already provided by S^t. The measure of the "order of importance" is based on conditional min-entropy. The construction of the sequence is assumed to be done interactively with a test on the accuracy achieved by the current subset, using one or more classifiers. This test will provide the stopping condition: once we obtain the desired level of accuracy, the algorithm stops and gives as result the current subset S^T. Of course, achieving a level of accuracy $1 - \varepsilon$ is only possible if $\mathcal{B}(C \mid F) \leq \varepsilon$.

Definition 1. *The series $\{S^t\}_{t\geq0}$ and $\{f^t\}_{t\geq1}$ are inductively defined as follows:*

$$S^0 \overset{\text{def}}{=} \emptyset$$
$$f^{t+1} \overset{\text{def}}{=} \operatorname{argmin}_{f \in F \setminus S^t} H_\infty(C \mid f, S^t)$$
$$S^{t+1} \overset{\text{def}}{=} S^t \cup \{f^{t+1}\}$$

The algorithms in [6,28] are analogous, except that they use Shannon entropy. They also define f^{t+1} based on the maximization of mutual information instead of the minimization of conditional entropy, but this is irrelevant. In fact $I_1(C; f \mid S^t) = H_1(C \mid S^t) - H_1(C \mid f, S^t)$, hence maximizing $I_1(C; f \mid S^t)$ with respect to f is the same as minimizing $H_1(C \mid f, S^t)$ with respect to f.

Our algorithm is locally optimal, in the sense stated by the following proposition, whose proof can be found in [20]:

Proposition 1. *At every step, the set S^{t+1} minimizes the Bayes risk of the classification among those which are of the form $S^t \cup \{f\}$, namely:*

$$\forall f \in F \ \mathcal{B}(C \mid S^{t+1}) \leq \mathcal{B}(C \mid S^t \cup \{f\})$$

In the following sections we analyze some extended examples to illustrate how the algorithm works, and also compare it with the ones of [6,28].

3.1 An Example in Which Rényi Min-Entropy Gives a Better Feature Selection Than Shannon Entropy

Let us consider the dataset in Fig. 1, containing ten records labeled each by a different class, and characterized by six features (columns $f_1, \ldots, f_5$). We note that f_0 separates the classes in two sets of four and six elements respectively, while all the other columns are characterized by having two values, each of which univocally identify one class, while the third value is associated to all the remaining classes. For instance, in column f_1 value A univocally identifies the record of class 0, B univocally identifies the record of class 1, and all the other records have the same value along that column, i.e. C.

The last five features combined are necessary and sufficient ton completely identify all classes, without the need of the first one. Note of the last five features can be replaced by f_0 for this purpose. In fact, each pair of records which are separated by one of the features $f_1, \ldots, f_5$, have the same value in column f_0.

If we apply the discussed feature selection method and we look for the feature that minimizes $H(Class|f_i)$ for $i \in \{0, \ldots, 5\}$ we obtain that:

- The first feature selected with Shannon is f_0, in fact $H_1(Class|f_0) \approx 2.35$ and $H_1(Class|f_{\neq 0}) = 2.4$. (The notation $f_{\neq 0}$ stands for any of the f_i's except f_0.) In general, indeed, with Shannon entropy the method tends to choose a feature which splits the classes in a way as balanced as possible. The situation after the selection of the feature f_0 is shown in Fig. 2(a).
- The first feature selected with Rényi min-entropy is either f_1 or f_2 or f_3 or f_4 or f_5, in fact $H_\infty(Class|f_0) \approx 2.32$ and $H_\infty(Class|f_{\neq 0}) \approx 1.74$. In general, indeed, with Rényi min-entropy the method tends to choose a feature which divides the classes in as many sets as possible. The situation after the selection of f_1 is shown in Fig. 2(b).

Class	f_0	f_1	f_2	f_3	f_4	f_5
0	A	C	F	I	L	O
1	A	D	F	I	L	O
2	A	E	G	I	L	O
3	A	E	H	I	L	O
4	B	E	F	J	L	O
5	B	E	F	K	L	O
6	B	E	F	I	M	O
7	B	E	F	I	N	O
8	B	E	F	I	L	P
9	B	E	F	I	L	Q

Fig. 1. The dataset

Going ahead with the algorithm, with Shannon entropy we will select one by one all the other features, and as already discussed we will need all of them to completely identify all classes. Hence at the end the method with Shannon entropy will return all the six features (to achieve perfect classification). On the

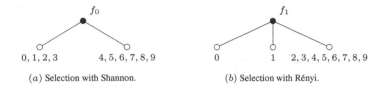

(a) Selection with Shannon.   (b) Selection with Rényi.

Fig. 2. Classes separation after the selection of the first feature.

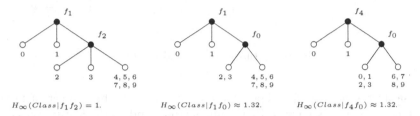

$H_\infty(Class|f_1 f_2) = 1.$ $H_\infty(Class|f_1 f_0) \approx 1.32.$ $H_\infty(Class|f_4 f_0) \approx 1.32.$

Fig. 3. Selection of the second feature with Rényi.

other hand, with Rényi min entropy we will select all the remaining features except f_0 to obtain the perfect discrimination. In fact, at any stage the selection of f_0 would allow to split the remaining classes in at most two sets, while any other feature not yet considered will split the remaining classes in three sets. As already hinted, with Rényi we choose the feature that allows to split the remaining classes in the highest number of sets, hence we never select f_0.

For instance, if we have already selected f_1, we have $H_\infty(Class|f_1 f_0) \approx 1.32$ while $H_\infty(Class|f_1 f_{\neq 0}) = 1$. If we have already selected f_4, we have $H_\infty(Class|f_4 f_0) \approx 1.32$ while $H_\infty(Class|f_4 f_{\neq 0}) = 1$. See Fig. 3.

At the end, the selection of features using Rényi entropy will determine the progressive splitting represented in Fig. 4. The order of selection is not important: this particular example is conceived so that the features $f_1, \ldots, f_5$ can be selected in any order, the residual entropy is always the same.

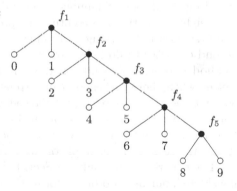

Fig. 4. Sequence of class splitting with Rényi.

Discussion. It is easy to see that, in this example, the algorithm based on Rényi min-entropy gives a better result not only at the end, but also at each step of the process. Namely, at step t (cfr. Definition 1) the set S^t of features selected with Rényi min-entropy gives a better classification (i.e., more accurate) than the set S'^t that would be selected using Shannon entropy. More precisely, we have $\mathcal{B}(C \mid S^t) < \mathcal{B}(C \mid S'^t)$. In fact, as discussed above the set S'^t contains

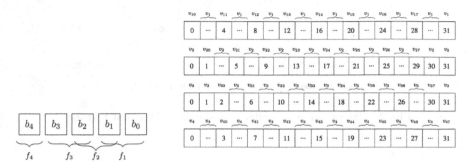

Fig. 5. Features F (left) and F' (right).

necessarily the feature f_0, while S^t does not. Let S^{t-1} be the set of features selected at previous step with Rényi min-entropy, and f^t the feature selected at step t (namely, $S^{t-1} = S^t \setminus \{f^t\}$). As argued above, the order of selection of the features $f_1, \ldots, f_5$ is irrelevant, hence we have $\mathcal{B}(C \mid S^{t-1}) = \mathcal{B}(C \mid S'^t \setminus \{f_0\})$ and the algorithm *could* equivalently have selected $S'^t \setminus \{f_0\}$. As argued above, the next feature to be selected, with Rényi, must be different from f_0. Hence by Proposition 1, and by the fact that the order of selection of $f_1, \ldots, f_5$ is irrelevant, we have: $\mathcal{B}(C \mid S^t) = \mathcal{B}(C \mid (S'^t \setminus \{f_0\}) \cup \{f^t\}) < \mathcal{B}(C \mid S'^t)$.

As a general observation, we can see that the method with Shannon tends to select the feature that divides the classes in sets (one for each value of the feature) as balanced as possible, while our method tends to select the feature that divides the classes in as many sets as possible, regardless of the sets being balanced or not. In general, both Shannon-based and Rényi-based methods try to minimize the height of the tree representing the process of the splitting of the classes, but the first does it by trying to produce a tree *as balanced as possible*, while the second one tries to do it by producing a tree *as wide as possible*. Which of the method is best, it depends on the correlation of the features. Shannon works better when there are enough uncorrelated (or not much correlated) features, so that the tree can be kept balanced while being constructed. Next section shows an example of such situation. Rényi, on the contrary, is not so sensitive to correlation and can work well also when the features are highly correlated, as it was the case in the example of this section.

The experimental results in Sect. 4 show that, at least in the cases we have considered, our method outperforms the one based on Shannon entropy. In general however the two methods are incomparable, and perhaps a good practice would be to construct both sequences at the same time, so to obtain the best result of the two.

3.2 An Example in Which Shannon Entropy May Give a Better Feature Selection Than Rényi Min-Entropy

Consider a dataset containing samples equally distributed among 32 classes, indexed from 0 to 31. Assume that the data have 8 features divided in 2 types

F and F', each of which consisting of 4 features: $F = \{f_1, f_2, f_3, f_4\}$ and $F' = \{f'_1, f'_2, f'_3, f'_4\}$. The relation between the features and the classes is represented in Fig. 5.

Because of space restriction we have omitted the computations, the interested reader can find them in the report version of this paper [20]. At step 3 one of the possible outcomes of the algorithm based on Shannon is the set of features $S_1^3 = \{f_1, f_3, f_4\}$, and one of the possible outcomes of the algorithm based on Rényi is $S_\infty^3 = \{f'_1, f'_2, f'_i\}$ where i can be, equivalently, 3 or 4. At this point the method with Shannon can stop, since the residual Shannon entropy of the classification is $H_1(C \mid S_1^3) = 0$, and also the Bayes risk is $\mathcal{B}(C \mid S_1^3) = 0$, which is the optimal situation in the sense that the classification is completely accurate. S_∞^3 on the contrary does not contain enough features to give a completely accurate classification, for that we have to make a further step. We can see that $S_\infty^4 = F'$, and finally we have $H_\infty(C \mid S_\infty^4) = 0$.

Thus in this particular example we have that for small values of the threshold on the accuracy our method gives better results. On the other hand, if we want to achieve perfect accuracy (threshold 0) Shannon gives better results.

4   Evaluation

In this section we evaluate the method for feature selection that we have proposed, and we compare it with the one based on Shannon entropy by [6,28].

To evaluate the effect of feature selection, some classification methods have to be trained and tested on the selected data. We used two different methods to avoid the dependency of the result on a particular algorithm. We chose two widely used classifiers: the Support Vector Machines (SVM) and the Artificial Neural Networks (ANN).

Even though the two methods are very different, they have in common that their efficiency is highly dependent on the choice of certain parameters. Therefore, it is worth spending some effort to identify the best values. Furthermore, we should take into account that the particular paradigm of SVM we chose only needs 2 parameters to be set, while for ANN the number of parameters increases (at least 4).

It is very important to choose values as robust as possible for the parameters. It goes without saying that the strategy used to pick the best parameter setting should be the same for both Shannon entropy and Rényi min-entropy. On the other hand for SVM and ANN we used two different hyper-parameter tuning algorithms, given that the number and the nature of the parameters to be tuned for those classifiers is different.

In the case of SVM we tuned the cost parameter of the objective function for margin maximization (C-SVM) and the parameter which models the shape of the RBF kernel's bell curve (γ). Grid-search and Random-search are quite time demanding algorithms for the hyper-parameter tuning task but they're also widely used and referenced in literature when it comes to SVM. Following the guidelines in [9,21], we decided to use Grid-search, which is quite suitable when we have to deal with only two parameters. In particular we performed Grid-search including a 10 folds CV step.

Things are different with ANN because many more parameters are involved and some of them change the topology of the network itself. Among the various strategies to attack this problem we picked Bayesian Optimization [27]. This algorithm combines steps of extensive search for a limited number of settings before inferring via Gaussian Processes (GP) which is the best setting to try next (with respect to the mean and variance and compared to the best result obtained in the last iteration of the algorithm). In particular we tried to fit the best model by optimizing the following parameters:

– number of hidden layers
– number of hidden neurons in each layer
– learning rate for the gradient descent algorithm
– size of batches to update the weight on network connections
– number of learning epochs

To this purpose, we included in the pipeline of our code the *Spearmint* Bayesian optimization codebase. Spearmint, whose theoretical bases are explained in [27], calls repeatedly an objective function to be optimized. In our case the objective function contained some *tensorflow* machine learning code which run a 10 folds CV over a dataset and the objective was to maximize the accuracy of validation. The idea was to obtain a model able to generalize as much as possible using only the selected features before testing on a dataset which had never been seen before.

We had to decide the stopping criterion, which is not provided by *Spearmint* itself. For the sake of simplicity we decided to run it for a time lapse which has empirically been proven to be sufficient in order to obtain results meaningful for comparison. A possible improvement would be to keep running the same test (with the same number of features) for a certain amount of time without resetting the computation history of the package and only stop testing a particular configuration if the same results is output as the best for k iterations in a row (for a given k).

Another factor, not directly connected to the different performances obtained with different entropies, but which is important for the optimization of ANN, is the choice of the activation functions for the layers of neurons. In our work we have used ReLU for all layers because it is well known that it works well for this aim, it is easy to compute (the only operation involved is the max) and it avoids the sigmoid saturation issue.

4.1   Experiments

As already stated, at the i-th step of the feature selection algorithm we consider all the features which have already been selected in the previous $i - 1$ step(s). For the sake of limiting the execution time, we decided to consider only the first 50 selected features with both metrics. We tried our pipeline on the following datasets:

- BASEHOCK dataset: 1993 instances, 4862 features, 2 classes. This dataset has been obtained from the 20 newsgroup original dataset.
- SEMEION dataset: 1593 instances, 256 features, 10 classes. This is a dataset with encoding of hand written characters.
- GISETTE dataset: 6000 instances, 5000 features, 2 classes. This is the discretized version of the NIPS 2003 dataset which can be downloaded from the site of Professor Gavin Brown, Manchester University.

We implemented a bootstrap procedure (5 iterations on each dataset) to shuffle data and make sure that the results do not depend on the particular split between training, validation and test set. Each one of the 5 bootstrap iterations is a new and unrelated experimental run. For each one of them a different training-test sets split was taken into account. Features were selected analyzing the training set (the test set has never been taken into account for this part of the work). After the feature selection was executed according to both Shannon and Rényi min-entropy, we considered all the selected features adding one at each time. So, for each bootstrap iteration we had 50 steps, and in each step we added one of the selected features, we performed hyper-parameter tuning with 10 folds CV, we trained the model with the best parameters on the whole training set and we tested it on the test set (which the model had never seen so far). This procedure was performed both for SVM and ANN.

We computed the average performances over the 5 iterations and the results are in Figs. 6, 7, and 8. In all cases the feature selection method using Rényi min-entropy usually gave better results than Shannon, especially with the BASE-HOCK dataset.

5   Related Works

In the last two decades, thanks to the growing interest in machine learning, many methods have been setup to tackle the feature reduction problem. In this section we discuss those closely related to our work, namely those which are based on information theory. For a more complete overview we refer to [5,6,28].

The approach most related to our proposal is that of [6,28]. We have already discussed and compared their method with ours in the technical body of this paper.

As far as we know, Rényi min-entropy has only been considered, in the context of feature selection, by [12] (although in the experiments they only show results for other Rényi entropies). The definition they consider, however, is that of [7] which, as already mentioned, has the unnatural characteristic that a feature may increase the entropy of the classification instead of decreasing it. It is clear, therefore, that basing a method on this notion of entropy could lead to strange results.

Two key concepts that have been widely used are *relevance* and *redundancy*. Relevance refers to the importance for the classification of the feature under consideration f^t, and it is in general modeled as $I_1(C; f^t)$. Redundancy represents

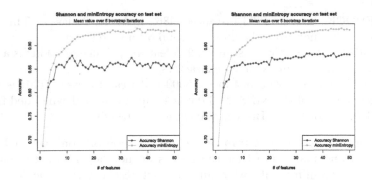

Fig. 6. Accuracy of the ANN and SVM classifiers on the BASEHOCK dataset

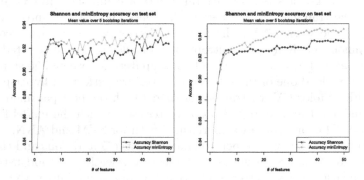

Fig. 7. Accuracy of the ANN and SVM classifiers on the GISETTE dataset

how much the information of f^t is already covered by S. It is often modeled as $I_1(f^t, S)$. In general, we want to maximize relevance and minimize redundancy.

One of the first algorithms ever implemented was the MIFS algorithm proposed by [4], based on a greedy strategy. At the first step it selects $f^1 = \text{argmax}_{f_i \in F} I_1(C; f_i)$, and at step t it selects $f^t = \text{argmax}_{f_i \in F \setminus S^{t-1}} [I_1(C, f_i) - \beta \sum_{f_s \in S^{t-1}} I_1(f_i, f_s)]$ where β is a parameter that controls the weight of the redundancy part.

The mRMR approach (redundancy minimization and relevance maximization) proposed by [22] is based on the same strategy as MIFS. However the redundancy term is now substituted by its mean over the elements of the subset S so to avoid its value to grow when new attributes are selected.

In both cases, if relevance outgrows redundancy, it might happen that many features highly correlated and so highly redundant can still be selected. Moreover, a common issue with these two methods is that they do not take into account the conditional mutual information $I_1(C, f^t \mid S)$ for the choice of the next feature to be selected f^t.

More recent algorithms involve the ideas of joint mutual entropy $I_1(C; f_i, S)$ (JMI, [5]) and conditional mutual entropy $I_1(C; f_i \mid S)$ (CMI, [13]). The step for choosing the next feature with JMI is $f^t = \text{argmax}_{f_i \in F \setminus S^{t-1}}$

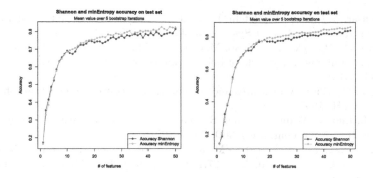

Fig. 8. Accuracy of the ANN and SVM classifiers on the SEMEION dataset

$\{min_{f_s \in S^{t-1}} I(C; f_i, f_s)\}$, while with CMI is $f^t = \text{argmax}_{f_i \in F \setminus S^{t-1}}$ $\{min_{f_s \in S^{t-1}} I(C; f_i \mid f_s)\}$. In both cases the already selected features are taken into account one by one when compared to the new feature f^t. The correlation between JMI and CMI is easy to prove [29]: $I_1(C; f_i, S) = H_1(C) - H_1(C \mid S) + H_1(C \mid S) - H_1(C \mid S) = I_1(C; S) + I(C; f_i \mid S)$.

6   Conclusion and Future Work

We have proposed a method for feature selection based on a notion of conditional Rényi min-entropy. Although our method is in general incomparable with the corresponding one based on Shannon entropy, in the experiments we performed it turned out that our methods always achieved better results.

As future work, we plan to compare our proposal with other information-theoretic methods for feature selection. In particular, we plan to investigate the application of other notions of entropy which are the state-of-the-art in security and privacy, like the notion of g-vulnerability [1], which seems promising for its flexibility and capability to represent a large spectrum of possible classification strategies.

References

1. Alvim, M.S., Chatzikokolakis, K., Palamidessi, C., Smith, G.: Measuring information leakage using generalized gain functions. In: Proceedings of CSF, pp. 265–279 (2012)
2. Amaldi, E., Kann, V.: On the approximability of minimizing nonzero variables or unsatisfied relations in linear systems. Theor. Comput. Sci **209**(1–2), 237–260 (1998)
3. Arimoto, S.: Information measures and capacity of order α for discrete memoryless channels. In: Topics in Information Theory, Proceedings of Colloquia Mathematica Societatis Janos Bolyai, pp. 41–52 (1975)
4. Battiti, R.: Using mutual information for selecting features in supervised neural net learning. IEEE Trans. Neural Netw. **5**(4), 537–550 (1994)

5. Bennasar, M., Hicks, Y., Setchi, R.: Feature selection using joint mutual information maximisation. Expert Syst. Appl. **42**(22), 8520–8532 (2015)
6. Brown, G., Pocock, A.C., Zhao, M.J., Luján, M.: Conditional likelihood maximisation: a unifying framework for information theoretic feature selection. JMLR **13**, 27–66 (2012)
7. Cachin, C.: Entropy measures and unconditional security in cryptography. Ph.D. thesis, ETH (1997)
8. Cai, J., Luo, J., Wang, S., Yang, S.: Feature selection in machine learning: a new perspective. Neurocomputing **300**, 70–79 (2018)
9. Chang, C.C., Lin, C.J.: LIBSVM: a library for support vector machines. ACM Trans. Intell. Syst. Technol. **2**, 27:1–27:27 (2011). http://www.csie.ntu.edu.tw/~cjlin/libsvm
10. Cover, T.M., Thomas, J.A.: Elements of Information Theory. Wiley, New York (1991)
11. Csiszár, I.: Generalized cutoff rates and Rényi's information measures. Trans. Inf. Theory **41**(1), 26–34 (1995)
12. Endo, T., Kudo, M.: Weighted Naïve Bayes classifiers by Renyi entropy. In: Ruiz-Shulcloper, J., Sanniti di Baja, G. (eds.) CIARP 2013. LNCS, vol. 8258, pp. 149–156. Springer, Heidelberg (2013). https://doi.org/10.1007/978-3-642-41822-8_19
13. Fleuret, F.: Fast binary feature selection with conditional mutual information. JMLR **5**, 1531–1555 (2004)
14. Guyon, I., Elisseeff, A.: An introduction to variable and feature selection. JMLR **3**, 1157–1182 (2003)
15. Jain, A.K., Duin, R.P.W., Mao, J.: Statistical pattern recognition: a review. IEEE Trans. Pattern Anal. Mach. Intell. **22**(1), 4–37 (2000)
16. Kohavi, R., John, G.: Wrappers for feature subset selection. AI **97**, 273–324 (1997)
17. Liu, H., Yu, L.: Toward integrating feature selection algorithms for classification and clustering. IEEE Trans. Knowl. Data Eng. **17**(4), 491–502 (2005)
18. Liu, J., Lin, Y., Wu, S., Wang, C.: Online multi-label group feature selection. Knowl.-Based Syst. **143**, 42–57 (2018)
19. Nakariyakul, S.: High-dimensional hybrid feature selection using interaction information-guided search. Knowl.-Based Syst. **145**, 59–66 (2018)
20. Palamidessi, C., Romanelli, M.: Feature selection with rényi min-entropy. Technical report, INRIA (2018). https://hal.archives-ouvertes.fr/hal-01830177
21. Pedregosa, F.: Scikit-learn: machine learning in Python. JMLR **12**, 2825–2830 (2011)
22. Peng, H., Long, F., Ding, C.H.Q.: Feature selection based on mutual information: criteria of max-dependency, max-relevance, and min-redundancy. IEEE Trans. Pattern Anal. Mach. Intell **27**(8), 1226–1238 (2005)
23. Rényi, A.: On measures of entropy and information. In: Proceedings of the 4th Berkeley Symposium on Mathematics, Statistics, and Probability, pp. 547–561 (1961)
24. Sheikhpour, R., Sarram, M.A., Gharaghani, S., Chahooki, M.A.Z.: A survey on semi-supervised feature selection methods. Pattern Recognit. **64**, 141–158 (2017)
25. Sibson, R.: Information radius. Z. Wahrscheinlichkeitsth. und Verw. Geb **14**, 149–161 (1969)
26. Smith, G.: On the foundations of quantitative information flow. In: de Alfaro, L. (ed.) FoSSaCS 2009. LNCS, vol. 5504, pp. 288–302. Springer, Heidelberg (2009). https://doi.org/10.1007/978-3-642-00596-1_21

27. Snoek, J., Larochelle, H., Adams, R.P.: Practical Bayesian optimization of machine learning algorithms. In: Proceedings of NIPS 2012, pp. 2960–2968 (2012)
28. Vergara, J.R., Estévez, P.A.: A review of feature selection methods based on mutual information. Neural Comput. Appl. **24**(1), 175–186 (2014)
29. Yang, H.H., Moody, J.: Feature selection based on joint mutual information. In: Proceedings of International ICSC Symposium on Advances in Intelligent Data Analysis, pp. 22–25 (1999)

Applications

Extracting Gamma-Ray Information from Images with Convolutional Neural Network Methods on Simulated Cherenkov Telescope Array Data

Salvatore Mangano$^{(\boxtimes)}$, Carlos Delgado, María Isabel Bernardos,
Miguel Lallena, Juan José Rodríguez Vázquez, and for the CTA Consortium

CIEMAT - Centro de Investigaciones Energéticas, Medioambientales y Tecnológicas,
Av. Complutense, 40, 28040 Madrid, Spain
salvatore.mangano@ciemat.es
http://cta.ciemat.es/

Abstract. The Cherenkov Telescope Array (CTA) will be the world's leading ground-based gamma-ray observatory allowing us to study very high energy phenomena in the Universe. CTA will produce huge data sets, of the order of petabytes, and the challenge is to find better alternative data analysis methods to the already existing ones. Machine learning algorithms, like deep learning techniques, give encouraging results in this direction. In particular, convolutional neural network methods on images have proven to be effective in pattern recognition and produce data representations which can achieve satisfactory predictions. We test the use of convolutional neural networks to discriminate signal from background images with high rejections factors and to provide reconstruction parameters from gamma-ray events. The networks are trained and evaluated on artificial data sets of images. The results show that neural networks trained with simulated data can be useful to extract gamma-ray information. Such networks would help us to make the best use of large quantities of real data coming in the next decades.

Keywords: Gamma-ray astronomy · Cherenkov Telescope Array
Reconstruction technique · Image recognition · Deep learning
Convolutional neural networks

1 Introduction

The ground-based observation of the very high energy gamma-ray sky ($E_{gamma} > 100\,\text{GeV}$) has greatly progressed during the last 40 years through the use of imaging atmospheric Cherenkov telescopes (IACTs). These telescopes aim to detect the air shower produced by the interaction of a primary cosmic gamma ray in the Earth's atmosphere. Charged air shower particles that travel

---

CTA website: https://www.cta-observatory.org/.

© Springer Nature Switzerland AG 2018
L. Pancioni et al. (Eds.): ANNPR 2018, LNAI 11081, pp. 243–254, 2018.
https://doi.org/10.1007/978-3-319-99978-4_19

at ultra-relativistic speed emit Cherenkov light. This Cherenkov light propagates to the ground producing a faint pool of Cherenkov light of about 120 m in radius. The optical mirrors of the telescopes reflect the collected Cherenkov light into the focal plane where photomultipliers convert light into an electrical signal that is digitized and transmitted to record the image.

The image in the camera represents the electromagnetic air shower and is used to identify the primary cosmic gamma-ray. However, Cherenkov light is not only produced by cosmic gamma-rays but also by the more abundant hadronic cosmic rays. These massive charged particles arriving from outer space are mostly protons, but they also include heavier nuclei, which are known atoms without their electron shells. The shape, intensity and orientation of the image provides information about the primary cosmic particle type, energy, direction of propagation and depth of first interaction.

Several different classification and reconstruction techniques exist which on one hand discriminate gamma-ray events from the more numerous hadron events and on the other hand infer the primary gamma-ray energy and direction. One of the first developed reconstruction methods [1], the so called Hillas parametrization, used direction and elliptical shape of the gamma-ray images as the main features to discriminate them against the hadronic cosmic ray background which produces wider and more irregular images. Later more advanced reconstruction methods with superior performance have been developed using machine learning algorithms as in the case of random forest [2] for the MAGIC [3] telescope and boosted decision trees [4] for the H.E.S.S. [5] and VERITAS [6] telescopes. A further reconstruction method is to fit the image to results of a fast simulation under the hypotheses that the image is an electromagnetic shower [7–9].

Recently, several gamma-ray observatories with Cherenkov telescopes started using convolutional neural networks (CNNs) for classification and regression problems [10–13]. CNNs belong to a class of supervised machine learning techniques that have achieved impressive results in image processing [14,15] with little need of human intervention in finding significant image features. With enough training data CNNs can find patterns in the data that when applied to images maximize the gamma-ray reconstruction performance or background rejection. In general the study of such machine learning follows always the same procedure: to start define a data set, then determine a cost function that has to be minimized, next design a neural network architecture where computationally efficient changes on adjustable parameters works, and in the end apply some sort of stochastic gradient descent to minimize the cost function. For an in depth treatment of the literature, see the following references [16–18].

The Cherenkov Telescope Array [19,20] (CTA) will be the next generation ground-based gamma-ray observatory to study very high energy processes in the Universe. The main goal of CTA is to identify and study high energy gamma-ray sources, including objects such as supernova remnants, pulsars, binary stars and active galaxies. The measured fluxes, energy spectra and arrival directions of gamma rays will help to find answers to the origin of these high energy particles and provide information on the morphology of the sources. Also some more speculative models are investigated, like theories which incorporate the violation of Lorentz invariance and predict unexpected cosmological effects on gamma-ray propagation, or the search of possible signals from annihilating dark matter particles. CTA is expected to have around one order of magnitude improvement in sensitivity in the energy range from $\sim$50 GeV to $\sim$50 TeV compared to currently operating IACTs. This is due to the fact that CTA has the capability to detect gamma-rays over larger areas than existing observatories. CTA will provide whole-sky coverage with an observatory in the Southern Hemisphere (Cerro Paranal, Chile) and an observatory in the Northern Hemisphere (La Palma, Spain). The Southern Hemisphere observatory has a total of 99 telescopes of three different sizes with an area of 4.5 km^2 and the Northern Hemisphere observatory has a total of 19 telescopes with an area of 0.6 km^2. These telescopes will provide a large amount of images that encode primary particle information and it is essential to develop efficient statistical tools to best extract such information. Moreover both observatories will be equipped with four large size telescopes [21], each with a mirror diameter of about 23 m and a focal length of 28 m. The large size telescope will dominate the performance of the observatory between 20 GeV and 200 GeV and will be equipped with a 1855 pixels camera with 4.6° full field of view. First real data from such a telescope should be available already in the end of 2018.

In this note, we aim to asses the use of CNNs to discriminate signal from background images and to provide reconstruction parameters from gamma-ray events for the CTA observatory. To evaluate the performance of the CNNs we use official simulated CTA data exploiting the pixel wise information of minimally treated images. In contrast to previous mentioned existing works, we apply for the first time CNNs to simulated CTA data to reconstruct the gamma-ray parameters. We focus only on large size telescopes with showers triggered in all four telescopes. The remainder of this note is organized as follows. Section 2 gives a short description of data simulation and data selection. In Sect. 3 we present details about specific networks, explain analysis strategy and discuss results, followed in Sect. 4 by concluding remarks.

2   Monte Carlo Simulation and Preselection

A Monte Carlo simulation has been used [22,23] to produce a large artificial data set[1] and to examine the performance of different CNN architectures. As presented in [22] the Monte Carlo generated gamma-ray data has been verified against real gamma-ray data from the existing Cherenkov telescopes. For this study the directions of the primary gamma rays and protons are distributed isotropically and extend well beyond the CTA field of view. In particular, for this diffuse emission, no previous knowledge about the true direction of the primary gamma-ray source position is assumed. The development of extensive air showers caused by primary gamma-rays and protons including emission of

---

[1] The simulation data used for this study were extracted from the so called CTA prod-3 data set.

Cherenkov light is simulated with CORSIKA [24]. The primary particles enter the atmosphere as diffuse emission within 10° of the field of view center with an average zenith angle of 20° and an average azimuth angle of 0°. These events have been produced in the energy range from 3 GeV to 330 TeV for gamma rays and from 4 GeV to 660 TeV for protons. The distinct energy range values is due to the fact that at the same energy, the Cherenkov photon intensity in a proton shower is smaller than the one produced in a gamma-ray shower. The ratio of Cherenkov photon yield between gamma-ray shower and proton shower is around two for the selected energy range. In proton showers a large fraction of the total energy is carried by hadrons and neutrinos which produce little or no amounts of Cherenkov photons [25]. The atmospheric conditions of the La Palma site have been reproduced and the response of the telescope is simulated by the sim_telarray [26] package. The generated camera images of telescopes consisting of calibrated integrated charge and pulse arrival time per pixel is extracted from the simulation using the MARS [27] package.

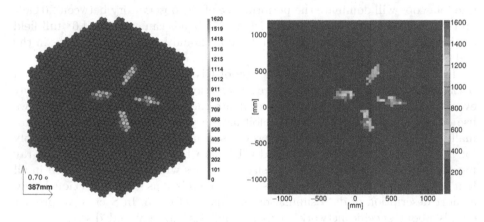

Fig. 1. Left: Camera pixel intensity of a four combined telescope image of a gamma-ray event in a hexagonal grid with hexagonal pixels. Right: Same event as in the left Figure but as a squared image with squared pixels produced by oversampling technique.

The main aim for IACTs is to fully reconstruct properties like type, energy, direction and depth of first interaction of the primary particle from the Cherenkov light produced by atmospheric shower. The use of more than one telescope significantly improves the ability to reconstruct these particle properties as the air shower can be recorded under different viewing angles, usually referred to as stereoscopic imaging. To incorporate this stereoscopic information and reduce the complexity of different numbers of telescopes we select only events that trigger four large size telescopes. To simplify the further analysis we combine four images into a single image by summing pixel values. As CTA images are arranged in a hexagonal grid like the one presented in Fig. 1 left, whereas the CNN framework is designed to process only rectangular pixels, some image processing is needed. A straightforward conversion from the hexagonal (1855 pixels) to squared image (64 × 64 pixels) is to use an oversampling technique. One such realization is presented in the Fig. 1 right. The CNN has been supplied with such preprocessed integrated charge per pixel images and with labels like primary particle type, energy, direction and depth of first interaction.

The following selection criteria have been used to simplify the reconstruction task. The incoming directions of the randomized primary particles were selected within a cone with four degrees radius centered on the pointing direction. The impact points of the uniformly distributed primary particles on the ground have been selected within a circle with a radius of 200 m around the coordinates of each single telescope. This ensures that the superimposed elliptical images do not overlap much. In principle such a selection can be done as a two class classification problem, distinguishing images with small overlap versus large overlap. However for this study we did not include such a classification selection and we leave this as a future development.

3 Convolutional Neural Networks for Simulated Cherenkov Telescope Array Data

We present results of one CNN that separately classifies signal and background events and a second one that reconstructs parameters of the primary gamma-ray particles. We use TensorFlow [28] to implement a network architecture handling as input the preprocessed images mentioned in the previous section. In the following, we give details about architecture and training of the CNNs and provide examples of applications to official simulated CTA data.

A typical CNN architecture consist of several successive convolutional layers followed by one or more fully connected layers. In the first convolutional layer the input image is convolved by a filter (also referred as kernel) over a restricted region (also referred as receptive field) producing activation maps. The restricted region is in general much smaller than the input images and allows to identify in the first layer simple features, like edges or curves. Applying filters on following layers obtain activation maps that represent more and more complex features producing an automated feature extractor. Such a feature extractor can possibly identify discriminative information in the images that is not fully exploited by existing reconstruction algorithms.

The goal of the CNN is not to achieve good predictions on training data examples, but to make good predictions for new examples that are not contained in the training set. This requires that the neural network finds the underlying main information in data and generalizes in a meaningful way. Various neural network architectures were trained tuning hyperparameters in order to optimize performance on the test set. The performance is given by energy and angular resolution. Once the architecture and hyperparameters are decided, the algorithm is fully automatic. Due to the large amount of possible parameter combinations, the currently used solution was obtained by random search. Several different sequential architectures with two, four and eight convolution layers combined with one or two fully connected layers with different activation functions and kernel sizes have been tested. Usually neural networks which have larger numbers of parameters may generalize better than neural networks with fewer parameters, but larger networks may have an increased overfitting problem and may require longer periods of time in order to complete a calculation than smaller networks. Even if training of CNNs can take huge computer resources, the finally trained network reconstructs a new event in a short time compared to the training time. Using simpler architectures over more complex architectures with similar performance reduce reconstruction times and so reduce computing costs. Quicker reconstruction means quicker scientific results, which is better for many scientific objectives such as for example transient phenomena and short-timescale variability searches.

The selected architecture, which gives reasonable performance in terms of function loss during testing, consists of four convolutional layers with a kernel of 5×5 pixels and with a feature sizes of 32 in each layer. The convolutional layers and fully connected layers both had exponential linear unit activation for the regression and classification problem. In order to account for rotational invariance, the data is augmented artificially with rotated examples (e.g. 0, 120 and 240°)[2]. However the rotational invariance is only an approximation, since the geomagnetic field actually breaks such a symmetry. Each convolution layer is followed by a batch normalization layer [30] and an average pooling layer [31], with pool size of two and stride length of two, which reduce the size of images to half in pixels. A dropout layer [32] with 80% to keep the neurons is used during training, whereas at the final test time dropout uses all neurons.

The flattened representations from the fourth convolution layer is then followed by a fully connected layer of 256 parameters with the same activation functions used in the previous layers. Finally we apply a sigmoid activation function for probabilistic predictions in the classification problem and no activation function in the regression problem.

The initialization scheme used for the parameters is commonly referred to as the Xavier initialization [33]. The cost function for the classification problem is cross entropy and for the regression problem is mean squared error. Backpropagation [34] is explicitly used to minimize the cost function by adapting parameters using a gradient descent optimization algorithm [35]. Training proceeds by optimizing the cost function with L2 regularization and learning rate decay using the Adam algorithm [36]. At each training step, we select a random sample of simulated data with batch size of 256 and use them to optimize the network parameters. The models were trained on a cluster with Tesla K80 GPUs. The data set for the classification problem consists of the same number of gamma-ray and proton events with about 24000 simulated events and for the regression problem consists of about 40000 gamma-ray events. The data was randomly divided into two sets: a training set (80%) and a test set (20%).

---

[2] One approach producing similar results as the one explained in the text was to use harmonic networks [29] to grasp the rotational invariance of the problem.

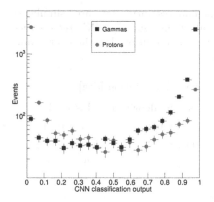

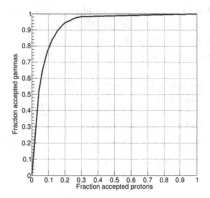

Fig. 2. The CNN gamma-ray and proton classification output for events using an independent test set.

Fig. 3. ROC curve for simulated true energy for gamma-ray and proton events above 10 GeV.

After having trained the CNN for the classification problem, the classifier is tested with an independent test set of gamma-ray and proton events. As an example, Fig. 2 shows the result of the classification of this test set with the trained CNN, representing the classification power of the CNN approach in terms of gamma-ray and proton separation for events above 10 GeV. To illustrate the general performance of our binary classification problem, we use the receiver operator characteristic (ROC) curve shown in Fig. 3. The ROC curve is a graphical plot that illustrates the true positive rate versus the false positive rate for each possible discrimination value.

We trained separate dedicated CNNs to estimate gamma-ray energies, direction and depth of first interaction. The trained networks for the regression problem are able to reproduce the simulated energy of the events as seen in the Fig. 4, where reconstructed energy as a function of true energy is presented. Figure 5 shows the energy resolution as a function of the true energy of our CNN for two different data sets. The on-axis and off-axis data set represent the energy resolution of diffuse gamma-ray events with angles with respect to the field of view center of less and more than two degrees, respectively. The energy resolution is defined as the one standard deviation of a Gaussian function fit of the distribution of the difference between true and reconstructed energy divided by true energy for a given energy range. The expected energy resolution performance of CTA [23] based on combination of Hillas parametrization and multivariate classification methods is slightly better with about an energy resolution of 9% at 300 GeV. Table 1 compares the energy resolution of the baseline algorithm with the results of this work for three distinct energy bins. However these numbers hava to be taken with care as such comparison are dependent on the differences in data sample like diffuse and point like emission, data selection, number of telescopes and selected strategy.

Table 1. Comparison of energy resolution and angular resolution for three simulated true energy bins for baseline CTA reconstruction algorithm and the CNN reconstruction presented in this work. Lower values implies better resolution. For the energy and angular resolution only statistical uncertainties are shown.

Simulated true energy	Energy resolution [%]		Angular resolution [deg]	
	Baseline algorithm	This work	Baseline algorithm	This work
30 GeV	25 ± 0.5	21 ± 0.4	0.26 ± 0.005	0.26 ± 0.01
300 GeV	9 ± 0.5	13 ± 0.4	0.09 ± 0.005	0.10 ± 0.005
3000 GeV	7 ± 0.5	11 ± 1.6	0.05 ± 0.005	0.08 ± 0.01

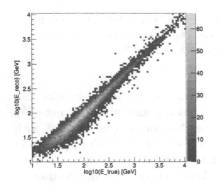

Fig. 4. Reconstructed energy as function of simulated true energy for only diffuse gamma-ray events using a separate CNN than the one used for classification.

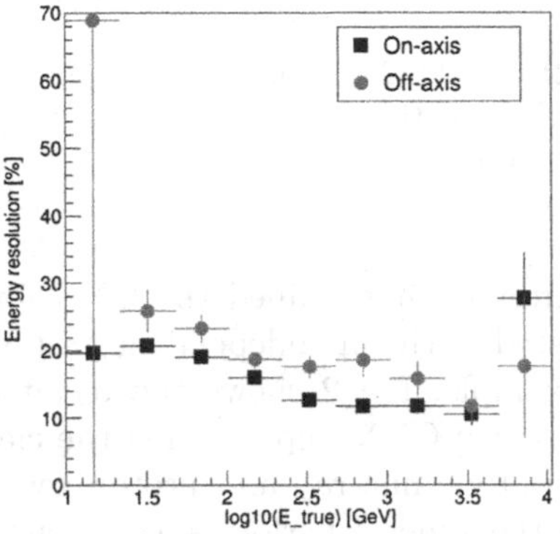

Fig. 5. Energy resolution as a function of simulated true energy for two different data sets of diffuse gamma-ray events. The on-axis (off-axis) points represent the resolution of diffuse gamma-ray events with angles with respect to the field of view center of less (more) than two degrees.

The directional reconstruction performance as a function of true energy of the CNN is given in the Fig. 6 for the two different on-axis and off-axis data sets. As can be seen from the Fig. 6 the on-axis angular resolution is better than the off-axis one. The angular resolution is defined as the angular offset, relative to the true gamma-ray direction, within which 68% of the gamma-ray events are reconstructed. The angular resolution for a point like emission for the CTA baseline algorithm is about 0.09° at 300 GeV. Table 1 shows the angular resolution of the baseline algorithm and the results of this work for three distinct energy bins.

Finally, in contrast to energy and directional primary particle reconstruction, the reconstruction of depth of first interaction of the primary particle is not used in many analyses, although depth of first interaction is useful to sepa-

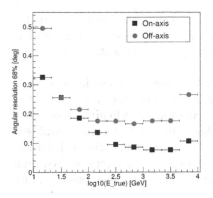

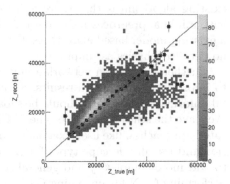

Fig. 6. Angular resolution as a function of simulated true energy for two different data sets of diffuse gamma-ray events. The on-axis (off-axis) points represent the resolution of diffuse gamma-ray events with angles with respect to the field of view center of less (more) than two degrees.

Fig. 7. Reconstructed depth of first interaction as a function of simulated true depth of first interaction using the same CNN architecture as for reconstruction of primary particle direction.

rate lepton from gamma-ray initiated showers. This quantity can be difficult to reconstruct if the number of triggered telescopes is small. Moreover, the algorithm [37] used to reconstruct this variable needs knowledge about the physics interaction and detector response. In contrast CNN algorithm needs no additional physical knowledge except what is in the simulation and we use the same CNN architecture as for directional reconstruction. Figure 7 shows the reconstructed depth of first interaction as a function of true depth of that interaction. A clear correlation is seen suggesting that the height of first interaction can be estimated automatically without any further changes on the CNN architectures.

In this study we did not exploit all the information as images should be separated according to individual telescopes. We should also take advantage of including all relevant telescope types and the timing information. It has been shown [38] that the primary particle information as well as the background rejection can be significantly improved by using timing knowledge.

4   Summary and Conclusion

The aim of this work is to investigate a deep learning technique for atmospheric Cherenkov telescopes classification and primary particle parameter estimation. The approach of the work is to treat gamma-ray detection as a two class classification problem (gamma-ray versus proton events) as well as to reconstruct gamma-ray shower parameters and solve it with supervised learning methods.

Promising CNN results have been found and a first comparison to previously published baseline algorithm can be made. The main advantages of CNN over existing algorithm is that there is little need of specialized physics knowledge with minimal preprocessing of data. Although the results are still not as good as a existing model based algorithms, CNN have simpler implementation requiring no detailed physics assumptions.

Further analysis on network architecture and image preprocessing is needed to improve reconstruction results. Specifically our method does not exploit the full information as images should be separated according to individual telescopes. We leave the study for a more general CNN taking into account of more sophisticated approaches, like use hexagonal symmetric features, include timing information and use all telescope types for the future work. All these steps are required to add more complexity and generalize our analysis in order to provide a more performing CNN for upcoming CTA data.

Acknowledgments. We gratefully acknowledge the support of the project (reference AYA2014-58350-JIN) funded by MINECO through the young scientist call (year 2014). This work is partially supported by the Maria de Maeztu Units of Excellence Program (reference MDM-2015-0509). We also gratefully acknowledge financial support from the agencies and organizations listed here: http://www.cta-observatory.org/consortium_ acknowledgments. This research shows in Table 1 the CTA energy and angular resolution values provided by the CTA Consortium and Observatory, see http://www. cta-observatory.org/science/cta-performance/ (version prod3b-v1) for more details. We want to thank K. Bernlöhr for the simulation CTA prod-3 data set, which was carried out at the Max Planck Institute for Nuclear Physics in Heidelberg. This work was conducted in the context of the CTA Analysis and Simulation Working Group and this paper has gone through internal review by the CTA Consortium.

References

1. Hillas, A.M.: Cherenkov light images of EAS produced by primary gamma rays and by nuclei. In: Proceedings of the 19th International Cosmic Ray Conference, La Jolla, vol. 3, pp. 445–448 (1985)
2. Albert, J., et al.: Implementation of the random forest method for the imaging atmospheric Cherenkov telescope MAGIC. Nucl. Instrum. Meth. A **588**, 424–432 (2008)
3. Aleksić, J., et al.: Performance of the MAGIC stereo system obtained with Crab Nebula data. Astropart. Phys. **35**, 435–448 (2012)
4. Ohm, S., et al.: Gamma-hadron separation in very-high-energy gamma-ray astronomy using a multivariate analysis method. Astropart. Phys. **31**, 383–391 (2009)
5. Aharonian, F., et al.: Observations of the Crab nebula with H.E.S.S. Astron. Astrophys. **457**(3), 899–915 (2006)
6. Holder, J., et al.: The first VERITAS telescope. Astropart. Phys. **25**, 391–401 (2006)
7. de Nauroisa, M., Rolland, L.: A high performance likelihood reconstruction of gamma-rays for imaging atmospheric Cherenkov telescopes. Astropart. Phys. **32**, 231–252 (2009)
8. Holler, M., et al.: Photon reconstruction for H.E.S.S. using a semi-analytical shower model. PoS (ICRC 2015) **980** (2015). arXiv:1509.02896

9. Parsons, R.D., Hinton, J.A.: A Monte Carlo template based analysis for air-Cherenkov arrays. Astropart. Phys. **56**, 26–34 (2014)
10. Feng, Q., Lin, T.T.Y.: The analysis of VERITAS muon images using convolutional neural networks. Proc. IAU **12**(S325), 173–179 (2016). arXiv:1611.09832
11. Holch, T.L., et al.: Probing convolutional neural networks for event reconstruction in γ-ray astronomy with Cherenkov telescopes. PoS (ICRC 2017) **795** (2017). arXiv:1711.06298
12. Nieto, D., et al.: Exploring deep learning as an event classification method for the Cherenkov telescope array. PoS (ICRC 2017) **809** (2017). arXiv:1709.05889
13. Shilon, I., et al.: Application of deep learning methods to analysis of imaging atmospheric Cherenkov telescopes data (2018). arXiv:1803.10698
14. LeCun, Y., et al.: Gradient-based learning applied to document recognition. IEEE Proc. **86**(11), 2278–2324 (1998)
15. Krizhevsky, A., et al.: ImageNet classification with deep convolutional neural networks. NIPS Proc. **25**, 1097–1105 (2012)
16. Bishop, C.M.: Pattern Recognition and Machine Learning. Springer, New York (2006)
17. LeCun, Y., et al.: Deep learning. Nature **521**, 436–444 (2015)
18. Goodfellow, I., Bengio, Y., Courville, A.: Deep Learning. MIT Press, Cambridge (2016)
19. Acharya, B.S., et al.: Science with the Cherenkov telescope array (2017). arXiv:1709.07997
20. Mangano, S.: Cherenkov telescope array status report (2017). arXiv:1705.07805
21. Mazin, D., et al.: Large size telescope report (2016). arXiv:1610.04403
22. Bernlöhr, K., et al.: Monte Carlo design studies for the Cherenkov telescope array. Astropart. Phys. **43**, 171–188 (2013)
23. Maier, G., et al.: Performance of the Cherenkov telescope array (2017). arXiv:1709.01381
24. Heck, D., et al.: CORSIKA: a Monte Carlo code to simulate extensive air showers. Technical report FZKA 6019, Forschungszentrum Karlsruhe (1998)
25. Ong, R.A.: Very high-energy gamma-ray astronomy. Phys. Rep. **305**, 93–202 (1998)
26. Bernlöhr, K.: Simulation of imaging atmospheric Cherenkov telescopes with COR-SIKA and sim_telarray. Astropart. Phys. **30**, 149–158 (2008)
27. Zanin, R., et al.: MARS, the MAGIC analysis and reconstruction software. In: Proceedings of 33rd ICRC, Rio de Janeiro, Brazil, ld. 773 (2013)
28. Abadi, et al.: TensorFlow: large-scale machine learning on heterogeneous distributed systems. In: 12th USENIX Symposium on OSDI 2016, pp. 265–283 (2016). arXiv:1603.04467
29. Worrall, D.E., et al.: Harmonic networks: deep translation and rotation equivariance (2016). arXiv:1612.04642
30. Ioffe, S., Szegedy, C.: Batch normalization: accelerating deep network training by reducing internal covariate shift. PMLR **37** (2015). arXiv:1502.03167
31. Boureau, Y., et al.: A theoretical analysis of feature pooling in visual recognition. In: ICML (2010)
32. Srivastava, N., et al.: Dropout: a simple way to prevent neural networks from overfitting. J. Mach. Learn. Res. **15**, 1929–1958 (2014)
33. Glorot, X., Bengio, Y.: Understanding the difficulty of training deep feedforward neural networks. PMLR **9**, 249–256 (2010)
34. Rumelhart, D.E., et al.: Learning representations by back-propagating errors. Nature **323**, 533–536 (1986)

35. Ruder, S.: An overview of gradient descent optimization algorithms (2017). arXiv:1609.04747
36. Kingma, D.P., Ba, J.: Adam: a method for stochastic optimization (2017). arXiv:1412.6980
37. Sitarek, J., et al.: Estimation of the height of the first interaction in gamma-ray showers observed by Cherenkov telescopes (in preparation)
38. Stamatescu, V., et al.: Timing analysis techniques at large core distances for multi-TeV gamma ray astronomy. Astropart. Phys. **34**, 886–896 (2011)

Automatic Hand Sign Recognition: Identify Unusuality Through Latent Cognizance

Pisit Nakjai⬤ and Tatpong Katanyukul[✉]⬤

Khon Kaen University, Khon Kaen, Thailand
mynameisbee@gmail.com, tatpong@kku.ac.th

Abstract. Sign language is a main communication channel among a hearing disability community. Automatic sign language transcription could facilitate better communication and understanding between a hearing disability community and a hearing majority.

As a recent work in automatic sign language transcription has discussed, effectively handling or identifying a non-sign posture is one of the key issues. A non-sign posture is a posture unintended for sign reading and does not belong to any valid sign. A non-sign posture may arise during a sign transition or simply from an unaware posture. Confidence ratio (CR) has been proposed to mitigate the issue. CR is simple to compute and readily available without extra training. However, CR is reported to only partially address the problem. In addition, CR formulation is susceptible to computational instability.

This article proposes alternative formulations to CR, investigates an issue of non-sign identification for Thai Finger Spelling recognition, explores potential solutions and has found a promising direction. Not only does this finding address the issue of non-sign identification, it also provide an insight behind a well-learned inference machine, revealing hidden meaning and new interpretation of the underlying mechanism. Our proposed methods are evaluated and shown to be effective for non-sign detection.

Keywords: Hand sign recognition · Thai Finger Spelling
Open-set detection · Novelty detection · Zero-shot learning
Inference interpretation

1 Introduction

Sign language is a main face-to-face communication channel in a hearing disability community. Like spoken languages, there are many sign languages, e.g., American Sign Language (ASL), British Sign Language (BSL), French Sign Language (LSF), Spanish Sign Language (LSE), Italian Sign Language (LIS), Chinese Sign Language (CSL), Indo-Pakistani Sign Language (IPSL), Thai Sign Language (TSL), etc. A sign language usually has two schemes: a semantic sign

© Springer Nature Switzerland AG 2018
L. Pancioni et al. (Eds.): ANNPR 2018, LNAI 11081, pp. 255–267, 2018.
https://doi.org/10.1007/978-3-319-99978-4_20

scheme and a finger spelling scheme. A semantic sign scheme uses hand gestures, facial expressions, body parts, and actions to communicate meaning, tone, and sentiment. A finger spelling scheme uses hand postures to represent alphabets in its corresponding language. Automatic sign language transcription would allow better communication between a deaf community and hearing majority. Sign language recognition has been subjects of various studies [2,7,11]. A recent study [7], investigating hand sign recognition for Thai Finger Spelling (TFS), has discussed issues and challenges in automatic transcription of TFS. Although the discussion is based on TFS, some issues are general across languages or even general across domains beyond sign language recognition. One of the key issues discussed in the study [7] is an issue of a non-sign or an invalid TFS sign, which may appear unintentionally during a sign transition or from unaware hand postures.

The appearance of non-signs may undermine the overall transcription performance. Nakjai and Katanyukul [7] proposed a light-weight computation approach to address the issue. Sign recognition is generally based on multi-class classification, whose output is represented in softmax coding. That is, a softmax output capable of predicting one of K classes is noted $y = [y_1 y_2 \ldots y_K]^T$, whose coding bit $y_i \in [0,1], i = 1, \ldots, K$ and $\sum_{i=1}^{K} y_i = 1$. A softmax output y represents predicted class k when y_k is the largest component: $k = \arg\max_i y_i$. Their approach is based on the assumption that the ratio between the largest value of the coding bit and the rest shows the confidence of the model in its class prediction. Softmax coding values have been normalized so that it can be associated to both probability interpretation and cross-entropy calculation. Despite the benefits of normalization, they use the penultimate values instead of the softmax values for rationale that some information might have been lost during the softmax activation. Penultimate values are inference values before going through softmax activation (i.e., a_k in Eq. 1). Specifically, to indicate a non-sign posture, they proposed a confidence ratio (CR), $cr = \frac{a}{b}$, where a and b are the largest and second largest penultimate values, respectively: $a = a_m$ and $b = a_n$ where $m = \arg\max_i a_i$ and $n = \arg\max_{i \neq m} a_i$. Their CR has been reported to be effective in identifying a posture that is likely to get a wrong prediction. However, on their evaluating environment, they reported that CR could hardly distinguish the cause of the wrong prediction whether it was a misclassified valid sign or it was a forced prediction on an invalid sign. In addition, generally each penultimate output is a real number, $a_i \in \mathbb{R}$. This nature poses a risk on CR formulation for when there is zero or a negative number, CR can be misleading or its computation can even collapse (when the denominator is zero).

Our study investigates development of an automatic hand sign recognition for Thai Finger Spelling (TFS), alternative formulations to CR, a non-sign issue and potential mitigations for a non-sign issue. TFS has 25 hand postures to represent 42 Thai alphabets using single-posture and multi-posture schemas [7]. Single-posture schema directly associates a hand posture to a corresponding alphabet. Multi-posture schema associates a series of 2 to 3 hand postures to a corresponding alphabet. Based on probability interpretation of an inference output, Bayes theorem, and examining an internal structure of a commonly adopted inference model, various formulations alternative to CR are investigated (Sect. 3). Sections 2, 4, and 5 provide related background, methodologies and experimental results, and discussion and conclusions, respectively.

2 Background

TFS Hand Sign Recognition. A recent visual-based state-of-the-art in TFS sign recognition A-TFS [7] frames hand sign recognition as a pipeline of hand localization and sign classification problem. A-TFS is an approach based on a color scheme and a contour area using Green's theorem for hand localization. Then, an image region dominated by a hand is scaled to a pre-defined size (i.e., 64×64) and passed through a classifier, implemented with a convolution neural network. The classifier predicts the most likely class out of the 25 pre-defined classes, each corresponding to a valid TFS sign.

Most visual-based TFS sign recognition studies [7,11] focus on static images. However, a practical system should anticipate video and streaming data, where unintended postures may be passed through the pipeline and cause confusion to the final transcription result. Unintended postures can accidentally match valid signs. This challenging case is worth a dedicated study and could be addressed through a language model. However, even when the unintended postures do not match any of the valid signs, a classifier is forced to predict one out of its pre-defined classes. No matter which class it predicts, the prediction is wrong. This could cause immediate confusion on its recognition result or undermine performance of its subsequence process when using this recognition as a part of a larger system. Confidence ratio (CR) [7] was proposed to address the issue, but reported to be marginally effective.

Novelty Detection. A conventional classifier specifies a fixed number of classes that it can predict and is forced to predict. This constraint allows it to be efficiently optimized to its classification task, but it has a drawback, which is more apparent when the assumption of all-inclusive classes is strongly violated. The concept of flagging out an instance belonging to a class that an inference machine has not seen at all in the training phase is a common issue and a general concern beyond sign language recognition. The issue has been extensively studied under various terms[1], e.g., novelty detection, anomaly detection, outlier detection, zero-shot learning, and open-set recognition.

Pimentel et al. [9] summarize a general direction in novelty detection. That is, a detection method usually builds a model using training data containing no examples or very few examples of the novel classes. Then, somehow depending on approaches, a novelty score s is assigned to a sample under question x and the

---

[1] Definition of novelty, anomaly, outlier, and zero-shot may be slightly different. Approaches may be various [9,13], but they are generally addressing a similar concern.

final novelty judgement is decided by thresholding, i.e., the sample x is judged a novelty (belonging to a new class) when $s(x) > \tau$ for τ is a pre-defined threshold.

To obtain the novelty score, various approaches have been examined. Pimentel et al. [9] categorize novelty detection into 5 approaches: probabilistic, distance-based, reconstruction-based, domain-based, and information-theoretic based techniques. A probabilistic approach relies on estimating a probability density function (pdf) of the data. A sample x is tested by thresholding the value of its pdf: $pdf(x) < \tau$ indicates x being novel. Training data is used to estimate the pdf. Although this approach has a strong theoretical support, estimating a pdf in practice requires a powerful generative model along with an efficient mechanism to train it. A generative model at its fullest potential could provide greater inference capabilities on data, such as expressive representation, reconstruction, speculation, generation, and structured prediction. Its applicability is much beyond novelty detection. However, high-dimension structured data, e.g., images, render this requirement very challenging. A computationally traceable generative model is a subject of highly active research. Another related issue is to determine a sensible value for τ, in which many studies [1,3] have resorted to extreme value theory (EVT) [8]. A distance-based approach is presumably [9] based on an assumption that data seen in a training process is tightly clustered and data of new types locate far from their nearest neighbors in the data space. Either a concept of nearest neighbors [14] or of clustering [6] is used. Roughly speaking, a novelty score is defined by a distance either between a sample x and its nearest neighbors or between x and its closest cluster centroids. The distance is often measured with Euclidean or Mahalanobis distance. The approach relies on a mechanism to identify the nearest neighbors or the nearest clusters. This usually is computationally intensive and becomes a key factor attributed to its scalability issue in terms of data size and data dimensions. A reconstruction-based approach involves building a re-constructive model, often called "auto-encoder," which learns to find a compact representation of input and reproduce it as an output. Then, to test a sample, the sample is put through a reconstruction process and a degree of dissimilarity between the sample and its reconstructed counterpart is used as a novelty score. Hawkins et al. [4] used a 3-hidden-layer artificial neural network (ANN) learned to reproduce its input. As an auto-encoder, a number of input nodes is equal to a number of output nodes and a number of nodes in at least one hidden layer is smaller than a number of input nodes in order to force ANN to learn a compressed representation of the data. Any sample that cannot be reconstructed well is taken for novelty, as this infers that its internal characteristics do not align with the compressed structure fine tuned to the training data. This approach may also resort to distance measurement for a degree of dissimilarity, but it does not require to search for the nearest neighbors. Therefore, once an auto-encoder is tuned, it is easier to scale up than a distance-based approach. A domain-based approach associates building a boundary of the data domain in a feature space. Any sample x is considered novelty if its location on the feature space lies outside the boundary. Schölkopf et al. [10] proposed one-class support vector machine (SVM) for novelty detection. SVM learns to build a boundary in a feature space to adequately cover most training examples, while having a user-defined parameter to control a degree to allow some training samples to be outside the boundary. This compromising mechanism is a countermeasure to outliers in the training data. The last approach—information-theoretic—involves measurement of information content in the data. It assumes that samples of novelty increase information content in the dataset significantly. As their task was to remove outliers from data, He et al. [5] used a decrease in entropy of a dataset after removal of the samples to indicate a degree of the samples being outliers. The samples were heuristically searched. Pimentel et al. [9] note that this approach often requires an information measure that is sensitive enough to pick up the effect of novelty samples, especially when a number of these samples is small. Noted that most approaches do not scale well to high-dimension structured data, like images. Novelty detection in high-dimension structured data is still in an early stage.

Based on this categorization [9], a probabilistic approach is closest to the direction we are taking. However, unlike many early works, firstly, rather than requiring a dedicated model, our proposed method builds upon a well-adopted classifier. It can be used with an already-trained model without requirement for re-training. Secondly, most works including a notable work of OpenMax [1]—whose performance achieves F-measure[2] of 0.595—determine a degree of novelty by how unlikely the sample belongs to any seen class. Another word, most previous works have to examine every probability of sample x being seen class i, $Pr[class = i|x]$, for $i = 1, \ldots, K$, when K is a number of all seen classes. Our work follows our interpretation of a softmax output, i.e., $y_i \equiv Pr[class = i|s, x]$, where s represents a state of being a seen class (not novelty). How likely sample x is novel then can be directly deduced.

3   Prediction Confidence and Non-sign Identification

Confidence Score (cs). To quantify confidence in classification output, our study investigates various candidates (shown in Table 1) based on that y_k associates to a probability of being class k and y_k is generally obtained through a softmax mechanism (Eq. 1). Formulation cs_1 is straightforward. Formulation cs_2 associates to a logarithm of probability. Formulation cs_3 is similar to confidence ratio [7] (CR), but with an attempt to link an empirical utility to a theoretical rationale. In addition, formulation cs_3 is preferable in terms of computational cost and stability. Formulation cs_4—a logit function—has a more direct interpretation of the starting assumption that the confidence is high when probability of the predicted class is much higher than the rest.

Latent Cognizance. Given the input image x, the predicted sign in softmax coding $y \in \mathbb{R}^K$, where K is a number of the pre-defined classes, is derived through a softmax activation: for $k = 1, \ldots, K$,

---

[2] Tested on 80,000 images (including 15,000 unknown images).

Table 1. Formulations under investigation for confidence score (cs). Softmax value $y_l = \frac{e^{a_l}}{\sum_{i=1}^{K} e^{a_i}}$, where K is a number of predefined classes; a_l is a penultimate value; k and j are indices of the largest and the second largest components, respectively

Confidence score	$cs_1 = y_k$	$cs_2 = a_k$	$cs_3 = \log\left(\frac{y_k}{y_j}\right) = a_k - a_j$	$cs_4 = \log\left(\frac{y_k}{1-y_k}\right)$

Table 2. Formulations under investigation for cognizance function ($\tilde{g}$). Term a represents a penultimate value

Cognizance function	$\tilde{g}_0(a) = a$	$\tilde{g}_1(a) = e^a$	$\tilde{g}_2(a) = a^2$	$\tilde{g}_3(a) = a^3$	$\tilde{g}_4(a) = |a|$

$$y_k = \frac{e^{a_k}}{\sum_{i=1}^{K} e^{a_i}}, \tag{1}$$

where a_k is the kth component of penultimate output. Each y_k can be interpreted as a probability that the given image belongs to sign class k, or more precisely a probability that the given valid input belongs to class k. That is, $y_k \equiv Pr[k|s, \boldsymbol{x}]$ where k indicates one of the K valid classes, $\boldsymbol{x}$ is the input under question, and s indicates that $\boldsymbol{x}$ is representing one of the valid classes (being a sign). For conciseness, conditioning on $\boldsymbol{x}$ may be omitted, e.g., $y_k \equiv Pr[k|s, \boldsymbol{x}]$ may be written as $y_k = Pr[k|s]$. Noted that, this insight is distinct to a common interpretation [1] that a softmax coding bit y_k of a well-learned inference model estimates probability of being in class k, i.e., $y_k = Pr[k|\boldsymbol{x}]$. This common notion does not emphasize its conditioning on an inclusiveness of all pre-defined classes.

Identifying a non-sign can be achieved through determining the probability of a sample $\boldsymbol{x}$ not belonging to any of the sign classes: $Pr[\bar{s}|\boldsymbol{x}] = 1 - Pr[s|\boldsymbol{x}]$. To deduce $Pr[s|\boldsymbol{x}]$, or concisely $Pr[s]$, consider Bayesian relation: $Pr[k|s] = \frac{Pr[k,s]}{\sum_{i=1}^{K} Pr[i,s]}$ where $Pr[k, s]$ is a joint probability. Given the Bayesian relation, the inference mechanism (Eq. 1), and our new interpretation of y_k, the following relation is found:

$$\frac{e^{a_k}}{\sum_{i=1}^{K} e^{a_i}} = \frac{Pr[k, s]}{\sum_{i=1}^{K} Pr[i, s]}. \tag{2}$$

Based on Eq. 2, it should be easier to find an appropriate mapping between e^{a_k} and $Pr[k, s]$ for the interpretability of the equation. Here, we draw the assumption that penultimate value a_k relates to joint probability $Pr[k, s]$ through an unknown function $u : a_k(\boldsymbol{x}) \mapsto Pr[k, s|\boldsymbol{x}]$. Theoretically, this unknown function is difficult to exactly characterize. In practice, even without exact characteristics of this mapping, a good approximate is enough to accomplish a task of identifying a non-sign. Supposed there exists an approximate mapping g, i.e., $g(a_k) \approx Pr[k, s]$, therefore given $g(a_i)$'s (for $i = 1, \ldots, K$), a non-sign can be identified by $Pr[s|\boldsymbol{x}] = \sum_i Pr[i, s|\boldsymbol{x}] \approx \sum_i g(a_i(\boldsymbol{x}))$. Further refining, to lessen burden on enforcing proper probability properties on g, define a "cognizance" function $\tilde{g}$ such that $\tilde{g}(a_i(\boldsymbol{x})) \propto g(a_i(\boldsymbol{x}))$. Consequently, define primary and secondary latent cognizance as the following relations, respectively:

$$\tilde{g}\left(a_i(\boldsymbol{x})\right) \propto Pr[i, s|\boldsymbol{x}], \tag{3}$$

$$\sum_i \tilde{g}\left(a_i(\boldsymbol{x})\right) \propto Pr[s|\boldsymbol{x}]. \tag{4}$$

Various formulations (Table 2) are investigated for an effective cognizance function. Identity $\tilde{g}_0$ is chosen for its simplicity. Exponential $\tilde{g}_1$ is chosen for its immediate reflection on Eq. 2. It should be noted that a study on a whole family of $\tilde{g} = m \cdot e^a$, where m is a constant, is worth further investigation. Other formulations are intuitively included on an exploratory purpose.

4   Experiments

Various formulations of confidence score and choices of cognizance function are evaluated on TFS sign recognition system. Our TFS sign recognition follows the current state-of-the-art in visual TFS sign recognition [7] with a modification of convolution neural network (CNN) configuration and its input resolution. Instead of a 64×64 gray-scale image, our work uses a 128×128 color image as an input for CNN. Our CNN configuration uses a VGG-16 [12] with the 2 fully-connected layers each having 2048 nodes, instead of 3 fully-connected layers in the original VGG-16. Figure 1 illustrates our processing pipeline.

Fig. 1. Processing pipeline of our TFS sign recognition.

Sign Data. The main dataset contains images of 25 valid TFS sign postures. Twelve signers[3] were employed to perform TFS signs. Each signer performed all 25 valid TFS signs for 5 times. That resulted in a total number of 1500 images (5 times ×25 postures ×12 signers), which were augmented to 15000 images. The augmentation process generated new images from the original images using different image processing methods, e.g., skewing, scaling, rotating, and translating. All augmented images were visually inspected for human readability and semantic integrity. Every image is a color image with a resolution of approximately 800×600 pixels.

---

[3] A signer is an individual person who performs TFS signs.

Experimentation. The data was separated based on signers into a training set and a test set, i.e., 11250 images from 9 signers for training set (75%) and 3750 images from the other 3 signers for test set (25%). The experiments were conducted for 10 repetitions in a 10-fold manner. Specifically, each repetition separated data differently, e.g., the 1st fold used data from signers 1, 2, and 3 for test and used the rest for training; the 2nd fold used test data from signers 2, 3, and 4; and so on till the last fold using test data from signers 10, 1, and 2.

The mean Average Precision (mAP), commonly used in object detection [7], is a key performance measurement. Area under curve (AUC) and receiver operating characteristic (ROC) are used to evaluate effectiveness of various formulations for confidence score and latent cognizance. AUC is often referred to as an estimate area under Precision-Recall curve, while ROC is usually referred to an estimate area under Detection-Rate–False-Alarm-Rate curve. However, generally both areas are equivalent. We use them to differentiate the purpose of our evaluation rather than taking them as different metrics. AUC is used for identification of samples not to be correctly predicted[4]. It is more direct to measure a quality of a replacement for confidence ratio (CR) [7]. ROC is used for identifying non-sign samples[5]. It is more direct to the very issue of non-sign postures.

Non-sign Data. In addition to the sign dataset, a non-sign dataset containing images of various non-sign postures is used to evaluate non-sign identification methods. All non-sign postures were carefully choreographed to be perceivably different from any valid TFS sign and performed by a signer before augmented to 1122 images. All augmented images had been visually inspected that they all were readable and did not accidentally match to any of the 25 valid signs.

Results. Table 3 shows TFS recognition performance of the previous studies and our work. The high performing mAP (97.59%) indicates that our model is well-trained. The results were shown to be non-normal distributed, based on Lilliefors test at 0.05 level. Wilcoxon rank-sum test was conducted on each treatment for comparing (1) difference between correctly classified samples (CP) and misclassified samples (IP), (2) difference between CP and non-sign samples (NS), and (3) difference between IP and NS. At 0.01 level, Wilcoxon rank-sum test confirmed all 3 differences in all treatments. Figure 2 shows boxplots of all treatments. Y-axes show the treatment values, e.g., the top left plot has its Y-axis depicting values of $\frac{a_k}{a_j}$. The values are observed in 10 cross-validations each testing on 4872 images (3750 sign images and 1122 non-sign images). Hence, each subplot depicts 48720

---

[4] Positive is defined to be a sample of either a non-sign or an incorrect prediction.
[5] Positive is defined to be a non-sign.

data points[6] categorized into 3 groups. Although the significance tests confirm that the 3 groups are distinguishable using any of the treatments, the boxplots show a wide range of degrees of difficulty to distinguish each individual sample, e.g., cubic cognizance ($\sum_i a_i^3$) seems to be easier than others on thresholding the 3 cases. To measure a degree of effectiveness, Tables 4 and 5 provide AUC and ROC. Noted that, since treatment $\tilde{g}_0$ gives results in a different manner than others: a higher value associates to a non-sign (c.f. a lower value in others), the evaluation logic is adjusted accordingly.

On finding an alternative to CR [7], maximal penultimate output a_k appears promising with the largest AUC (0.934) and it is simple to obtain (no extra computation, thus no risk of computational instability). On addressing a non-sign issue, cubic cognizance a^3 gives the best ROC (0.929). Its smoothed estimate densities[7] of non-sign samples (NS) and sign samples (combining CP and IP) are shown on Fig. 3a. Plots of detection rate versus false alarm rate of the 4 strongest candidates and CR are shown in Fig. 3b. Table 6 shows non-sign detection performance of the 4 strongest cognizance functions compared to a baseline, CR. Non-sign detection performance is measured with accuracy—a ratio of correctly classified sign/non-sign samples to all test samples—and F-measure—a common performance index for novelty detection [1]—at thresholds selected so that every treatment has its False Alarm Rate closest to 0.1.

Table 3. Performance of visual-based TFS sign recognition.

Method	TFS coverage	Data size (# images)	Key factors	Performance
Chansri and Srinonchat [2]	16 signs	320	Kinect 3D camera, HOG and ANN	83.33%
Silanon [11]	21 signs	2100	HOG and ANN	78.00%
A-TFS [7]	25 signs	1500	Hand Extraction and CNN	91.26%
Our work (V-TFS)	25 signs	15000	Hand Extraction and VGG-16	97.59%

Table 4. Evaluation of confidence score formulations.

	$cr = \frac{a_k}{a_j}$	$cs_1 = y_k$	$cs_2 = a_k$	$cs_3 = a_k - a_j$	$cs_4 = \log\left(\frac{y_k}{1-y_k}\right)$
AUC	0.814	0.919	0.934	0.900	0.919
ROC	0.740	0.879	0.921	0.847	0.879

---

[6] Extreme values—under 0.25 quantile and over 0.75 quantile—were removed.

[7] A normalized Gaussian-smoothing version of histogram produced through smoothed density estimates of ggplot2 (http://ggplot2.tidyverse.org) with default parameters.

Table 5. Evaluation of various $\tilde{g}$ formulations on $\sum_i \tilde{g}(a_i) \propto Pr[s]$.

	Identity $\tilde{g}_0(a) = a$	Exponential $\tilde{g}_1(a) = e^a$	Quadratic $\tilde{g}_2(a) = a^2$	Cubic $\tilde{g}_3(a) = a^3$	Absolute $\tilde{g}_4(a) = |a|$
AUC	0.437	0.930	0.855	0.934	0.737
ROC	0.419	0.920	0.845	0.929	0.726

Table 6. Non-sign detection performance of the cognizance functions c.f. CR. Thresholds were selected so that every treatment has its False Alarm Rate closest to 0.1.

Treatment	Threshold	Accuracy	F-measure
CR [7]	1.02	0.769	0.029
e^a	100000.00	0.919	0.807
a^2	26.70	0.866	0.627
a^3	1700.63	0.926	0.831
$|a|$	50.88	0.825	0.425

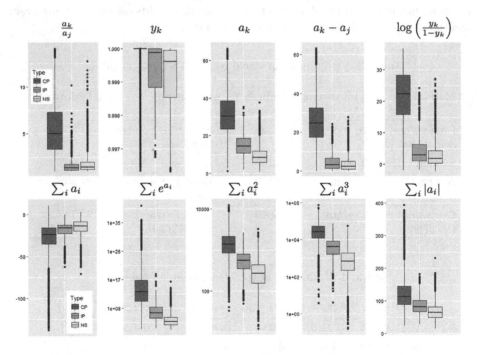

Fig. 2. Upper row: boxplots of confidence ratio and candidates for confidence score. A Y-axis shows values of confidence score in linear scale. The confidence score formulations are indicated in the subplot titles. Lower row: boxplots of 5 candidates for a cognizance function. A Y-axis shows $\sum_i \tilde{g}(a_i)$ values ($\sum_i a_i$ and $\sum_i |a_i|$ in linear scale; the rest in log scale). The confidence score or cognizance values are shown in 3 groups: CP for correctly classified samples; IP for misclassified samples; NS for non-sign samples.

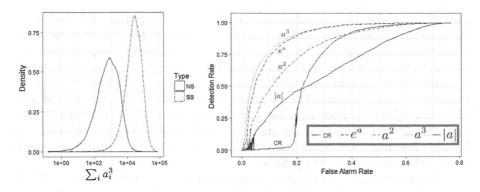

Fig. 3. (a) Left: illustration of smoothed estimated densities of sign (denoted SS) and non-sign (denoted NS) data over $\sum_i a_i^3$. (b) Right: detection rate versus false alarm rate curves of the 4 strongest candidates and the confidence ratio (CR).

5   Discussion and Conclusions

The cubic function has shown to be the best cognizance function among other candidates, including the exponential function. In addition, the cubic cognizance has ROC par to the max-penultimate confidence score. On the other hand, the max-penultimate confidence score also provide a competitive ROC and could be used to identify non-sign samples as well. Noted that OpenMax [1]— a state-of-the-art in open-set recognition—uses penultimate output as one of its crucial parts. Our finding could contribute to the development of OpenMax. A study of using cubic cognizance in OpenMax system seems promising, since it is shown to be more effective than a penultimate output. Another point worth noting is that the previous work [7] evaluated confidence score on identifying non-signs and could not confirm its effectiveness with the significance tests. Their results agree with our early experiments when using a lower resolution image, a smaller CNN structure, and training and testing on smaller datasets. In our early experiment, only a few of the treatments could be confirmed for non-sign identification. Those that were confirmed are consistent with ROC presented here. This observation implies a strong relation between state of the inference model and non-sign-identification effectiveness. This relation deserves a dedicated systematic study. Regarding applications of the techniques, thresholding can be used and a proper value for the threshold has to be determined. This can be simply achieved through tracing Fig. 3b with the corresponding threshold values. Alternatively, the proper threshold can be determined based on Extreme Value Theory, like many previous studies [1,3]. Another interesting research direction is to find a similar solution for other inference families. Our techniques target a softmax-based classifier, which is well-adopted especially in artificial neural network. However, Support Vector Machine (SVM), another well-adopted classifier, is built on a different paradigm. Application of latent cognizance to SVM might not work or might be totally irrelevant. Investigation into the issue on other inference paradigms could provide a unified insight of the underlying inference mechanism and benefits beyond addressing the novelty issue. Regarding starting assumptions, high ROC values of exponential and cubic cognizances support our new interpretation and its following assumptions. However, the penultimate output, according to our new interpretation, has relation $a_k(x) = \log(Pr[k|s, x]) + C$, where $C = -\log \sum_i a_i(x)$. This relation only partially agrees with our results. High value of AUC agrees with $\log(Pr[k|s, x])$ that a class is confidently classified, but $Pr[k|s, x]$ alone is not enough to determine a non-sign, which needs $Pr[\bar{s}|x]$. This implies that our research is on a right direction, but it still needs more studies to complete the picture.

In brief, our study investigates (1) alternatives to confidence ratio (CR) [7] and (2) methods to identify a non-sign. The max-penultimate output is shown to be a good replacement for CR in terms of detection performance and simplicity. Its large value associates to a sample likely to be correctly classified and vice versa. The cognizance $\sum_i a_i^3$ is shown to be a good indicator for a non-sign such that $\sum_i a_i^3(x) \propto Pr[s|x]$, i.e., a low value of $\sum_i a_i^3(x)$ associates to a non-sign sample. To wrap up, our findings give an insight into a softmax-based inference machine and provide a tool to measure a degree of confidence in the prediction result as well as a tool to identify a non-sign. The implications may go beyond our current scope of TFS hand-sign recognition and contribute to open-set recognition or other similar concepts. Latent cognizance is viable for its simplicity and effectiveness in identifying non-signs. These would help improve an overall quality of the translation, which in turn hopefully leads to a better understandingc among people of different physical backgrounds.

References

1. Bendale, A., Boult, T.E.: Towards open set deep networks. CoRR abs/1511.06233 (2015). http://arxiv.org/abs/1511.06233
2. Chansri, C., Srinonchat, J.: Hand gesture recognition for thai sign language in complex background using fusion of depth and color video. Procedia Comput. Sci. **86**, 257–260 (2016). https://doi.org/10.1016/j.procs.2016.05.113
3. Clifton, D., Hugueny, S., Tarassenko, L.: Novelty detection with multivariate extreme value statistics. J. Signal Process. Syst. **65**, 371–389 (2011)
4. Hawkins, S., He, H., Williams, G., Baxter, R.: Outlier detection using replicator neural networks. In: Kambayashi, Y., Winiwarter, W., Arikawa, M. (eds.) DaWaK 2002. LNCS, vol. 2454, pp. 170–180. Springer, Heidelberg (2002). https://doi.org/10.1007/3-540-46145-0_17
5. He, Z., Deng, S., Xu, X., Huang, J.Z.: A fast greedy algorithm for outlier mining. In: Ng, W.-K., Kitsuregawa, M., Li, J., Chang, K. (eds.) PAKDD 2006. LNCS (LNAI), vol. 3918, pp. 567–576. Springer, Heidelberg (2006). https://doi.org/10.1007/11731139_67
6. Kim, D., Kang, P., Cho, S., Lee, H., Doh, S.: Machine learning-based novelty detection for faulty wafer detection in semiconductor manufacturing. Expert. Syst. Appl. **39**(4), 4075–4083 (2011)
7. Nakjai, P., Katanyukul, T.: T. J Sign Process Syst (2018). https://doi.org/10.1007/s11265-018-1375-6

8. Pickands, J.I.: Statistical inference using extreme order statistics. Ann. Stat. **3**(1), 119–131 (1975). https://doi.org/10.1214/aos/1176343003
9. Pimentel, M., Clifton, D., Clifton, L., Tarassenko, L.: A review of novelty detection. Signal Process. **99**, 215–249 (2014)
10. Schölkopf, B., Williamson, R., Smola, A., Shawe-Taylor, J., Platt, J.: Support vector method for novelty detection. In: Proceedings of the 12th International Conference on Neural Information Processing Systems, NIPS 1999, pp. 582–588. MIT Press (1999)
11. Silanon, K.: Thai finger-spelling recognition using a cascaded classifier based on histogram of orientation gradient features. Comput. Intell. Neurosci. **11** (2017). https://doi.org/10.1155/2017/9026375
12. Simonyan, K., Zisserman, A.: Very deep convolutional networks for large-scale image recognition. CoRR abs/1409.1556 (2014). http://arxiv.org/abs/1409.1556
13. Xian, Y., Schiele, B., Akata, Z.: Zero-shot learning - the good, the bad and the ugly. In: IEEE Computer Vision and Pattern Recognition (CVPR) (2017)
14. Zhang, J., Wang, H.: Detecting outlying subspaces for high-dimensional data: the new task, and performance. Knowl. Inf. Syst. **10**(3), 333–355 (2006)

Cascade of Ordinal Classification and Local Regression for Audio-Based Affect Estimation

Maxime Sazadaly[1], Pierre Pinchon[1], Arthur Fagot[1],
Lionel Prevost[1(✉)], and Myriam Maumy-Bertrand[2]

[1] Learning, Data and Robotics Lab, ESIEA, Ivry-sur-Seine, France
`lionel.prevost@esiea.fr`
[2] Centre National de la Recherche Scientifique, Institut de Recherche
Mathématique Avancé, Université de Strasbourg, Strasbourg, France

Abstract. Affective dimensions (i.e. valence, arousal, etc.) are continuous, real variables, bounded on [−1,+1]. They give insights on people emotional state. Literature showed that regressing these variables is a complex problem due to their variability. We propose here a two-step process. First, an ensemble of ordinal classifiers predicts the optimal range within [−1, +1] and a discrete estimate of the variable. Then, a regressor is trained locally on this range and its neighbors and provides a finer continuous estimate. Experiments on audio data from AVEC'2014 and AV+EC'2015 challenges show that this cascading process can be compared favorably with state of art and challengers results.

Keywords: Affective computing · Ensemble of classifiers · Random forests

1 Introduction

Nowadays, vocal recognition of emotions has multiple applications in domains as diverse as medicine, telecommunications or transport [1]. For example, in telecommunications, it would become possible to priorities the calls from individuals in imminent danger situations over less relevant ones. In general, emotion recognition enables the improvement of human/machine interfaces, which justifies the unexpected increase of research on this field, due to the progresses in artificial learning.

Human interactions rely on multiple sources: body language, facial expressions, etc. A vocal message carries a lot of information that we translate implicitly. This information can be expressed or perceived verbally, but also non-verbally, through the tone, the volume or the speed of the voice. The automatic analysis of such information gives insights on the speaker emotional state.

The conceptualization of emotions is still a hot topic in psychology. Opinions do not converge towards a unique model. In fact, we can mainly differentiate three approaches [9]: (1) the basic emotions (Anger, Disgust, Fear, Happiness, Sadness, Surprise) described by Ekman [6], (2) the circumplex model of affect and (3) the appraisal theory. In the second model, the affective state is generally described, at least, by two dimensions: the valence which determines the positivity of the emotion and the arousal which determines the activity of the emotion [18, 23]. These two values, bounded on [−1,+1], describe much more precisely the emotional state of an individual

© Springer Nature Switzerland AG 2018
L. Pancioni et al. (Eds.): ANNPR 2018, LNAI 11081, pp. 268–280, 2018.
https://doi.org/10.1007/978-3-319-99978-4_21

than the basic emotions. However, it has been shown that other dimensions were necessary to report more accurately this state during an interaction [8].

The choice of one model or the other restrains the kind of machine learning algorithms used to estimate the emotional state. In case of basic emotions, the variable to be predicted is qualitative and nominal. Classification methods must be used. On the contrary, affective dimensions are quantitative, continuous, and bounded variables. So, regression predictor will be needed. To take advantage of the best of both worlds, we propose in this study a method that combines classification and regression. To predict a continuous and bounded variable, we first quantize the affect variable into bounded ranges. For example, a 5 ranges valence quantization would give the following boundaries $\{-1, -0.5, -0.2, +0.2, +0.5, +1\}$. It could be interpreted as "very negative", "negative", "neutral", "positive" and "very positive". Then, we proceed into 3 steps:

- Train an ensemble of classifiers to estimate if the affect variable associated to an observation is higher than a given boundary;
- Combine the ensemble decisions to predict the optimal range;
- Regress locally the affect variable on this range.

The proposed method is therefore a cascade of ordinal classifiers and local regressors (COCLR). We will see in the following state of the art that similar proposals have been made. But in this paper, we perform a thorough study on the key parameter of this method: the number of ranges to be separated by the ensemble of ordinal classifiers. We show experimentally that:

- On small and numerous ranges, ordinal classification performs well;
- On large ranges, the COCLR cascade performs better;
- On challenging databases (AVEC'2014 [23] and AV+EC'2015 [17], described in Sect. 4), the COCLR cascade can be compared favorably with challengers' and winner's proposals with an acceptable development and computational cost.

This paper is organized as follows. Section 2 focuses on the state of the art on affect prediction on audio data. In Sect. 3, we will present the COCLR flowchart. In Sect. 4, we will introduce the datasets used to train and evaluate our system and the different pre-processing realized. Then, in Sect. 5, we will expose and discuss our results. Finally, Sect. 6 offers some conclusions.

2 State of Art

The Audio-Visual Emotion recognition Challenges (AVEC), that takes place every year since 2011, enables to assess the systems proposed on similar datasets. The main objective of these challenges is to ensure a fair comparison between research teams by using the same data. Particularly, the unlabeled test set is released to registered participants some days before the challenge deadline. Moreover, the organizers provide to the competitors a set of audio and video descriptors extracted by approved methods.

The prosodic features such as the height, the intensity, the speech rate, and the quality of the voice, are important to identify the different types of emotions. Low level acoustic descriptors like energy, spectrum, cepstral coefficients, formants, etc. enable an accurate description of the signal [23]. Furthermore, it has recently been demonstrated that features learned by the first layers of deep convolutional networks were quite similar to some acoustic descriptors [22].

2.1 Emotion Classification and Prediction

The classification of emotion is done through classical methods like support vector machines (SVM) [2], Gaussian mixture models (GMM) [20] or random forests (RF) [15]. For regression tasks, numerous models have been proposed: support vector regressors (SVR) [5], deep belief networks (DBN) [13], bidirectional long-short term memory networks (BLSTM) [14], etc. As all these models having their own pros and cons, recent works focus on model combinations to improve overall accuracy. Thus, in [11], authors propose to associate BLSTM and SVR to benefit from the treatment of the past/present context of the BLSTM and the generalization ability of the SVR.

AV+EC'2015 challenge winners proposed in [12] a hierarchy of BLSTM. They deal with 4 information channels: audio, video (described by frame-by-frame geometric features and temporal appearance features), electrocardiogram and electro dermal activity. They combine the predictions of single-modal deep BLSTM with a multi-modal deep BLSTM that perform the final affect prediction.

2.2 Ordinal Classification and Hierarchical Prediction

The standard approach to ordinal classification converts the class value into a numeric quantity and applies a regression learner to the transformed data, translating the output back into a discrete class value in a post-processing step [7]. Here, we work directly on numerical values of affect variables but quantify them into several ranges. Recently, a discrete classification of continuous affective variables through generative adversarial networks (GAN) has been proposed [3]. Five ranges are considered.

The idea of a combining regressors and classifiers has already been applied to deal with age estimation from images. In [10], a first "global" regression is done with a SVR on all ages. Then, it is refined by locally adjusting the age regressed value by using an SVM. In [21] authors propose another hierarchy on the same issue. They define 3 age ranges (namely "child", "teen" an "adult"). An image is classified by combining the results of a pool of classifiers (SVC, FLD, PLS, NN and naïve Bayes) in a majority rule. Then, a second stage uses the appropriate relevant vector machine regression model (trained on one age range) to estimate the age.

The idea of such a hierarchy is not new, but its application to affect data, have not been proposed yet. Moreover, we show in the following experiments that the number of boundaries to be considered impacts the performance of the whole hierarchy.

3   Cascade of Ordinal Classifiers and Local Regressors

The cascade of ordinal classifiers and local regressors proposed here is a hybrid combination of classification and regression systems. Let us note X, the observation (feature vector), y the affective variable to be predicted (valence or arousal) and $\hat{y}$, the prediction. The variable y is continuous and defined on the bounded interval $[-1, +1]$. Therefore, it is possible to segment this interval into a set of n smaller sub-intervals called "ranges" in the following, bounded by the boundaries bi and bi+1 with $i \in \{1, n + 1\}$. For example, n = 2 define 2 ranges: $[-1, 0[$ ("negative") and $[0, +1]$ ("positive") and 3 boundaries $bi \in \{-1, 0, +1\}$. Each boundary bi (except -1 and $+1$) may define a binary classification issue: given the observation X, the prediction $\hat{y}$ is lower (resp. higher) than bi. By combining the outputs of the $(n - 1)$ binary classifiers, we get an ordinal classification. Given the observation X, the prediction $\hat{y}$ is probably (necessarily in case of perfect classification) located within the range $[bi, bi + 1[$. Once this range obtained, a local regression is run on it along to its direct neighbors to predict y. Figure 1 illustrates the full cascade. The structure of this system is modular and compatible with any kind of classification and regression algorithms. Moreover, it is generic and may be adapted to other subjects than affective dimension prediction.

3.1 Ordinal Classification

The regression of an affect value y on an observation X can be bounded by the minimum and the maximum this value might take. The interval on which y is defined, $I = [min(y), max(y)]$, can be divided in n ranges.

The first stage of the cascade is an ensemble of $(n - 1)$ binary classifiers. Each classifier decides if, given the observation X, the variable to be predicted is higher than the lower boundary b_i of a range or not. Training samples are labeled -1 if their y value is lower than b_i and +1 otherwise. Considering the sorted nature of the boundaries b_i, we build here an ensemble of ordinal classifiers [7].

We combine the decisions of these classifiers to compute the lower and upper bounds of the optimal range $[b_i, b_{i+1}[$. Consider an observation X with $y = 0.15$. Suppose the number of ranges $n = 6$ and linearly distributed boundaries b_i. The following ranges are defined: $[-1.0, -0.5, -0.25, 0, 0.25, 0.5, 1.0]$. In case of perfect classification, the output vector of the ensemble of classifiers would be: $\{1, 1, 1, 1, -1, -1\}$ where -1 means "y is lower than b_i" while +1 means "y is higher than b_i". Obviously, b_i is the bound associated to the "latest" classifier with a positive output and b_{i+1} the first classifier with a negative output. By combining the local decisions of these binary classifiers, we get the (optimal) range $[b_i, b_{i+1}[$. This range C_i will be used in the second stage to locally predict y. In the example, this range is $[0, 0.25[$. However, indecision between two classifiers can happen [16]. This indecision will be handled by the second stage of the cascade.

The performance measure of the ordinal classifiers, the accuracy, is directly linked to the definition of the ranges. The choice of the number of ranges n is a key parameter of our system and can be seen as a hyper-parameter. The n ranges and their corresponding boundaries b_i can be defined in several ways. If they are linearly distributed, they will define a kind of affective scale as in [3]. But the choice of the boundaries b_i could also prevent strong imbalances between classes. In case of highly imbalanced classes, the application of a data augmentation method is strongly recommended [4].

From now on, we can evaluate the accuracy (ranges detection rates) of the classifier combination. It can also be used to compute a discrete estimate of y, by using the center

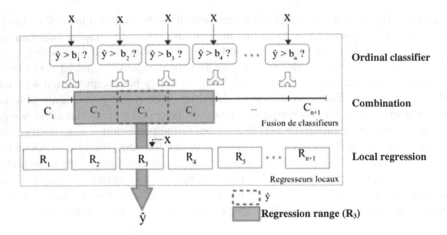

Fig. 1. COCLR: a two-stage cascade. The first stage is a combination of binary classifiers which aim is to estimate y's range. The observation X is handled by the corresponding local regressor which will evaluate the value of y on this range and its neighbors.

of the predicted range as the value $\hat{y}$. Finally, we can estimate the correlation of $\hat{y}$ to the ground truth y.

3.2 Local Regression

The aim of the second stage of the cascade is to compute the continuous value of y. Thus, each range i is associated to a regressor R_i that locally regresses y on $[b_i, b_{i+1}]$. So, each regressor is specialized in the regression on a specific range. However, as explained previously, indecisions between nearby classes throughout the ordinal classification may induce an improper prediction of the range. *De facto*, the wrong regressor can be activated, causing a drop of the correlation. The analysis of the first stage results, illustrated by the confusion matrix (Fig. 2), indicates that prediction mistakes are close enough or even connected to the optimal range which y belongs to. Thus, we can expand the regression range to $[b_{i-1}, b_{i+2}]$, if they exist.

Widening the local regression ranges helps to solve the indecision issue between the nearby boundaries. Moreover, it frees us from the obligation to strongly optimize the first stage. In fact, the use of a perfect classifier instead of a classifier that reaches an accuracy of 90% on the first stage won't have a significant impact on the result of the whole cascade.

4   Databases

4.1   AVEC'2014

The AVEC'2014 database is an ensemble of audio and video recordings of human/machine interaction [23]. This base is composed of 150 recordings, each of them containing the reactions of only one person, realized from 84 German subjects.

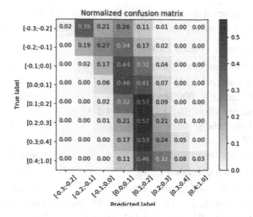

Fig. 2. Confusion matrix of the first stage of the cascade

The age of the subjects varies between 18 and 63 years old. In order to create this dataset, a part of the subjects has been recorded many times with a break of two weeks between each recording session. The distribution of the records is arranged as following: 18 subjects have been recorded three times, 31 of them have been recorded twice and the 34 lefts have been recorded only once. In these recordings, the subjects had to realize two tasks:

- NORTHWIND – The participants read out loud an extract of "Die Sonne und der Wind" (The North Wind and the Sun)
- FREEFORM – The participants answer numerous questions such as: "What is your favorite meal?"; "Tell us a story about your childhood."

Then the recordings are split in 3 parts: learning set, validation set, and test set, in which 150 couples of Freeform-Northwind are equally distributed. Low-level descriptors are described in Table 1.

4.2   AV+EC'2015/RECOLA

The second dataset we used to measure the performances of our system is the affect recognition challenge AV+EC'2015 [17]. The AV+EC'2015 relies on the RECOLA base. This one is composed of a set of 9.5 h of audio, video and physiologic recordings (ECG, EDA) from 46 records of French people from different origins (Italian, German, and French) and different genders. The AV+EC'2015 relies on a sub-set of 27 recordings completely labelled. In our case, we only used the audio recordings and only worked on the valence which is known as the most complex to be predicted.

The learning, development and testing partitions contain 9 recordings each. The diversity of origins and genders of the subjects has been preserved in these. The different audio features used are available in the AV+EC'2015 presentation paper [17].

Table 1. 42 acoustic low-level descriptors (LLD); 1 computed on voiced and unvoiced frames, respectively; 2 computed on voiced, unvoiced and all frames, respectively [17].

1 energy related LLD	Group
Sum of auditory spectrum (loudness)	Prosodic
25 spectral LLD	**Group**
α ratio (50–1000 Hz/1–5 kHz)	/ Spectral
Energy slope (0–500 Hz, 0.5–1.5 kHz)	/ Spectral
Hammarberg index	/ Spectral
MFCC 1–4	Cepstral
Spectral flux	Spectral
16 voicing related LLD	**Group**
F_o (linear and semi-tone)	Prosodic
Formants 1, 2, 3 (freq., bandwidth, ampl.)	Voice qual.
Harmonic difference H1-H2, HI-A3	Voice qual.
Log. HNR, jitter (local), shimmer (local)	Voice qual.

4.3 Data Augmentation

The study of the valence on a bounded interval allows the identification of several intensity thresholds of the felt emotion. Then, we can qualify this as very negative, negative, neutral, positive, and very positive, depending on this value. However, for the AVEC'2014 and the AV+EC'2015/RECOLA bases, these intensity thresholds are no equally represented. Figure 3 shows a clear unbalance of classes, favoring the representation of observations corresponding to a neutral valence (between −0.1 and 0.1). Considering the fact that some systems poorly support strong unbalances of classes [19], we increased the volume of data using the Synthetic Minority Over-sampling Technique [4].

5 Experimental Results

5.1 Performance Metrics

The cascade performances are directly linked to those of both stages. Thus, the performances of the ensemble of ordinal classifiers are measured by the accuracy. It measures the ratio of examples for which the interval has been correctly predicted. We use the confusion matrix in order to analyze the behavior of this system in a more precise way.

The performances of the ensemble of local regressors are measured using Pearson's correlation (PC), gold standard metric of the challenge AVEC'2014 [23] on which we base our study. However, as these data are not normally distributed, we decided to measure the performances of our system with Spearman's correlation (SC) and the concordance correlation coefficient (CCC) as well.

The experimental results presented in the following are computed on the development/validation set of the different databases. Due to the temporal nature of

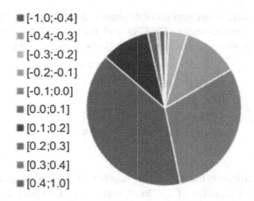

Fig. 3. Valence distribution on the training set with 10 ranges of $[-1; +1]$.

audio data, we have also decided to analyze the outputs of both stages on complete sequences and applied temporal smoothing to refine results.

5.2 Preliminary Results

As previously stated, our architecture is modular and adapted to any kind of classification or regression method. Throughout our experiments, we tried to use support vector machines (C-SVM with RBF kernels) and random forests (RF with 300 decision trees[1], attribute bagging on $\sqrt{nfeatures}$) as classifiers. The Table 2 presents the ordinal classification rate obtained by these two systems on the development sets of AVEC'2014 and AV+EC'2015, for the prediction of valence. We choose this affect variable because it is known to be particularly hard to predict (see baseline results in row 1). By taking the center of the predicted intervals as values of $\hat{y}$, we have been able to compute the correlations of these two systems. These correlations enable to compare the performances of our classifier ensemble to those of a unique "global" random forest regressor dealing with the whole interval $[-1, +1]$.

The results obtained on both databases encouraged us to continue with random forests rather than support vector machines. Indeed, the results returned by these are significantly sharper than the SVM ones, independently of the choice of the sub-intervals. For the same reasons, we have decided to use random forests to perform local regression.

5.3 Results on AVEC'2014

The Table 3 compares the performances of the different systems presented on the development base of the AVEC'2014, while using several number of ranges n.

First, the interval I has been split here in 10 ranges: $[-1.0, -0.4, -0.3, -0.2, -0.1, 0.0, 0.1, 0.2, 0.3, 0.4, 1.0]$. The most performant system in term of correlation is here, without a doubt, the ordinal classifier ensemble, where the values are the centers of the

---

[1] Sensitivity analysis on the number of decision trees is presented in Fig. 5.

Table 2. Valence prediction: comparison of different ordinal classifiers (SVM-OC and RF-OC), one global random forest regressor (RF-GR) and the challenge baseline [11, 22] on the development set. The performance measure is the Pearson correlation coefficient.

	AVEC'2014	AV+EC'2015
Baseline	0.38	0.17
RF-GR	0.45	–
SVM-OC	0.61	0.56
RF-OC	**0.77**	**0.65**

Table 3. Valence prediction: impact of the number of ranges on performances of global regressor (GR), ordinal classifier (OC), local regressors (LR) and cascade (COCLR). LR performances are computed considering the classification as "perfect" (Accuracy = 1).

n	Model	Accuracy	Pearson C	Spearman C	CCC
1	GR	–	0.45	0.47	0.27
10	OC	0.78	0.69	0.70	0.60
	LR	–	*0.91*	*0.90*	*0.89*
	COCLR	–	0.51	0.53	0.37
6	OC	0.83	0.63	0.66	0.54
	LR	–	*0.85*	*0.85*	*0.76*
	COCLR	–	0.54	0.53	0.39
4	OC	0.89	0.47	0.48	0.29
	LR	–	*0.80*	*0.81*	*0.77*
	COCLR	–	**0.77**	**0.77**	**0.65**

predicted ranges. It is as well relevant to point out that, despite the very high correlation of the local regressors alone, the COCLR system does not seem efficient.

Then, the interval I has been split into 6 ranges: [−1.0, −0.4, −0.2, 0.0, 0.2, 0.4, 1.0].

We compare the performances of the different systems on the AVEC 2014 development base. The most performant system, as far as the correlation is concerned, is still the ordinal classifier ensemble. However, the performance gap between the COCLR and the ordinal classifier ensemble has tightened. It is also noteworthy that the accuracy of the classification system has risen and the correlation of the local regressors alone, has slightly dropped.

Finally, the interval I here has been split into $n = 4$ ranges, [−1.0, 0.3, 0.0, 0.3, 1.0]. Previous conclusions on ordinal classifiers and local regressors remain the same. But this time, the COCLR cascade turned out to be significantly the most efficient one. The correlation related to this system is the highest obtained for every choice of intervals of any sort. These different results highlighted the importance of the choice of the number of ranges on which the COCLR system stands. It seems, as well, that the correlation of the local regressors alone decreases when we increase the size of the ranges, contrary to the accuracy of the classification system.

5.4   Results on AV+EC'2015/ RECOLA

As we did previously, we measured the performances of our system according to the different sub-intervals. Affect value varies within [−0.3, 1.0] so we discard classifier and regressors trained on]−1.0,−0.3[. Throughout our tests, we used 3 groups of different sub-intervals. The biggest, composed of 8 ranges, is: [−0.3, −0.2, −0.1, 0.0, 0.1, 0.2, 0.3, 0.4, 1.0]. The second one, composed of 5 ranges, is: [−0.3, −0.1, 0.0, 0.1, 0.3, 1.0]. Finally, the last one, composed of 3 ranges, is: [−0.3, 0.0, 0.3, 1.0]. The Table 4 presents a summary of these results.

Table 4. Valence prediction: best obtained models for each number of ranges on AV+EC'2015 development set. Challenge results [12] on the audio channel (AC) and their multimodal system (MM). The performance measure is the Pearson correlation coefficient.

	Proposal			Baseline AC	Winner AC	Winner MM
n	1	5	3	–	–	–
Best model	GR	OC	COCLR	–	–	–
Pearson C	0.463	0.521	0.675	0.167	0.529	**0.725**

We can observe that the results derived from the RECOLA database are similar to the ones the AVEC'2014. In fact, the most performant system remains the COCLR, when we chose a small number of ranges. The correlation obtained by the cascade of ordinal classifiers and local regressors for the valence on the development base is worth 0.67. As previously, we have observed a decline of the correlation of the local regressors and a rise of the accuracy of the first stage of the cascade when the size of the sub-intervals increased. Comparisons with challenge winner's results [12] are encouraging. Though our cascade get lower results (0.675) than their multimodal system (0.725), it gets better result than those obtained on the audio channel only (0.529). These latter are similar to those of the first stage ordinal classifier (0.521).

Last but not least, our proposal is fast to train (<10 mn for 3 ranges) and evaluate (<0.1 ms) on an Intelcore I7-8 cores-3.4 GHz and doesn't require a great amount of memory space (<1Go for 3 ranges).

5.5   Temporal Smoothing

As previously stated, the AVEC'2014 and AV+EC'2015 are based on audio recordings. As a result, the observations provided are temporally linked. Because the system we trained do not consider this characteristic, we have analyzed our results and the ground truth in a temporal way. The Fig. 4 presents the ground trough valences and the ones predicted by the ordinal classification system, according to the time on a sequence of the development base. It is outstanding that the system seems to only miss punctually. Indeed, it is exceptional that the system is majorly failing on a w time window wide enough. By rendering a temporal smoothing operation with a sliding window of size 5, we have been able to increase the performances of our system, as shown on Fig. 5.

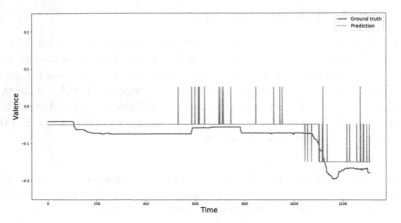

Fig. 4. Comparison between the ground truth and the ordinal classification on a sequence of the AVEC'2014 database.

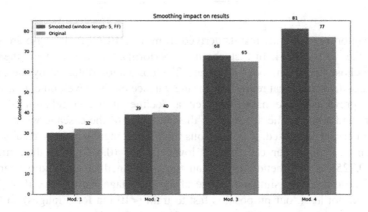

Fig. 5. Comparison of the results obtained before (right) and after running a temporal smoothing (left) on the output of RF-ordinal classifiers (resp. 25, 50, 150 and 300 trees).

6   Conclusions

We propose in this article an original approach for the regression of a continuous, bounded variable, based on a cascade of ordinal classifiers and local regressors. We chose to applicate it to the estimation of affective variables such as the valence. The first stage allows us to predict a trend, depending to the chosen interval. Thus, taking into account, for example, four intervals, the emotional state of a person will be qualified as very negative, negative, positive or very positive. We have been able to observe that this trend is more accurately estimated while the number of interval is increasing. The second stage enable a sharper prediction of the variable by regressing locally, on its interval and its direct neighbors. It seems even more efficient when the number of considered interval is low. Indeed, it allows to reduce the influence of the first stage on the prediction. Finally, we showed that the performances of this cascade can be compared favorably to the ones of the winner of the challenge AV+EC'2015.

Despite these satisfying results, there are still room to improve it (others than applying it to the prediction of the *arousal* and the – running – assessment of the performances on the challenges test data). The COCLR is a cascade which first stage is an ensemble of classifiers. The decision here is sanctioned by the least performant classifier. A more adapted combination rule would impact advantageously the global performances. The outputs (binary or probabilistic) of the ordinal classifier might also enrich the descriptors used by the local regressors.

To conclude, the research introduces a cascading architecture which obtains promising results on a challenging dataset. Several hypotheses have been issued concerning the impact of the different parameters involved, but none of them has been generalized yet. Testing this architecture on other datasets would help us to validate these hypotheses and justify the general interest of this proposal.

References

1. Basu, S., Chakraborty, J., Bag, A., Aftabuddin, M.: A review on emotion recognition using speech. In: International Conference on Inventive Communication and Computational Technologies (ICICCT), pp. 109–114 (2017)
2. Bitouk, D., Verma, R., Nenkova, A.: Class-level spectral features for emotion recognition. Speech Commun. **52**(7), 613–625 (2010)
3. Chang, J., Scherer, S.: Learning representations of emotional speech with deep convolutional generative adversarial networks. In: ICASSP, pp. 2746–2750 (2017)
4. Chawla, N.V., Bowyer, K.W., Hall, L.O., Kegelmeyer, P.W.: SMOTE: synthetic minority oversampling technique. J. Artif. Intell. Res. **16**, 321–357 (2002)
5. Drucker, H., Burges, C.J., Kaufman, L., Smola, A.J., Vapnik, V.: Support vector regression machines. In: Advances in Neural Information Processing Systems, pp. 155–161 (1997)
6. Ekman, P.: Basic emotions. In: Handbook of Cognition and Emotion, pp. 45–60. Wiley (1999)
7. Frank, E., Hall, M.: A simple approach to ordinal classification. In: De Raedt, L., Flach, P. (eds.) ECML 2001. LNCS (LNAI), vol. 2167, pp. 145–156. Springer, Heidelberg (2001). https://doi.org/10.1007/3-540-44795-4_13
8. Fontaine, J.R., Scherer, K.R., Roesch, E.B., Ellsworth, P.C.: The world of emotions is not two-dimensional. Psychol. Sci. **18**(12), 1050–1057 (2007)
9. Grandjean, D., Sander, D., Scherer, K.R.: Conscious emotional experience emerges as a function of multilevel, appraisal-driven response synchronization. Conscious. Cognit. **17**(2), 484–495 (2008)
10. Guo, G., Fu, Y., Huang, T.S., Dyer, C.R.: Locally adjusted robust regression for human age estimation. In: WACV (2008)
11. Han, J., Zhang, Z., Ringeval, F., Schuller, B.: Prediction-based learning for continuous emotion recognition in speech. In: ICASSP, pp. 5005–5009 (2017)
12. He, L., Jiang, D., Yang, L., Pei, E., Wu, P., Sahli, H.: Multimodal affective dimension prediction using deep bidirectional long short-term memory recurrent neural networks. In: AVEC, pp. 73–80 (2015)

13. Hinton, G.E., Osindero, S., Teh, Y.: A fast learning algorithm for deep belief nets. Neural Comput. **18**(7), 1527–1554 (2006)
14. Nicolaou, M.A., Gunes, H., Pantic, M.: Continuous prediction of spontaneous affect from multiple cues and modalities in valence-arousal space. IEEE Trans. Affect. Comput. **2**(2), 92–105 (2011)
15. Noroozi, F., Sapinski, T., Kaminska, D., Anbarjafari, G.: Vocal-based emotion recognition using random forests and decision tree. Int. J. Speech Technol. **20**(2), 239–246 (2017)
16. Qiao, X.: Noncrossing Ordinal Classification. arXiv:1505.03442 (2015)
17. Ringeval, F., et al.: AV+EC 2015: the first affect recognition challenge bridging across audio, video, and physiological data. In: AVEC, pp. 3–8 (2015)
18. Russell, J.: A circumplex model of affect. J. Pers. Soc. Psychol. **39**(6), 1161–1178 (1980)
19. Saranya, R., Yamini, C.: Survey on ensemble alpha tree for imbalance Classification problem
20. Sethu, V., Ambikairajah, E., Epps, J.: Empirical mode decomposition based weighted frequency feature for speech-based emotion classification. In: ICASSP, pp. 5017–5020 (2008)
21. Thukral, P., Mitra, K., Chellappa, R.: A hierarchical approach for human age estimation. In: ICASSP, pp. 1529 –1532 (2012)
22. Trigeorgis, G., et al.: Adieu features? End-to-end speech emotion recognition using a deep convolutional recurrent network. In: ICASSP, pp. 5200–5204 (2016)
23. Valstar, M.F., et al.: Avec 2014: 3D dimensional affect and depression recognition challenge. In: AVEC (2014)

Combining Classical and Deep Learning Methods for Twitter Sentiment Analysis

Mohammad Hanafy[(✉)], Mahmoud I. Khalil, and Hazem M. Abbas

Computers and Systems Engineering Department, Faculty of Engineering,
Ain-Shams University, Cairo 11772, Egypt
mohammad.hanafy@hotmail.com

Abstract. Millions of tweets are published every day which contain
massive amount of opinions and sentiments. Thus, twitter is used heavily in research and business areas. Twitter is a global platform that is
accessed from all the globe. Users express their opinions freely, using
informal language, without any rules and with different languages. We
propose a unified system that could be applied on any raw tweets and
could be applied without any man-made intervention. We use emoticons
as heuristic labels for our system and extract features statistically or
with unsupervised techniques. We combine classical and deep learning
algorithms with an ensemble algorithm to make use of different features
of each model and achieve better accuracy. The results show that our
approach is reliable and achieves accuracy near the state-of-the-art with
a smaller set of labeled tweets.

Keywords: Sentiment analysis · Twitter · Deep learning
NLP · Ensemble classifiers

1 Introduction

Sentiment analysis or opinion mining is one of the hottest fields in research area
nowadays, as knowing the sentiments and the opinions of people is crucial in
every industry. Millions of reviews are posted everyday. Manually tracking those
reviews is a very hard task so sentiment analysis plays an important role in
different applications such as reviewing movies, products, etc. [6]. It also could
be used to predict the price of stocks from users' reviews in social media [2], even
presidential campaigns could use sentiment analysis to measure the popularity
of candidates and predict the winners [23].

Extracting the opinions from twitter[1] is a challenging task, as users use
different languages. Users tend to express their opinions in informal language
and the used language is evolving so fast with new abbreviations and acronyms.
That created the need for a unified method that does not need any manual tasks
to deal with such massive amount of evolving non-structured data. However,

---

[1] https://twitter.com/.

© Springer Nature Switzerland AG 2018
L. Pancioni et al. (Eds.): ANNPR 2018, LNAI 11081, pp. 281–292, 2018.
https://doi.org/10.1007/978-3-319-99978-4_22

most of the research was done for a specific language, sometimes for a specific topic or context and depended on manually created corpus and lexicons which need hard manual work.

Fortunately, twitter has some interesting features that could help in building models that are language-independent and self-learning. One of those features is emoticons/emoji. Even though there is a difference between emoticons and emoji, in this paper, we use them interchangeably as both of them are used to express sentiments and they could be mapped to each other. Emoticons are popular and are commonly used on text messages, as they are able to express the same sentiments among different users, wherever they come from.

Another feature is the geo-location data that is associated with each tweet. This data could enable us to define a model from scratch with emoticons that is specified for a particular region assuming that people from a certain area use the same language.

In this paper, we build a semi-supervised model that uses unlabeled tweets that are gathered by location from USA, then, we use these data to classify English tweets as we know that USA speaks English. Then, we auto-label the tweets using emoticons to generate our training data for our models. Then, we extract features from labeled data by statistical and unsupervised approaches i.e., tf-idf and word2vec. Finally, we apply different classical and deep learning algorithms and combine them to make use of their features. Our model is a unified model, that could be applied on any raw tweets without any prior knowledge of its language or the need for any manual tasks.

2   The Previous Work

In last decades, sentiment analysis held the interest of many researchers. [20] provided a pioneering paper of how to apply machine learning methods such as naive Bayes (NB), Support Vector Machine (SVM) and Maximum Entropy (MaxEnt) on the text classification problem. They used Bag of words (BOW) with unigram, bigram and part of speech (POS) as features. Researchers subsequently attempted to improve the accuracy [1,11,24], but most of the work depended on a specific language, certain context or needed different manual tasks.

To overcome those limitations, researchers tried to develop semi-supervised methods i.e. [13] built a model that needed only 3 words of any language and some unlabeled data to auto-generate training labels then classify texts into positive or negative. [7] proved that, depending on emoticons as heuristic data to label raw tweets could be reliable. [17] used emoticons to auto label tweets and applied the approach on four different languages and on multilingual tweets. Although these papers achieved good accuracies, their work used classical methods and new ones were emerged with better accuracies.

Other researchers worked with large neural networks. [4] tried to build a model that can extract most of Natural language processing tasks such as POS, chunking (CHUNK) and Named Entity Recognition (NER) almost from scratch, by building a multi-layer neural network and with using large amount of unlabeled data. In 2013, [16] introduced a revolutionary paper which introduced a word embedding model called word2vec that could represent the semantic meaning of words from their context in an unsupervised way.

Many researchers used word2vec with deep learning algorithms to achieve better accuracy. [10] used a deep Convolutional Neural Network (CNN) by representing each sentence as a list of its word embedding values, then he applied different filters with the width of word2vec representation and with different heights. Other researchers combined different deep learning algorithm to achieve better performance [18], but deep learning methods are very time consuming, vague and require a large amount of data.

3   Theory and Algorithms

The proposed model is a unified model, that could be applied on any raw tweets without any limitations on the used language or how tweets were written. The model is composed of four independent classical and deep learning algorithms that are combined using a voting ensemble. All models are semi-supervised models that use emoticons as a heuristic data to generate the training data. Then, the features get extracted either statistically or with unsupervised techniques i.e. BOW or word2vec. Then, different classifiers are applied i.e. MaxEnt, SVM, Long Short-Term Memory (LSTM) and CNN with the proper extracted features.

This section gives an overview of the steps that we followed to construct each of our models and the techniques that were used in each step.

3.1   Data Processing

The first step in our model is processing the data. Data processing is a crucial task in each text classification problem, as any subsequent step depends on it [14]. In contrast to typical approaches that use language specific operations such as removing stop words and CHUNK, we use only those operations that are common between all languages and those that are related to twitter itself:

- Identifying emoticons and replacing them with their scores that will be used in the next step to auto-label tweets.
- Replacing hashtags with their separate words.
- Replacing twitter's reserved words such as RT for retweeting and @ for mentions, with place holders.
- Replacing URLs with placeholders.
- Reduction of words by allowing only duplicate characters as users tend to repeat characters to emphasize the meanings.

3.2   Auto-Labeling

Auto-Labeling is the second task in our model. In this step, we generate our training data from unlabeled ones. We depend on the sentiment carriers i.e.

Table 1. Samlpe of Emoji and the equivalent emoticons with their Ranks

Emoji	Emoticons	Neg	Neut	Pos	Rank
👍	(Y)	0.115	0.248	0.637	0.521
😃	:D	0.072	0.299	0.629	0.557
😎	8)	0.106	0.297	0.597	0.491
🙀	:(	0.544	0.281	0.175	-0.368

emoticons or emoji to label the raw data by scoring each tweet with certain score based on the scores of the sentiment carriers that it contains. This task is similar to the work done by [7,17], but we use the scores of emoji that were provided by [19], not just dividing them into positive and negative ones. Table 1 shows a sample of emoji and their equivalent emoticons with their scores. Each tweet is scored by the average score of all its sentiment carriers, i.e. if the tweet has 2 emoticons and their scores are .9 and .5, then the score of the tweet will be $\frac{.5 + .9}{2} = .7$. The tweet that has only one sentiment or its $|score| > .7$ will be considered to be positive or negative, otherwise it will be neglected. By applying our approach on CIKM dataset, we were able to generate 170 K labeled tweets that were used as our training data.

3.3 Feature Extraction

The next step is extracting the important features from the available data. As we build multiple models, we use different feature representation techniques. The following section shows an overview of the used feature representation techniques.

TF-idf. Tf-idf is a statistical approach to find the important words in a corpus. It depends on the frequency of a word in a document and its frequency in all the corpora. Tf-idf is calculated as follows:

$$tf\text{-}idf(t, d, D) = tf(t, d) \times idf(t)$$
$$idf(t) = \log \frac{n_d}{1 + df(d, t)}$$

where, t is the word, d is the tweet, tf is the term frequency in the document, n_d is the number of tweets and df is document frequency where word exists.

BOW. Bag-of-words is one of the most used feature representation techniques [14] in sentiment analysis. The basic idea is to select a set of important words, then each document is represented as a vector of the number of the occurrences of the selected words. BOW does not consider the order of words thus it is often used with n-gram model. Selecting the important words could be done using a language-related features i.e. POS and semantic words or using a statistical approach i.e. Tf-idf as in the proposed model.

Word2Vec. Word2Vec is a word embedding technique that was developed by [16]. It depends on that the meaning of words defined by their company and context [8]. Word2Vec is implemented by training a 2-layer neural network to represent each word as a vector of certain length based on its context. The resulted vectors have some unique features and can solve the analogy problem by performing an arithmetic operation such as $king - man = queen - woman$.

3.4   Classifiers

The final step is using the classifiers. In our model, we use different classifiers, then we combine them using ensemble classifier to make use of their unique features. In this subsection, we demonstrate the different classifiers that we used.

SVM. Support Vector Machine has shown reliable results in the sentiment classification problem. SVM is one of the first methods applied in this field and is still being used. We found that using this model performs very well with BOW and word2vec summation.

MaxEnt. Maximum Entropy is one of the most used models in a wide range of applications. MaxEnt. follows this equation to find the probability of the output, given a certain input:

$$P_{ME}(c|d, \lambda) = \frac{\exp\left[\sum_i \lambda_i f_i(c, d)\right]}{\sum_{lc}\left[\exp \sum_i \lambda_i f_i(c, d)\right]}$$

In our model, c is a positive or negative, d is the tweet and λ is learned parameters of the model. We found that using this model performs very well with BOW and word2vec summation.

LSTM. Recurrent Neural Network was introduced in early 80s, it was designed to capture patterns in sequential data. RNN typically consists of multiple connected units of the same-structured single-layer neural network. LSTM [9] is an enhancement over RNN to overcome its selectivity and vanishing/exploding gradient problems. LSTM extends the architecture of RNN by replacing the multiple connected single-layer units with a more complex architecture consists of a memory cell, and three gates that control the flow of the state.

CNN. Convolutional Neural Network has attracted many researchers in recent years due to its out-performance, especially in image recognition and text classification [15]. The idea of CNN is to construct a convolutional layer from a set of filters that traverse through the input layer, then, this convolutional layer is followed by a pool layer and a deep neural network. CNN has three variations based on how the model updates the word-embedding values during the training stage: CNN-static, CNN-non-static and CNN-rand. In recent years, CNN has applied in text classification and achieved remarkable results.

Voting Ensembles. Ensemble classifiers have proved their reliability in research and applications [21]. Voting ensembles combine the different classifiers by considering the output of each classifier as a vote. Then, the ensemble takes its final decision based on those votes. Voting ensembles have different implementations i.e., majority voting and Weighted voting ensembles. Majority voting ensemble makes its final decision as the majority of the output of the base classifiers. Weighted voting ensemble is as enhancement over the Majority voting, instead of treating the different classifiers equally, each classifier has its weight/power when it votes. Voting ensemble classifiers often used when the base classifiers are different in architecture and nature. In our model, we combine the output of our base classifiers using majority and weighted voting classifiers that showed a remarkable improvement in the final accuracy.

4   The Proposed Models

The previous section showed all the general steps and techniques to build the models. This section shows how each of the models was built in details, the motivation behind each model and finally, how and why combining the different models generates better results.

4.1   Model 1: BOW with tf-idf

Model. In this model, we use tf-idf to construct our bag-of-words then we apply different classical models i.e. MaxEnt. and SVM. There were different parameters that affected our model such as the loss function, regularization parameter of the classifier, size of the BOW and n-gram size.

Motivation. Bag-of-words is one of the oldest techniques, but it still proves its reliability. BOW depends heavily on the selected n-gram features so it has some limitations such as it does not consider the order of the words and it is limited to the short context which is the size of n-gram. On the other hand, BOW can thrive when tweets are short and have direct meaning or when they contain powerful words such as "won" and "good".

4.2   Model 2: Aggregation of Word2Vec

Model. This model is similar to work done by [12]. It depends on the fact that each word is represented as a vector, and applying arithmetic operations on those vectors i.e. "mean" and "summation" will give the semantic meaning of the whole sentence. We then apply different classical classifiers such as MaxEnt. and SVM on the output of the arithmetic operations.

Motivation. Word2vec represents words as vectors that represent the semantic meaning of the words. The original paper [16] showed that applying arithmetic operations can solve the word-analogy problems such as $king - man + woman$ yields to $queen$. The original paper also showed that, words that have similar meaning are close to each other. So applying arithmetic operations is effective. On the other hand, aggregation of word2vec has its drawback as it does not consider the order of sentence which can affect its ability of the classification.

4.3   Model 3: LSTM

Model. In this model, we represent each document as a list of word-embedding values of its words, then we feed the document representation to the sequential units of LSTM model. Finally, we tune the LSTM model.

Motivation. LSTM is very powerful with sequential data. It can solve the problem of the order of the sentences and the long context due to its sequential architecture and memory units. On the other hand, LSTM is a deep learning algorithm so it needs massive amount of training data and needs much time to train the model.

4.4   Model 4: CNN

Model. We build a model similar to [10], each document is represented as a list of word-embedding values of its words, then we apply a CNN with different filters that has the same width as word-embedding and different heights. In this model, we use word2vec as our word-embedding representation.

Motivation. CNN have proved their validity when using different datasets for training and testing [22]. CNN provides the ability to define the window size of the filters which represent the context in our model. The drawbacks of CNN are same as LSTM as they both are deep learning algorithms.

4.5   Model 5: The Proposed Model

Motivation. We have built four different models. Each model has its own advantages and unique features that could solve the problem of the others. Figure 1 shows the confusion matrix of the number of common correctly classified tweets by the different models against the test dataset. The dark and light colors indicate the relation between the different models. The figure shows that each model can overcome the others on some tweets and all models are independent of each other. Beside the direct output of our models, there are some other features that could give more insights. Figure 2 shows the relation between the classification probability of different models and the actual accuracy at this probability. The figure shows that for all the models, the probability of the classification has a positive linear relationship with the accuracy so it could be considered as a confidence of prediction. From the previous notes, building a classifier that can combine the different models and make use of the different features could improve the final accuracy of the models.

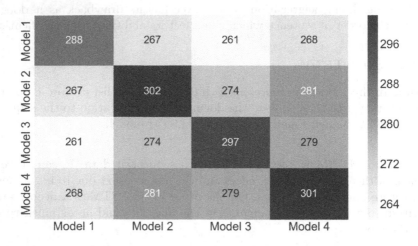

Fig. 1. Number of common correctly classified tweets between models

Model. To make use of the different models, a weighted voting ensemble classifier is used. The classifier combines the output of the different models and their classification probabilities. Then, each model is assigned a different weight when voting.

5   Experiments

This section describes the different datasets that were used, how the experiments were built and finally, the final results that achieved by the different models.

5.1   The Datasets

CIKM Datasets. This dataset was provided by [3]. It consists of 8M raw tweets and associated geo-location data. The tweets were collected by location from different locations in US with no restriction on the used language, thus it contains some tweets in French, Spanish, etc. We used this dataset to generate our training data by applying our auto-labeling technique.

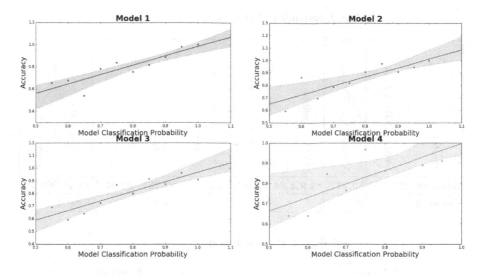

Fig. 2. Classification probability vs accuracy

STS-Test Datasets. STS-test set is one of the most popular datasets that is used in sentiment analysis. It was provided by [7] and it contains 498 manually annotated tweets. The dataset is divided as 177 negative, 139 neutral and 182 positive tweets. STS-test is a part bigger corpus containing additional 1.5M auto-labeled tweets gathered in 2009. The corpus was limited to English language and some categories e.g. products, companies, events, etc.

5.2   The Experimental Protocol

We use CIKM dataset to generate our training data by using auto-labeling technique that is described in Sect. 3.2 and we use STS-test set to test our models. During training and testing cycles, we deal with each tweet as an independent document that express even positive or negative sentiment [14].

In feature extraction step, we found that BOW achieved better results using 8K window size and word2vec with embedding size of 200 was enough, thus all our experiments used these values.

During classification step, SVM and MaxEnt were tuned for the best regularization value. CNN was tuned for different hidden dimension, dropout probability and number of filters. LSTM was tuned for different hidden dimension and dropout probability. Weighted voting ensembles were tuned for different weights of each model in range from 1:5.

Table 2. Summary of the accuracy of all models

Model	SVM + BOW 1-2 gram	MaxEnt + BOW 1-2 gram	SVM + mean(wor2vec)	MaxEnt + mean(wor2vec)	SVM + sum(wor2vec)	MaxEnt + sum(wor2vec)	LSTM	CNN-static	CNN-non-static	CNN-rand
Accuracy	79.66%	80.22%	84.9%	84.4%	84.4%	83.0%	82.73%	83.87%	81.2%	80.2%
Models	Model 1		Model 2				Model 3	Model 4		

Table 3. Proposed model accuracy

Model	Accuracy
BOW with TF-IDF(MaxEnt)	80.22%
Aggregation of Word2Vec (MaxEnt + Avg)	84.4%
Aggregation of Word2Vec (SVM + Avg)	84.9%
LSTM	82.73%
CNN-static	83.87%
Majority voting classifier	85.00%
Weighted voting classifier	85.79%
Weighted voting classifier(probabilities)	86.07%

5.3 The Experimental Results

Table 2 shows the summary of the results achieved by each of the four models. Table 3 compares the results achieved by each of the four models and the results achieved by our proposed model. We can notice that the last model that uses

Table 4. Comparing the proposed model with the previous models

Model	Accuracy
MaxEnt [7]	82.7%
SVM [7]	82.2%
SCNN [5]	85.2%
CharSCNN [5]	86.4%
(DeepCNN + Bi-LSTM) + SR + Glove [18]	86.6%
(DeepCNN + Bi-LSTM) + SR + Word2Vec [18]	86.35%
Our proposed model	86.07%

the probability has outperformed all the models. The results show that we have improved the accuracy of the individual models by more than 1% using a simple voting ensemble. Table 4 shows the previous work done on the same STS-test set. By comparing our results with previous work, we were able to achieve almost the state-of-the-art accuracy with only 170 K labeled tweets, i.e. only 10% of the others' training dataset, which saves much time and resources.

6   Conclusions and Future Work

We built a unified model that could be applied without any manual tasks and does not require any information about the used language. We used emoticons as a heuristic data to auto generate our training data. Then, we built multiple classical and deep learning algorithms then combined them to achieve better accuracy. We achieved accuracy near the-state-of-the-art results by using only 170 K of training data i.e. using only 10% of the others' models that used 1.5M training tweets.

Combining classical and deep learning algorithms improved the accuracy of both. Deep learning algorithms can infer the long and complex sentences correctly and thrive when we have a big amount of data. But training deep learning algorithms consumes time and resources. On the other hand, classical algorithms are simple and could be trained easily and fast. It also can overcome deep algorithms in some situations. Our model combines different models in a parallel manner as it uses voting ensembles. Thus, each model could be independent of the others and we can add more models easily to achieve better accuracy. In the future, we will work on improving the efficiency of ensemble classifier by adding more base models and by finding more parameters that could improve the accuracy of the overall accuracy.

References

1. Agarwal, A., Xie, B., Vovsha, I., Rambow, O., Passonneau, R.: Sentiment analysis of Twitter data. In: Proceedings of the workshop on languages in social media, pp. 30–38. Association for Computational Linguistics (2011)
2. Bollen, J., Mao, H., Zeng, X.: Twitter mood predicts the stock market. J. Comput. Sci. **2**(1), 1–8 (2011)
3. Cheng, Z., Caverlee, J., Lee, K.: You are where you tweet: a content-based approach to geo-locating Twitter users. In: Proceedings of the 19th ACM International Conference on Information and Knowledge Management, pp. 759–768. ACM (2010)
4. Collobert, R., Weston, J., Bottou, L., Karlen, M., Kavukcuoglu, K., Kuksa, P.: Natural language processing (almost) from scratch. J. Mach. Learn. Res. **12**(Aug), 2493–2537 (2011)
5. Dos Santos, C.N., Gatti, M.: Deep convolutional neural networks for sentiment analysis of short texts. In: The 25th International Conference on Computational Linguistics, COLING 2014, pp. 69–78 (2014)
6. Feldman, R.: Techniques and applications for sentiment analysis. Commun. ACM **56**(4), 82–89 (2013)

7. Go, A., Bhayani, R., Huang, L.: Twitter sentiment classification using distant supervision. CS224N Project Report, Stanford, vol. 1, no. 12 (2009)
8. Harris, Z.S.: Distributional structure. Word 10(2–3), 146–162 (1954)
9. Hochreiter, S., Schmidhuber, J.: Long short-term memory. Neural Comput. 9(8), 1735–1780 (1997)
10. Kim, Y.: Convolutional neural networks for sentence classification. In: Proceedings of the 2014 Conference on Empirical Methods for Natural Language Processing, pp. 1746–1751 (2014)
11. Kouloumpis, E., Wilson, T., Moore, J.D.: Twitter sentiment analysis: the good the bad and the omg!. Icwsm 11(538–541), 164 (2011)
12. Lilleberg, J., Zhu, Y., Zhang, Y.: Support vector machines and word2vec for text classification with semantic features. In: 2015 IEEE 14th International Conference on Cognitive Informatics & Cognitive Computing (ICCI* CC), pp. 136–140. IEEE (2015)
13. Lin, Z., Tan, S., Cheng, X.: Language-independent sentiment classification using three common words. In: Proceedings of the 20th ACM International Conference on Information and Knowledge Management, pp. 1041–1046. ACM (2011)
14. Liu, B., Zhang, L.: A survey of opinion mining and sentiment analysis. In: Aggarwal, C., Zhai, C. (eds.) Mining text data, pp. 415–463. Springer, Boston (2012). https://doi.org/10.1007/978-1-4614-3223-4_13
15. Liu, W., Wang, Z., Liu, X., Zeng, N., Liu, Y., Alsaadi, F.E.: A survey of deep neural network architectures and their applications. Neurocomputing 234, 11–26 (2017)
16. Mikolov, T., Sutskever, I., Chen, K., Corrado, G.S., Dean, J.: Distributed representations of words and phrases and their compositionality. In: Advances in Neural Information Processing Systems, pp. 3111–3119 (2013)
17. Narr, S., Hulfenhaus, M., Albayrak, S.: Language-independent Twitter sentiment analysis. In: Knowledge Discovery and Machine Learning (KDML), LWA, pp. 12–14 (2012)
18. Nguyen, H., Nguyen, M.L.: A deep neural architecture for sentence-level sentiment classification in Twitter social networking. arXiv preprint arXiv:1706.08032 (2017)
19. Novak, P.K., Smailović, J., Sluban, B., Mozetič, I.: Sentiment of emojis. PloS one 10(12), e0144296 (2015)
20. Pang, B., Lee, L., Vaithyanathan, S.: Thumbs up?: sentiment classification using machine learning techniques. In: Proceedings of the ACL-02 Conference on Empirical Methods in Natural Language Processing-Volume 10, pp. 79–86. Association for Computational Linguistics (2002)
21. Polikar, R.: Ensemble based systems in decision making. IEEE Circuits Syst. Mag. 6(3), 21–45 (2006)
22. Razavian, A.S., Azizpour, H., Sullivan, J., Carlsson, S.: CNN features off-the-shelf: an astounding baseline for recognition. In: 2014 IEEE Conference on Computer Vision and Pattern Recognition Workshops (CVPRW), pp. 512–519. IEEE (2014)
23. Tumasjan, A., Sprenger, T.O., Sandner, P.G., Welpe, I.M.: Predicting elections with Twitter: what 140 characters reveal about political sentiment. Icwsm 10(1), 178–185 (2010)
24. Wilson, T., Wiebe, J., Hoffmann, P.: Recognizing contextual polarity in phrase-level sentiment analysis. In: Proceedings of the Conference on Human Language Technology and Empirical Methods in Natural Language Processing, pp. 347–354. Association for Computational Linguistics (2005)

PHoG Features and Kullback-Leibler Divergence Based Ranking Method for Handwriting Recognition

Taraggy M. Ghanim[1,2]($\boxtimes$), Mahmoud I. Khalil[2], and Hazem M. Abbas[2]

[1] Faculty of Computer Science, Misr International University, Cairo, Egypt
taraggy.ghanim@miuegypt.edu.eg
[2] Faculty of Engineering, Ain Shams University, Cairo, Egypt
{mahmoud.khalil,hazem.abbas}@eng.asu.edu.eg

Abstract. Handwriting recognition is a research topic with a lot of challenges and complications. One of the main complications is big databases that affect classifier complexities and their ability to perform correctly. This paper introduces a new ranking approach that is proposed as a solution to this point of research. Per input test image, the approach sorts database classes from the nearest to the furthest based on the calculated ranks. Accordingly, the classification process is applied on only subset of best nearest neighbor classes rather than the whole database classes. The approach starts with computing simple regional-type features to group similar competitive database classes together using decision trees. This grouping process aims to split big database to multiple smaller ones. Decision trees match between test image and one of the constructed smaller databases. Finally, Kullback-Leibler divergence is measured between the pyramid histogram of gradients (PHoGs) features extracted from the test image and the members of the matched smaller database. This measurement sorts the matching classes to select smaller subset from them. This subset represents best nearest neighbors of test image that can be used for final classification. Reducing database size and focusing classification on subset of best nearest neighbor classes reduce the classifier complexity and increase the overall system classification accuracy. The proposed approach was applied on IFN-ENIT database, and its effect was tested on the SVM classifier.

Keywords: Handwriting recognition
Pyramid histogram of gradients · Kullback-Leibler divergence

1 Introduction

Handwriting recognition is one of the computer vision fields with many challenging aspects. It can be performed by two ways; online and offline. Online recognition process uses sequential writing order [4] and instant temporal information. Offline recognition process bases on visual features and pixel information

© Springer Nature Switzerland AG 2018
L. Pancioni et al. (Eds.): ANNPR 2018, LNAI 11081, pp. 293–305, 2018.
https://doi.org/10.1007/978-3-319-99978-4_23

only. Off-line recognition is more challenging due to lack of temporal information and absence of instant cursing.

Our work proposes a new ranking process that can be used as a guide to automatic offline handwriting classifiers. Additional to the mentioned challenges that affects this type of classifiers, being applied on big database increases difficulty to classification process, and affects overall accuracy. Accordingly, classification seems as a challenging competition that is responsible for matching test image with the correct class from database. Considering all the database classes to be competitive in such a challenging process is the main complication that affects classifier performance. The target of the proposed approach is to facilitate this challenging competition by selecting subsets of database classes; smaller in size than the whole database, to apply classifier on. This reduction in size of database classes affects classification performance and complexity positively. Now each test image with a selected subset of database classes are participant in the classification process.

The proposed approach starts by extracting five regional features [5] from input test image and database classes. Based on the extracted features, decision trees are applied to match between input test image and a set of matching database classes. These matching classes are grouped in one set. A new proposed ranking algorithm is then applied based on pyramid histogram of gradients (PHoGs) [29] and Kullback-Leibler divergence measure [17] to sort these matching classes relative to each input test image. Sorting classes from the nearest to the furthest enables the selection of smaller subset. These subsets represents best nearest neighbor classes of test image. Our objective is to use these subsets for final classification stage instead the whole database to vote for the correct test image class. By the end of the testing stage, the approach splits the big database to smaller sets of similar classes to construct a data-mined database version.

We tested our system on the Arabic handwriting IFN/ENIT database [26]. Arabic is one of the major languages [4] with a lot of challenges as mentioned briefly in [1]. Next sections are organized to summarize related work, proposed approach, experiments and finally conclusion respectively.

2   Related Work

Different recognition systems were proposed in the field of Arabic handwriting recognition. This section summarizes similar offline handwritten recognition systems. The pyramid histogram of gradients (PHoG) were extracted as features by Saïdani and Echi [29] to recognize Arabic/Latin and Machine-printed Handwritten Databases. A survey was provided by Lawgali [18]. Different classifiers were applied in this field.

Some systems used sliding window to compute different types of features. Hicham et al. [11] proposed a solution without segmenting words into PAWs and achieved 80.26% success rate on Arabic-Numbers database and 78.95% accuracy on IFN/ENIT database [26] using local densities and statistical features. AlKhateeb et al. [3] computed intensity and structure-like features. Re-ranking method was applied to improve the accuracy to 89.24% using Gaussian-like function. Jayech et al. [12] extracted statistical and structural features and applied synchronous Multi-Stream Hidden Markov Model (MSHMM) and achieved 91.1% recognition rate on IFN/ENIT database. MSHMM was also applied by Mezghani et al. [22], Maqqor et al. [21], Pechwitz and Maergner [27] and Kessentini et al. [14]. Mezghani et al. [22] automatically recognizes handwritten and printed Arabic and French words using Gaussian Mixture Models (GMMs). Kessentini et al. [14] worked offline using another multi-stream approach by combining density and contour based features. The system succeeded by 79.8% on the IFN/ENIT database. A confidence- and margin-based discriminative approach by Dreuw et al. [6] was applied using maximum mutual information (MMI) criterion and minimum phone error (MPE) criteria [9,10,25]. The MMI outperformed ML by 33% enhancement in word error rate on IFN/ENIT database.

Some other approaches recorded effective results using SVM like Al-Dmour and Abuhelaleh [2]. The system was applied on a subset of IFN/ENIT database with 85% recognition rate using SURF features. Khaissidi et al. [15] proposed an unsupervised segmentation-free method using Histograms of Oriented Gradients (HOGs) as features. The system achieved average precision 68.4% on Ibn Sina data-set [23]. The SVM classifier and HOG features was also applied using Sobel edge detectors by Elfakir et al. [7].

Mozaffari and Bahar [24] discriminate between Farsi/Arabic handwritten from machine-printed words using histogram features. The system achieved success rate 97% on a database of about 3000 words, printed/handwritten. Leila et al. [19] computed invariant pseudo-Zernike moments features and applied Fuzzy ARTMAP neural network. The system achieved 93.8% success rate on a database that includes 96 different classes written by 100 different writers.

3   The Proposed System

The paper proposes a new ranking algorithm that can precede final classification stages in pattern recognition process. The proposed system composes of two separate stages. The first stage applies decision trees as shown in Fig. 1 to match test image with a set of database classes. In the second stage, pyramid histogram of gradients (PHoG) [29] features are computed followed by Kullback-Leibler (KL) divergence measure [17] to introduce a new ranking algorithm. Ranking process aims to sort the classes of the chosen matching set. Sorting process helps to reduce computation complexity by shrinking the database part concerned in the final classification stage. This concerned part is a subset of best nearest neighbor classes chosen from the matching set after sorting.

3.1   Preprocessing

Input images during this stage is prepared for later valuable calculations. Our images are binary images of white background and black handwriting Arabic words. Negation of images [8] was necessary to concentrate computations on minor white pixels. Normalization of words' size is essential to unify computation environment of all database images.

3.2 Decision Trees Construction

This stage builds a tree-like model of decisions. It constructs a separate decision tree per test image. It is a commonly known data mining process [28] to derive a strategy to determine a particular set of matching classes for each test image. Its performance is evaluated by degree of matching test image with the correct set of classes. Matching process is considered to be correct when the decided set includes the correct class of the input test image. Finally, this is measured by true positive rate which is known as sensitivity [20]. In our proposed solution, the searching strategy bases on five computed regional values proposed by [5]. These features are number of piece of Arabic words PAWs, number of holes, extent, eccentricity and word orientation [5]. A sample of computed number of holes and PAWs are shown in Fig. 2.

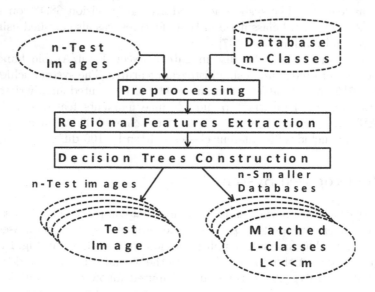

Fig. 1. The decision trees construction stage.

(a) Image with 4 holes   (b) Image with 4 PAWs

Fig. 2. A sample image after computing number of holes and PAWs.

Eccentricity [13] measures the aspect ratio as shown in Fig. 3. Extent [5] calculates the ratio between the area of the object and the area of its bounding box. Finally, the calculated word orientation measures the angle between the x-axis and the major axis of the word bounding ellipse.

$$ecc = \frac{2c}{b}, c^2 = a^2 - b^2 \tag{1}$$

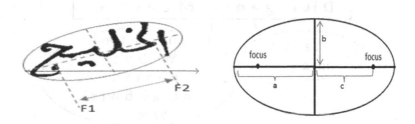

Fig. 3. Eccentricity, F1 and F2 are the two foci of the ellipse.

Equation (1) defines eccentricity *ecc* in terms of distances *a*, *b* and *c* shown in Fig. 3. The ranges of the five computed feature values are used to find the matching set of classes per test image using decision trees. This process excludes a large amount of training samples; which are not included in the matching set, where their inclusion in the final classification increases the recognition complexity and affects the accuracy negatively.

After applying this stage on the whole database, our database is data-mined. The database is split into smaller sets of similar classes characterized by their computed common features.

3.3 The Ranking Stage

The overview of the ranking stage is shown in Fig. 4. This stage begins by applying the PHoGs features followed by applying Kullback-Leibler Divergence Measure to sort the members of the matching set relative to the input image. The output of this stage is a subset of the best nearest neighbors that can be used for the final classification as shown in Fig. 4.

3.3.1 Pyramid Histogram of Gradients

Histogram of oriented gradients (HOG) [29,30] is a type of feature descriptor that represents edge information. It bases on counting gradient orientation occurrences in images' region of interest (ROI). Pyramids histogram of gradients (PHoG) is more powerful type of gradient features relative to the ordinary HOG features. PHoG feature extraction technique relies on computing histogram of gradients in localized portions of the image's ROI. It is an extension that captures fine and more discriminative details than the ordinary HoGs. Each level

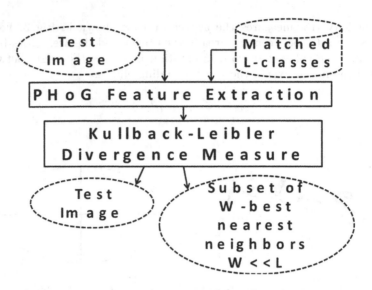

Fig. 4. The ranking stage.

is constructed by contracting edges in the level below. First level description is related to the original input data, representing the original (HoG). Higher level represents partitioning of the image into connected sub-graphs, which are subsets of pixels in each image. In every partition, features are computed independently relative to others on the same level. The union of sequential feature levels propagates information up and down laterally in the form of a pyramid as shown in Fig. 5.

Considering L number of levels and angle range from 0–360 quantized into discrete eight values, the following algorithm introduced by Saidani et al. [30] is applied,

1. Divide the image into smaller cells (connected components).
2. For each cell, compute histogram of oriented gradients.
3. Discretize into pre-defined angular bins.
4. Normalization of 8-bin histogram of adjacent cells.

In the experiments, we applied the PHoGs at different levels L as shown in Fig. 5a, b and c to study the effect of varying levels on the final classifier performance.

3.3.2   Kullback-Leibler Divergence Measure

Kullback-Leibler divergence [17] is a way to measure the disparity between any two probability distribution functions. It is an information-based measure of divergence between any given distributions $p(x)$ and $q(x)$ defined over the variable x. In our proposed approach, we considered the PHoGs to be distribution functions over gradients values.

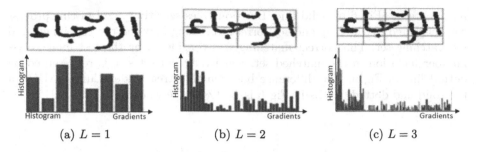

(a) $L = 1$ (b) $L = 2$ (c) $L = 3$

Fig. 5. Pyramid histogram of gradients at different levels.

$$D_{KL}(p(x) \parallel q(x)) = \sum_{x \in X} p(x) \ln \left(\frac{p(x)}{q(x)}\right) \tag{2}$$

The measure of divergence D_{KL} between the PHoG feature vector of each test image and the PHoGs of the training samples is defined by Eq. (2) considering $p(x)$ is the feature vector of test sample while $q(x)$ is the feature vector per training sample. The gradients are represented by the variable x. The smallest divergence measures vote for the most probable membership classes corresponding to the input test image. This measurement can be considered as the difference between the cross-entropy for $q(x)$ on $p(x)$, and the self-entropy [31] of $p(x)$. Cross entropy is always a bigger value than self-entropy except if $p(x)$ and $q(x)$ are identical. In this case, the cross entropy equals the self-entropy which causes a difference to equal zero which represents the maximum divergence.

The classes of the same group are ranked from the nearest to the furthest according to the computed divergence measure. This ranking process aims to exclude another part from the matching set corresponding to each test image, by choosing only subset of best nearest neighbors to pass finally for later classification.

4   Experiments

The proposed approach is applied on IFN/ENIT database [26], a database of Arabic handwriting words written by different 411 writers. The database is divided into four sets a, b, c and d which is the standard division stated in the database official website [26]. We considered sets a, b, c as training sets, then the approach was tested on set d as done previously by [3,12] and [6]. Set d contains 859 different classes. Each class represents one city name. The number of samples varies from one class to another.

4.1   Decision Trees Sensitivity

Decisions trees were applied firstly to match test image with a set of matching classes. The matching sets' sizes differ per sample. Sets' sizes frequencies is represented in Fig. 6 by solid line. Decision trees sensitivity [20] of this process was measured by the frequency of correct matching between the test image and its matching set. This correct matching was considered by the test image class number inclusion in the matched set. Sensitivity per set size is represented by dotted line in Fig. 6. The difference between the Area Under Curve (AUC) of the solid and dotted graphs in Fig. 6 is 13.4% of the whole AUC.

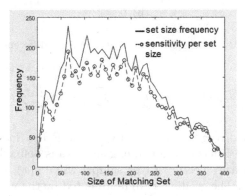

Fig. 6. Matching process sensitivity.

The IFN/ENIT database [26] has different ratios between the number of training to testing samples, which affects sensitivity of the matching. The probability to match correct set versus ratio of training to testing samples is shown in Fig. 7. The larger the ratio is, the better matching results occur which affect the system performance positively.

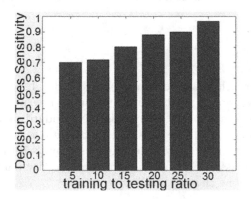

Fig. 7. Effect of number of training samples on sensitivity.

4.2   The Ranking Process

Figure 8 is a relation between ranks of correct classes inside their matching set after sorting and the possible success rates that can be achieved. The rank of the correct class comes to early positions after sorting. Saturation in graph means that the expected cumulative success rate approximates 100%. This ranking result allows later classification stage to be applied on a smaller subset of the database. Shrinking the length of classes in matching set reduces the computation complexity passing only a subset of best nearest neighbors to the last classification stage.

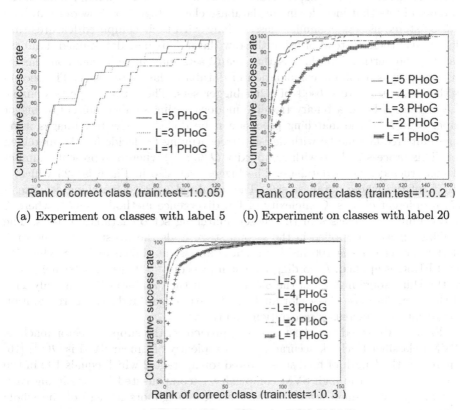

(a) Experiment on classes with label 5   (b) Experiment on classes with label 20

(c) Experiment on classes with label 30

Fig. 8. Effect of ranking on the expected success rate.

Expected success rate is measured from rate of desired class ranks at different PHoG levels after applying the ranking approach. Labels 30, 20, 5 in Fig. 8, parts a, b, c represent the ratio of training to testing samples. Increasing the PHoG level affects the ranking process positively, i.e. the rank of the correct class comes to earlier positions and graph saturates faster. Studying the relation between the expected success rate and correct classes' ranks indicates the minimum subset length that should be passed to the final classification stages. We applied different levels of PHoG and early saturation is achieved by higher levels of PHoGs. Higher ratio between number of training to testing samples outperforms smaller ratios. This was shown in Fig. 8c.

4.3 Complexity Analysis

The complexity of decision trees construction is $O(h)$ where h is the height of the tree; equals 5 in the proposed system according to the 5 computed features in this stage. During testing, the stage of decision tree construction results in the number of sets that include similar database classes together. Now our database can be stored as sets of similar classes along with their computed features that distinguish them instead of using the raw data. This new data-mined database has two properties. The first is that many sets may include common classes, but each set is a unique combination of database classes together. The second is that some sets are subsets of other bigger sets. These two properties enable the system to build a binary tree-like model relating smaller subsets to bigger super sets. Now the matching process can be done for new test images using our built tree-like model with decision tree pruning to decide for the matching set. This process is done with complexity $O(log(v))$ where v represents number of constructed sets in database. The biggest set size in Fig. 6 is 423 which is 49.2% of the whole database size. The weighted average matching set size is 172.6 different classes. Complexity of KL divergence method is $O(L)$, where L is the number of classes in the chosen matching set of database classes. The ranking algorithm applied in the second stage finds the worst case is the rank of the correct class is 100 inside its matching set as shown in Fig. 8 when 5th level PHoG is applied. Accordingly, we can pass only 100 best nearest neighbors to the third stage instead of the 859 classes of the database. Passing only 11% of the total database classes for final classification again reduces the recognition complexity and increases recognition accuracy.

Finally, we tested the effect of our approach on the support vector machine (SVM) classifier to check accuracy. The complexity of Linear SVM is $O(z^3)$ [16] where z is the length of final subset voted for in stage 2 which equals 100 in the worst case. Now the high SVM complexity is compensated by classifying each test image using small set of the best nearest neighbors instead of the whole data-set.

4.4 Comparison with Similar Systems

Table 1 shows the comparison of our approach with similar systems. Similar systems proposed different Holistic approaches, using the whole word for training and testing without segmentation to PAWs, the same as our approach. The system is trained by sets a, b and c and tested on set d and achieves a recognition rate of an average 96.4% for level 5-PHoG, 92.3% for level 3-PHoG 80% for level 1-PHoG using the Linear SVM without considering the false set matching. Considering the false set matching cause word error rate (WER) to be 16.6%

Table 1. A comparison with similar systems on IFN-ENIT database.

System	[IFN-ENIT] training-testing	Algorithm	WER
Dreuw 2011 [6]	abc-d	GMM-HMM	8.9%
Alkhateeb 2011 [3]	abc-d	Normalized intensity	10.76%
Jayech 2015 [12]	abc-d	MSHMM	16.9%
Al-Dmour 2016 [2]	18 classes only	Surf-SVM	15%
Proposed solution	abc-d	PHoG-KL-SVM	3.6%

when the system is applied on 859 classes of set d, which is relatively better than [2] who achieve 15% WER on 18 classes only of the same database.

5 Conclusions

Automatic Handwriting Recognition systems suffer from different complexities and variations concerning different characters' shapes and handwriting styles. A ranking algorithm is proposed as a solution to some complications. Ranking is done by measuring KL-divergence on PHoGs features. It was concluded that higher level L of PHoG and bigger number of available training samples relative to testing samples affects the applied ranking method effectively.

References

1. AbdelRaouf, A., Higgins, C.A., Pridmore, T., Khalil, M.: Building a multi-modal arabic corpus (MMAC). Int. J. Doc. Anal. Recognit. (IJDAR) **13**(4), 285–302 (2010)
2. Al-Dmour, A., Abuhelaleh, M.: Arabic handwritten word category classification using bag of features. J. Theor. Appl. Inf. Technol. **89**(2), 320 (2016)
3. AlKhateeb, J.H., Ren, J., Jiang, J., Al-Muhtaseb, H.: Offline handwritten Arabic cursive text recognition using hidden markov models and re-ranking. Pattern Recognit. Lett. **32**(8), 1081–1088 (2011)
4. Amin, A.: Off-line Arabic character recognition: the state of the art. Pattern Recognit. **31**(5), 517–530 (1998)
5. Dileep, D.: A feature extraction technique based on character geometry for character recognition. Department of Electronics and Communication Engineering, Amrita School of Engineering, Kollam, India (2012)
6. Dreuw, P., Heigold, G., Ney, H.: Confidence-and margin-based MMI/MPE discriminative training for off-line handwriting recognition. Int. J. Doc. Anal. Recognit. **14**(3), 273–288 (2011)
7. Elfakir, Y., Khaissidi, G., Mrabti, M., Chenouni, D.: Handwritten Arabic documents indexation using HOG feature. Int. J. Comput. Appl. **126**(9), 14–18 (2015)
8. Gonzalez, R.C., Woods, R.E.: Digital Image Processing. Pearson (2007)
9. Heigold, G.: A log-linear discriminative modeling framework for speech recognition. Ph.D. thesis, RWTH Aachen University (2010)

10. Heigold, G., Dreuw, P., Hahn, S., Schluter, R., Ney, H.: Margin-based discriminative training for string recognition. IEEE J. Sel. Top. Sig. Process. **4**(6), 917–925 (2010)
11. Hicham, E.M., Akram, H., Khalid, S.: Using features of local densities, statistics and HMM toolkit (HTK) for offline Arabic handwriting text recognition. J. Electr. Syst. Inf. Technol. **3**(3), 99–110 (2016)
12. Jayech, K., Mahjoub, M.A., Amara, N.E.B.: Arabic handwriting recognition based on synchronous multi-stream HMM without explicit segmentation. In: Onieva, E., Santos, I., Osaba, E., Quintián, H., Corchado, E. (eds.) HAIS 2015. LNCS (LNAI), vol. 9121, pp. 136–145. Springer, Cham (2015). https://doi.org/10.1007/978-3-319-19644-2_12
13. Kenna, L.: Eccentricity in ellipses. Math. Mag. **32**(3), 133–135 (1959)
14. Kessentini, Y., Paquet, T., Hamadou, A.B.: Off-line handwritten word recognition using multi-stream hidden Markov models. Pattern Recognit. Lett. **31**(1), 60–70 (2010)
15. Khaissidi, G., Elfakir, Y., Mrabti, M., Lakhliai, Z., Chenouni, D., El yacoubi, M.: Segmentation-free word spotting for handwritten Arabic documents. Int. J. Interact. Multimed. Artif. Intell. **4**, 6–10 (2016)
16. Koenderink, J.J., Rabins, J.M.: Solid shape. Appl. Opt. **30**, 714 (1991)
17. Kullback, S.: Information Theory and Statistics. Courier Corporation (1997)
18. Lawgali, A.: A survey on Arabic character recognition. Int. J. Sig. Process. Image Process. Pattern Recognit. **8**(2), 401–426 (2015)
19. Leila, C., Maâmar, K., Salim, C.: Combining neural networks for Arabic handwriting recognition. In: 2011 10th International Symposium on Programming and Systems (ISPS), pp. 74–79. IEEE (2011)
20. Macmillan, N.A., Creelman, C.D.: Detection Theory: A User's Guide. Psychology Press (2004)
21. Maqqor, A., Halli, A., Satori, K.: A multi-stream HMM approach to offline handwritten Arabic word recognition. Int. J. Natural Lang. Comput. (IJNLC) **2**(4), 21–33 (2013)
22. Mezghani, A., Slimane, F., Kanoun, S., Märgner, V.: Identification of Arabic/French handwritten/printed words using GMM-based system. In: CORIA-CIFED, pp. 371–374 (2014)
23. Moghaddam, R.F., Cheriet, M., Adankon, M.M., Filonenko, K., Wisnovsky, R.: IBN SINA: a database for research on processing and understanding of Arabic manuscripts images. In: Proceedings of the 9th IAPR International Workshop on Document Analysis Systems, Proceeding DAS 2010, pp. 11–18 (2010)
24. Mozaffari, S., Bahar, P.: Farsi/Arabic handwritten from machine-printed words discrimination. In: 2012 International Conference on Frontiers in Handwriting Recognition (ICFHR), pp. 698–703. IEEE (2012)
25. Nguyen: Vietnamese Characters Recognition based on supervised learning. Ph.D. thesis, A master's project Computer Science University of Colorado (2013)
26. Pechwitz, M., Maddouri, S.S., Märgner, V., Ellouze, N., Amiri, H., et al.: IFN/ENIT-database of handwritten Arabic words. In: Proceedings of CIFED, vol. 2, pp. 127–136 (2002). http://www.ifnenit.com
27. Pechwitz, M., Maergner, V.: HMM based approach for handwritten Arabic word recognition using the IFN/ENIT-database. In: 2003 Seventh International Conference on Document Analysis and Recognition, Proceedings, pp. 890–894. IEEE (2003)
28. Rokach, L., Maimon, O.: Data Mining with Decision Trees: Theory and Applications. World Scientific, Singapore (2014)

29. Saïdani, A., Echi, A.K.: Pyramid histogram of oriented gradient for machine-printed/handwritten and Arabic/Latin word discrimination. In: 2014 6th International Conference of Soft Computing and Pattern Recognition (SoCPaR), pp. 267–272. IEEE (2014)
30. Saidani, A., Kacem, A., Belaid, A.: Arabic/Latin and machine-printed/handwritten word discrimination using HOG-based shape descriptor. ELCVIA Electr. Lett. Comput. Vis. Image Anal. **14**(2), 1–23 (2015)
31. Shannon, C.E.: A mathematical theory of communication. ACM SIGMOBILE Mob. Comput. Commun. Rev. **5**(1), 3–55 (2001)

Pattern Recognition Pipeline
for Neuroimaging Data

Maxim Sharaev[1]([✉]), Alexander Andreev[2], Alexey Artemov[1],
Evgeny Burnaev[1], Ekaterina Kondratyeva[1], Svetlana Sushchinskaya[1],
Irina Samotaeva[3], Vladislav Gaskin[3], and Alexander Bernstein[1]

[1] Skolkovo Institute of Science and Technology, Skolkovo, Moscow Region, Russia
`m.sharaev@skoltech.ru`
[2] Deloitte Analytics Institute, Moscow, Russia
[3] Moscow Research and Clinical Center for Neuropsychiatry, Moscow, Russia

Abstract. As machine learning continues to gain momentum in the neuroscience community, we witness the emergence of novel applications such as diagnostics, characterization, and treatment outcome prediction for psychiatric and neurological disorders, for instance, epilepsy and depression. Systematic research into these mental disorders increasingly involves drawing clinical conclusions on the basis of data-driven approaches; to this end, structural and functional neuroimaging serve as key source modalities. Identification of informative neuroimaging markers requires establishing a comprehensive preparation pipeline for data which may be severely corrupted by artifactual signal fluctuations. We propose a new unified data analysis pipeline for neuroimaging-based diagnostic classification problems using various different feature extraction techniques, Machine Learning algorithms and processing toolboxes for brain imaging. We illustrate the approach by discovering potential candidates for new biomarkers for diagnostics of epilepsy and depression presence in simple and complex cases based on clinical and MRI data for patients and healthy volunteers. We also demonstrate that the proposed pipeline in many classification tasks provides better performance than conventional ones.

Keywords: Pattern recognition · Neuroimaging data · Biomarkers
Neurology · Psychiatry · MRI · fMRI

1 Introduction

Nowadays, Pattern Recognition (PR), Machine Learning (ML), and Intelligent data analysis techniques are used in medical research for diagnostic biomarkers discovery and treatment outcomes prediction with the use of neuroimaging data collected for the targeted groups of patients or healthy volunteers. To unify these

This study was performed in the scope of Skoltech SBI Gelfand/Bernstein and Skoltech NGP Solomon/Burnaev projects.

L. Pancioni et al. (Eds.): ANNPR 2018, LNAI 11081, pp. 306–319, 2018.
https://doi.org/10.1007/978-3-319-99978-4_24

processes, we offer a common data preprocessing and analysis pipeline for structural and functional magnetic resonance imaging (MRI) data which dramatically reduces research time and allows a researcher to try many preprocessing/data cleaning/feature extraction/classification options and compare results. In future, it is planned to add a visualization step (of informative features) together with classification results, which is highly desirable by medical community.

One challenge for the successful applications of automated diagnostics based on pattern recognition approaches in clinical settings is to ensure the highest possible quality of source signal employed for decision making. Cleaning the artifactual (irrelevant to the process in question) noise incidental to scanning deems necessary, as such fluctuations drastically hurt recognition performance, blocking the way to the identification of neuroimaging markers for mental disorders [4]. To this end, denoising schemes must be proposed, which involve the extensive examination of spatiotemporal constituents of the source signal and identification of the relevant components against the artifactual noise [3,28]. In the present work, we investigate a pattern classification pipeline for mental disorders featuring a denoising step, and observe consistent performance improvements w.r.t. the baseline approach.

A different but important challenge is to design highly sensitive and robust predictive models. Research indicates that the pattern of brain activity changes associated with disorders such as depression might have limited discriminative power, leading to performance drops for common machine learning algorithms. Current accuracy of around 75%, thus, does not allow direct clinical application of these models [18].

Data used in the pilot study consists of structural and functional MR-images from 100 subjects: 25 healthy volunteers (H), 25 patients with major depressive disorder in an acute depressive episode (D), as well as 25 epilepsy patients (E) and 25 epilepsy patients with major depressive disorder (ED).

In summary, our contributions in this work are the following:

- We provide a short review of current fMRI applications and data cleaning methods prior to feature extraction;
- We propose a principled noise-aware pattern recognition pipeline for neuroimaging tailored to pattern classification;
- Using a real MRI/fMRI dataset, we demonstrate the effectiveness of our methodology, searching for epilepsy-specific patterns. We aim to discriminate between healthy controls, patients with epilepsy, and patients with both epilepsy and depression.

The potential importance of the work arises from the fact that the majority of commercial and clinical scanners are 1.5T compared to high quality 3T and 7T research scanners. Here we prove that both structural and (low quality and short duration) resting-state data could be enough for epilepsy/depression diagnostics.

2   MR Data Properties, Preprocessing and Feature Extraction

Magnetic resonance imaging (MR imaging, MRI) and its part — functional magnetic resonance imaging (fMRI) — use strong magnetic fields to create images of biological tissues and physiological changes in them.

Upon acquisition, MRI data should be cleaned to eliminate the noise associated with the scanning procedure (low-level hardware artefacts such as magnetic field inhomogeneity, radiofrequency noise, surface coil artefacts and others) and signal processing (chemical shift, partial volume, etc.); besides there are artefacts associated with the scanned patient (physiological noise such as blood flow, movements, etc.). The artefacts should be considered in accordance with the noise origin and filtered out [9].

In addition to MRI data cleaning problem, there is another common challenge of the brain imaging analysis related to the large dimensionality of the measured data, which mostly depends on resolution parameters of the scanner inductive detection coil. For instance, standard voxel sizes are within 0.5–2 mm^3 in case of structural imaging (resulting in 10^7 voxels for the whole brain volume). Thus, an MRI image, composed of huge number of small sized voxels, has higher spatial resolution and, hence, high dimensionality. To avoid the curse of the dimensionality phenomenon, ML methods are usually applied to lower dimensional features extracted from original scans by feature selection procedures. These procedures are also included into the preprocessing stage.

2.1   Structural MRI Preprocessing and Feature Extraction

Preprocessing stage has two main goals: MRI data cleaning and avoiding the curse of the dimensionality phenomenon caused by high dimensionality of initial MRI data. The latter goal can be achieved by constructing lower dimensional and biomedically interpretable brain characteristics from the initial data.

The goal of this sub-stage is to extract informative features (biomedically significant brain characteristics, clinically meaningful features) with lower dimensionality. The approach is typically realized in several steps:

- selection of an appropriate brain atlas [17] which splits the brain into the anatomical areas (e.g. Hippocampi, cortical areas and etc.),
- 3D MRI images segmentation into disjoint sets (sub-images), consisting of voxels, corresponding to different brain regions (Regions of Interest, ROIs),
- various characteristics calculation for each ROI.

The examples of such characteristics could be structural morphometric parameters (volumes, thicknesses, curvatures) of the selected anatomical areas from the MRI-image, which together form a volumetric vector. For example, MRI processing toolbox [12] parcels MRI images into regions corresponding to the chosen Desikan-Killiany atlas; calculates 7 volumetric characteristics for each cortical region (NVoxels, Volume_mm3, normMean, normStdDev, normMin, normMax, normRange) and 9 geometric characteristics of subcortical regions (NumVert, SurfArea, GrayVol, ThickAvg, ThickStd, MeanCurv, Gaus-Curv, FoldInd, CurvInd).

For constructed objects different characteristics reflecting meaningful properties of these objects, can be computed for further use in Machine learning studies like segments of MRI-image consisting of 3D MRI-voxels from chosen brain areas (to be used as inputs for deep learning procedures [23]). Most often domain-specific lower dimensional features (morphometric or functional connectivity features) could be extracted from original data in specialized MRI processing toolboxes [2].

2.2 Functional MRI Preprocessing and Feature Extraction

fMRI-Related Noise. Not depending on equipment, fMRI signal is very noisy. As T2*-weighted image (BOLD-contrast) is a mixture of signals from many sources, the desired signal from the neuronal activity only represents a relatively small percentage of the variance of the signal [3]. Non-neuronal contributions to the BOLD fMRI time series include receiver coil thermal noise, instrumental drifts, spike-like artifactual signals induced by the hardware instabilities, rapid and high-amplitude spikes due to the head motion. The physiological noise of non-neuronal origin (which is essentially BOLD-signal, but of no interest) comprises of cardiac and respiratory noise, changes in arterial carbon dioxide concentration associated with varying respiration rate, vasomotor effects, changes in blood pressure and cerebral autoregulation mechanisms [20].

Noise Identification and Suppression. Three significantly different general approaches for noise identification and removal in fMRI data can be highlighted [5, 28]: the first is based on using additional sensors measuring physiological activities exploitation (model-based approach, [14]), the second is noise elimination specific for each type of noise (e.g. motion correction or thermal noise cleaning), finally the third one is data-driven using only fMRI data itself and prior information about fMRI signal and noise. The first approach is limited as it covers only physiological nature of the noise and can't handle e.g. scanner artifacts. Moreover, large amounts of data have already been collected (and are being collected) without additional "noise" information, so the aforementioned method cannot be useful here. Data collection with additional equipment introduces additional challenges from increased experimental time to equipment cost, MR-compatibility, and instability.

Independent Component Analysis (ICA) based technique could be viewed as a second step as it might be applied to components extracted by PCA [26]. The resulting independent components are assumed to be either noise-related or signal-related, each representing one of the sources in a source separation task solved by ICA. ICA transforms source fMRI signal into a set of components with distinct spatial and temporal structures, which further could be classified as noise or signal. Three possible approaches to this classification can be highlighted.

The first one is an expert-based technique: an individual with expertise in fMRI processing must examine every component (time courses, spatial distribution, and spectrum) and manually label it as either signal or noise [15,19]. [15] present a detailed guideline for evaluating and categorizing independent components and provide examples of each component class. Expert-based categorization may be tedious and error-prone for data with low SNR ratios, e.g. for ubiquitous medical 1.5T scanners.

Another option is to utilize a pre-trained classifier such as the one provided by the FIX package of FSL that achieves 99% classification accuracy based on the annotation created by human experts [28]. In this work, ICA components are used to extract features for the machine learning methods (supervised learning classifiers) that aim to classify noise components from signal automatically based on labeled training data. 46 temporal and 131 spatial features are extracted and a feature selection procedure is performed during classifier training.

Finally, the third approach is to combine the first two approaches, i.e. to calibrate the existing model pre-trained on vast amounts of data with different characteristics (such as the one provided by the FIX package) for the particular problem. This requires creating a new task-oriented labeled dataset using the expert knowledge and using transfer learning techniques known from machine learning to 'fine-tune' the classification model on the newly labeled data. This approach seems to be the most promising when data quality is low and number of patients is relatively high.

Described approaches to signal-noise separation for fMRI data might prove useful for classification tasks in the medical domains described above, specifically epilepsy, schizophrenia, and depression diagnostics. The crucial point here is that physiological noise having no direct relation to the neuronal activity (i.e. signal), might still carry valuable discriminating information for the classification task. For instance, cerebral blood flow fluctuations might reveal unobservable brain states, which correlate with the target variable (disease/no disease). Broadly speaking, two ways to approaching the classification problem exist. The first assumes building classifier based on the hand-crafted features extracted from the independent components (such as, for instance, described in [28]), that could prove effective for discriminating between patients vs. healthy controls. An alternative approach may be based on the reconstruction of the 4D fMRI signal itself after noise elimination and its utilization as a source data for training (i.e. data might be denoised, or its signal and noise parts investigated separately).

3   The Proposed Pipeline

The literature review has allowed us to identify established and prospective building blocks and organize them into a unified and highly automated fMRI processing pipeline, see [13] for detailed discussion. Our pipeline accepts raw functional and structural scans of a subject and outputs the predicted task-specific scores, whose meaning vary according to the application. For instance, for a depression vs. healthy control classification task, our pipeline should score the patient according to a probability of depression diagnosis for him. The entire chain of steps can be implemented in a modular way via existing or prospective software by respecting interfaces between the modules. We note that as some of the modules may carry computationally intensive processing, the runtime of the pipeline may vary from minutes to several hours. We briefly describe our proposed pipeline below.

The input to our pipeline comprises of functional and structural MRI scans. The raw scans are passed through a standard preprocessing step, an established low-level MRI handling stage (involving slice-time correction, motion correction, filtering in space and time, anatomical alignment across individuals) [7], yielding preprocessed data on the same scale and format. A second stage accepts pre-processed scans and runs a manual denoising procedure analogous to the one discussed in [15], producing a scan with irrelevant components excluded and an increased SNR. The third stage of our pipeline performs correlation-based and graph-based feature extraction for fMRI and structural morphometric parameters (volumes, thicknesses, curvatures) extraction for structural MRI data, which together form a volumetric vector. Lastly, our pipeline performs pattern recognition by making use of available implementations of conventional machine learning approaches such as SVMs [29], neural networks [25], and decision trees combined with imbalanced classification approaches [24], to name a few. Additionally, for each selected combination of analysis steps it is possible to select and visualize most informative structural/functional features (i.e., potentical candidates for biomarkers) and evaluate True Positive Rate (TPR) with any fixed False Positive Rate (FPR), which is extremely useful in medical practice.

4   Illustrative Example: Pattern Recognition for Epilepsy Detection

4.1   fMRI-based Pattern Recognition

The purpose of this example study is to demonstrate the possible advantages of using novel sophisticated artifact removal procedures prior to feature extraction in clinical diagnostics. The data at our disposal consisted of functional MRI scans of four groups of subjects: 25 patients with epilepsy, 25 patients with depression, 25 patients with both epilepsy and depression, and 25 healthy controls. We aimed at finding patterns connected to epilepsy, thus discriminating patients with epilepsy against the rest of the sample: the diagnostic question is whether a particular subject has epilepsy (otherwise he might be healthy or have depression). Resting-state functional MRI was collected at 1.5T EXCEL ART VantageAtlas-X Toshiba scanner at Z.P.Solovyev Research and Clinical Center for Neuropsychiatry[1].

Raw data were preprocessed according to two different protocols:

---

[1] Skoltech biomedical partner. Website (in Russian): http://npcpn.ru/.

1. SPECTRUM: standard preprocessing pipeline implemented in SPM12 including slice-timing correction, bad-slices interpolation, motion-correction (bad volumes interpolation), coregistration with T1 images and spatial normalization.
2. MANUAL: a combination of standard pipeline with manual ICA classification into signal and noise components by fMRI experts.

For the parcellation of the brain, we used an Automatic Anatomic Labeling (AAL) atlas consisting of 117 regions. For each region corresponding time series were assigned, and then a correlation matrix was computed from them. Significant correlation values ($p = 0.05$, Bonferroni corrected) were set to ones, all other values were set to zeros, yielding a binary adjacency matrix of dimensionality 117×117. An example of raw and binarized matrix is in Fig. 1.

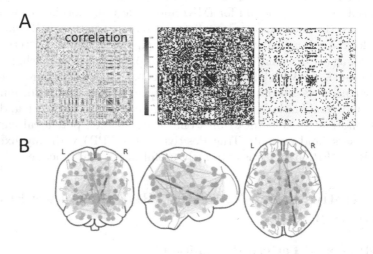

Fig. 1. Visualization of fMRI patterns used for the classification task. **A** depicts the correlation matrix, from left to right: raw, binarized (threshold = 0.15), binarized (threshold = 0.4). **B** depicts the functionally connected brain areas according to the elements of the correlation matrix in **A**.

The python library NetworkX was used to calculate features corresponding to each region of interest, explaining the level of functional activity of the region in terms of graph nodes. We calculated 5 metrics corresponding to each region of the brain: clustering coefficient, degree centrality, closeness centrality, betweenness centrality, average neighbor degree, and 2 metrics describing the graph in general: local efficiency and global efficiency [22]. Then the standard machine learning classifiers were applied: Support Vector Machine (SVM), Random Forests (RF), and Logistic Regression (LR).

Each model was validated using leave-one-out approach. Four comparisons were performed:

Table 1. *EvsH.* Results on 25 epilepsy only patients vs. 25 healthy controls classification task.

FPR	SPECTRUM TPR	MANUAL TPR
10%	8%	**36%**
15%	8%	**44%**
20%	24%	**68%**
30%	32%	**68%**

Table 2. *EDvsE.* Results on 25 epilepsy only patients vs. 25 epilepsy + depression patients classification task.

FPR	SPECTRUM TPR	MANUAL TPR
10%	**20%**	16%
15%	**20%**	20%
20%	**40%**	38%
30%	**44%**	48%

Table 3. *EvsNE.* Results on 50 subjects with epilepsy vs. 50 subjects without epilepsy classification task.

FPR	SPECTRUM TPR	MANUAL TPR
10%	10%	**36%**
15%	22%	**52%**
20%	32%	**60%**
30%	36%	**66%**

Table 4. *EDvsD.* Results on 25 depression only patients vs. 25 epilepsy + depression patients classification task.

FPR	SPECTRUM TPR	MANUAL TPR
10%	16%	**20%**
15%	24%	**32%**
20%	28%	**60%**
30%	40%	**60%**

- *EvsH.* 25 epilepsy only patients vs. 25 healthy controls,
- *EDvsE.* 25 epilepsy only patients vs. 25 epilepsy + depression patients,
- *EvsNE.* 50 subjects with epilepsy vs. 50 subjects without epilepsy (including healthy controls and subjects with depression).
- *EDvsD.* 25 depression only patients vs. 25 epilepsy + depression patients.

The results for each classification task are in Tables 1, 2, 3 and 4. Summary of classification results in terms of prediction accuracy for the two competing preprocessing pipelines is provided in Table 5.

Firstly, it can be seen that SPECTRUM preprocessing does not perform well when working on simple features (functional connectivity). Next, MANUAL shows relatively high performance in terms of accuracy and true positive rate, which means that additional sophisticated data cleaning could be beneficial prior to feature extraction for fMRI classification tasks. In all except EvsNE the classifier performance is rather high and stable after cross-validation meaning that epilepsy-specific fMRI pattern could be found, thus in it might be hard to distinguish between complex epilepsy + depression patients possibly due to similar brain disruptions. An additional work is needed here to construct new features sensitive to subtle differences between patients.

Table 5. Summary of classification results in terms of prediction accuracy for the two competing preprocessing pipelines.

Task	SPECTRUM	MANUAL
EvsH	$57 \pm 16\%$	$\mathbf{76 \pm 13\%}$
EDvsE	$67 \pm 15\%$	$66 \pm 16\%$
EvsNE	$54 \pm 16\%$	$\mathbf{73 \pm 16\%}$
EDvsD	$58 \pm 20\%$	$\mathbf{66 \pm 16\%}$

4.2 Structural MRI-based Pattern Recognition

Structural MRI data were cleaned, preprocessing and their features were extracted using MRI processing toolboxes [1,2] as described below (see also Fig. 2).

Structural morphometric features were calculated from $T1w$ images using [12]; for more than 100 brain regions corresponding features explaining brain structure (volumes, surface areas, thicknesses, etc.) were computed producing a vector with 894 features for each subject.

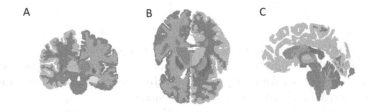

Fig. 2. An example of brain parcellaton used for feature extraction: structural brain MRI split into anatomical regions performed in Freesurfer 6.0 according to the Desikan-Killiany atlas. A: coronal view, B: axial view, C: saggital view.

Next, the ML exploratory pipeline was implemented on Ipython using scikit-learn library[2] and organized as follows:

- We considered two geometrical methods for dimensionality reduction: (1) Locally Linear Embedding, (2) Principal Component Analysis (see description of a weighted version in [6]).
- We considered two methods of feature selection: (1) feature selection based on Pearson's chi-squared test and ANOVA scoring, implemented via SelectKBest function in scikit-learn; (2) selection of relevant features based on a particular classification model used with Logistic Regression (LR), K-Nearest Neighbors (KNN) and Random Forest Classifier (RFC), implemented via the SelectFromModel function in scikit-learn.

---

[2] http://scikit-learn.org/.

Table 6. Results on EvsH task (best model: SVM).

FPR	10%	15%	20%	30%
TPR	32%	48%	56%	68%

Table 7. Results on EDvsE task (best model: SVM).

FPR	10%	15%	20%	30%
TPR	4%	8%	28%	32%

Table 8. Results on EvsNE task (best model: Logistic Regression).

FPR	10%	15%	20%	30%
TPR	42%	46%	52%	66%

Table 9. Results on EDvsD task (best model: Logistic Regression.

FPR	10%	15%	20%	30%
TPR	40%	52%	80%	80%

- We performed grid search for a number of selected features and for a number of components in dimension reduction procedure in the sets $\{10, 20, 50, 100\}$ and $\{5, 10, 15, 20\}$, correspondingly.

Data was whitened before training. Feature reduction was performed without double-dipping [21], therefore training and testing datasets are separated before feature selection/dimensionality reduction. Hyper-parameters grid search was based on cross-validation with stratification, repeated 10 times for each person being in test.

As mentioned above, the dataset explored with the proposed pipeline contains four groups of subjects. Patients from *epilepsy* and *epilepsy with depression* groups represent cohorts with several types of epilepsy localization: temporal, frontal, parietal, mixed and unknown cases. That allows exploration of general patterns of epilepsy, independently to the localization, which are known to be tacked particular subcortical regions as hippocampus. Then in other research, the ML methods are applied to classify epilepsy with a known epilepsy localization. There are two types of epilepsy being extensively explored: temporal lobe epilepsy (TLE) or even precisely TLE with mesial temporal sclerosis (TLE-MTS), and focal cortical dysplasia patients with TLE (TLE-FCD) [8,10,16]. In some cases, TLE patients from these selected groups are further separated in groups by the loci localization as right and left [11].

The results for each classification task are in Tables 6, 7, 8 and 9. Summary of classification results in terms of prediction accuracy is provided in Table 10. Thus, the obtained results on the in-homogenized cohort are the firstly reported and we consider the results with the classification accuracy more than 0.7 statistically significant with .05 level of confidence on the explored cohort of 50 patients. We provide the results for *EvsH* classification task in Table 6. Note that we obtain statistically significant results on structural MRI features. This could be explained by the fact that epilepsy leads to significant changes in brain structure, which can yield more accurate classification. Table 9 presents results for *EDvsE* classification task. We conclude that model performance is not statistically significant, highlighting the difficulty of isolation of depression against the complex illness picture based on MRI data alone. The results for *EvsNE* are in Table 8; here the model performance is comparable to ones from preselected cohort of patients [8,10,11,16] as for *EvsH* case, cf. table, where TPR reaches 80% with FPR fixed at 30%.

Table 10. Summary of classification results in terms of prediction accuracy.

Task	EvsH	EDvsE	EvsNE	EDvsD
Accuracy	$74.0 \pm 18.0\%$	$65.0 \pm 15.0\%$	$73.0 \pm 18.0\%$	$82.0 \pm 11.0\%$

The most by score important features from models *EvsH* and *EvsNE* are: Ventral Diencephalon Right, Hippocampus Left, Thalamus Left, Putamen Left, Angular gyrus Right, Frontal gyrus Right (Superior), Paracentral lobule Left, Postcentral gyrus Right, Precentral gyrus Left, Supra Marginal gyrus Left, Temporal Pole Left. These findings are in line with current knowledge of epileptogenic zones and brain areas mostly affected from epileptic seizures [27] as well as provide some new information on possible targets in epilepsy diagnostics and treatment.

5   Conclusions

In the current work, we reviewed some approaches to neuroimaging data cleaning aimed at the elimination of artifacts harmful for further pattern recognition. Based on well established and novel approaches, we proposed a principled noise-aware pattern recognition pipeline for neuroimaging tailored to pattern classification and showed the potential effectiveness of our proposed methodology in a pilot classification study. Our general data preprocessing and analysis pipeline for structural and functional MRI data could dramatically reduce research time, thus allowing a researcher to investigate a larger variety of preprocessing, data cleaning, feature extraction, and classification options as well as compare results based on desired metrics. From top performing combinations of analysis steps we obtained a number of stable structural and functional features (i.e., potential candidates for biomarkers), some of which are known and well established, whereas some of which are new and could possibly provide new medical knowledge on epilepsy mechanisms and its robust detection. We also evaluated True Positive Rates (TPR) with different fixed False Positive Rates (FPR), which is much more useful for clinicians than classification accuracy alone. According to the proposed pipeline, we found the ICA-based cleaning step to be crucial for further pattern recognition task: denoised data provides clearer and more informative features for machine learning-based diagnostics, and yields significant improvements in finding epilepsy-specific pattern in a group of patients with only epilepsy versus epilepsy + depression patients and healthy controls. We strongly believe that application of pattern recognition in functional neuroimaging is promising for clinical diagnostics of psychiatric disorders such as depression and neurological diseases such as epilepsy, though the classification performance achieved in our study may be not enough for immediate medical applications. We don't consider end-to-end (deep learning-based) pipeline, as we believe deep learning-based approaches would face strict limitations with raw data without any preprocessing due to (1) high dimensionality, (2) very low SNR ratios, and (3) very small sample size in comparison to usual deep learning problems.

The analyzed dataset has several obvious drawbacks: it is small and well balanced, which is not the usual case in clinical practice. As the number of public sources of epilepsy/depression clinical data is limited and the access to these datasets is difficult to obtain, it is hard to compare our results with other research groups, though several studies with similar objectives are discussed above.

We nevertheless obtained statistically significant results for *EvsH*, *EvsNE*, and *EDvsD* models. The epilepsy classification on mixed cohort *EvsNE* reached FPR 30% (model sensitivity 70%), and TNR 80% (specificity 80%) is comparable to the research conducted on pre-selected groups of patients [8,10,11,16], allowing the exploration of generalized disease biomarkers which were not analyzed with ML methods before.

Despite the limitations, the proposed approach is universal and has the potential to be implemented into clinical practice, as it is not based on high-quality data and sophisticated ML algorithms, but could be useful in real applications or serve as a starting point tutorial for building new MR processing pipelines.

References

1. Behroozi, M., Daliri, M.: Software tools for the analysis of functional magnetic resonance imaging. Basic Clin. Neurosci. **3**(5), 71–83 (2012)
2. Bernstein, A., Akzhigitov, R., Kondrateva, E., Sushchinskaya, S., Samotaeva, I., Gaskin, V.: MRI brain imagery processing software in data analysis. In: Perner, P. (ed.) Advances in Mass Data Analysis of Images and Signals in Medicine, Biotechnology, Chemistry and Food Industry. Proceedings of 13th International Conference on Mass Data Analysis of Images and Signals (MDA 2018). Springer (2018)
3. Bianciardi, M.: Sources of functional magnetic resonance imaging signal fluctuations in the human brain at rest: a 7 T study. Mag. Reson. Imaging **27**(8), 1019–1029 (2009)
4. Birn, R.M., Murphy, K., Handwerker, D.A., Bandettini, P.A.: fMRI in the presence of task-correlated breathing variations. Neuroimage **47**(3), 1092–1104 (2009)
5. Caballero-Gaudes, C., Reynolds, R.C.: Methods for cleaning the bold fMRI signal. Neuroimage **154**, 128–149 (2017)
6. Chernova, S., Burnaev, E.: On an iterative algorithm for calculating weighted principal components. J. Commun. Technol. Electron. **60**(6), 619–624 (2015)
7. Cohen, J.D., et al.: Computational approaches to fMRI analysis. Nature Neurosci. **20**(3), 304 (2017)
8. Del Gaizo, J.: Using machine learning to classify temporal lobe epilepsy based on diffusion MRI. Brain Behav. **7**(10), e00801 (2017)

9. Erasmus, L., Hurter, D., Naude, M., Kritzinger, H., Acho, S.: A short overview of MRI artefacts. SA J. Radiol. **8**, 13–17 (2004)
10. Fang, P., An, J., Zeng, L.L., Shen, H., Qiu, S., Hu, D.: Mapping the convergent temporal epileptic network in left and right temporal lobe epilepsy. Neurosci. Lett. **639**, 179–184 (2017)
11. Focke, N.K., Yogarajah, M., Symms, M.R., Gruber, O., Paulus, W., Duncan, J.S.: Automated MR image classification in temporal lobe epilepsy. Neuroimage **59**(1), 356–362 (2012)
12. FreeSurfer: Freesurfer toolbox - an open source software suite for processing and analyzing (human) brain MRI images (2018). https://surfer.nmr.mgh.harvard.edu/
13. Friston, K.J., Ashburner, J.T., Kiebel, S.J., Nichols, T.E., Penny, W.D.: Statistical Parametric Mapping: The Analysis of Functional Brain Images, vol. 8 (2007)
14. Glover, G.H., Li, T.Q., Ress, D.: Image-based method for retrospective correction of physiological motion effects in fMRI: retroicor. Mag. Reson. Med. **44**(1), 162–167 (2000)
15. Griffanti, L., et al.: Hand classification of fMRI ica noise components. Neuroimage **154**, 188–205 (2017)
16. Hong, S.J., Kim, H., Schrader, D., Bernasconi, N., Bernhardt, B.C., Bernasconi, A.: Automated detection of cortical dysplasia type II in MRI-negative epilepsy. Neurology **83**(1), 48–55 (2014)
17. Jean, T., Tournoux, P.: Co-Planar Stereotaxic Atlas of the Human Brain: 3-D Proportional System: An Approach to Cerebral Imaging (1988)
18. Kambeitz, J., et al.: Detecting neuroimaging biomarkers for schizophrenia: a meta-analysis of multivariate pattern recognition studies. Neuropsychopharmacology **40**(7), 1742 (2015)
19. Kelly Jr., R.E., et al.: Visual inspection of independent components: defining a procedure for artifact removal from fmri data. J. Neurosci. Methods **189**(2), 233–245 (2010)
20. Murphy, K., Birn, R.M., Bandettini, P.A.: Resting-state fMRI confounds and cleanup. Neuroimage **80**, 349–359 (2013)
21. Mwangi, B., Tian, T.S., Soares, J.C.: A review of feature reduction techniques in neuroimaging. Neuroinformatics **12**(2), 229–244 (2014)
22. Networkx: Networkx - software for complex networks (2018). https://networkx.github.io/
23. Notchenko, A., Kapushev, Y., Burnaev, E.: Large-scale shape retrieval with sparse 3D convolutional neural networks. In: van der Aalst, W.M., et al. (eds.) Analysis of Images, Social Networks and Texts, pp. 245–254. Springer International Publishing, Cham (2018)
24. Papanov, A., Erofeev, P., Burnaev, E.: Influence of resampling on accuracy of imbalanced classification. In: Verikas, A., Radeva, P., Nikolaev, D. (eds.) Proceedings of SPIE 9875, Eighth International Conference on Machine Vision, Barcelona, Spain, 8 December 2015, vol. 9875. SPIE (2015)
25. Prikhod'ko, P.V., Burnaev, E.V.: On a method for constructing ensembles of regression models. Autom. Remote Control **74**(10), 1630–1644 (2013). https://doi.org/10.1134/S0005117913100044
26. Rasmussen, P.M., Abrahamsen, T.J., Madsen, K.H., Hansen, L.K.: Nonlinear denoising and analysis of neuroimages with Kernel principal component analysis and pre-image estimation. NeuroImage **60**(3), 1807–1818 (2012)
27. Richardson, E., et al.: Structural and functional neuroimaging correlates of depression in temporal lobe epilepsy. Epilepsy Behav. **10**(2), 242–249 (2007)

28. Salimi-Khorshidi, G., Douaud, G., Beckmann, C.F., Glasser, M.F., Griffanti, L., Smith, S.M.: Automatic denoising of functional MRI data: combining independent component analysis and hierarchical fusion of classifiers. Neuroimage **90**, 449–468 (2014)
29. Smolyakov, D., Erofeev, P., Burnaev, E.: Model selection for anomaly detection. In: Verikas, A., Radeva, P., Nikolaev, D. (eds.) Proceedings of SPIE 9875, Eighth International Conference on Machine Vision, Barcelona, Spain, 8 December 2015. SPIE, vol. 9875 (2015)

Anomaly Pattern Recognition with Privileged Information for Sensor Fault Detection

Dmitry Smolyakov[1], Nadezda Sviridenko[1], Evgeny Burikov[2], and Evgeny Burnaev[1(✉)]

[1] Skolkovo Institute of Science and Technology, Moscow Region, Russia
{Dmitrii.Smoliakov,Nadezda.Sviridenko}@skolkovotech.ru,
E.Burnaev@skoltech.ru
[2] PO–AO "Minimaks-94", Moscow, Russia
Burikov@mm94.ru
http://adase.group

Abstract. Detection of malfunction sensors is an important problem in the field of Internet of Things. One of the classical approaches to recognize anomalous patterns in sensor data is to use anomaly detection techniques based on One Class Classification like Support Vector Data Description or One Class Support Vector Machine. These techniques allow to build a "geometrical" model of a sensor regular operating state using historical data and detect broken sensors based on a distance to the regular data patterns. Usually important signals/warnings, which can help to identify broken sensors, arrive only after their failures. In this paper, we propose the approach to utilize such data by using the privileged information paradigm: we incorporate signals/warnings, available only when training the anomaly detection model, to refine the location of the boundary, separating the anomalous region. We demonstrate the approach by solving the problem of broken sensor detection in a Road Weather Information System.

Keywords: Anomaly detection · Road weather information system
Learning using privileged information · Internet of Things

1 Introduction

Complex Internet of Things sensor systems can be used to control houses, monitor cars and airplanes. In particular, sensor systems can help in road condition monitoring. Road Weather Information Systems (RWIS) are located on the slide

The research, presented in Sect. 3 of this paper, was partially supported by the Russian Foundation for Basic Research grants 16-01-00576 A and 16-29-09649 ofi m. The research, presented in other sections, was supported solely by the Ministry of Education and Science of Russian Federation, grant No. 14.606.21.0004, grant code: RFMEFI60617X0004.

L. Pancioni et al. (Eds.): ANNPR 2018, LNAI 11081, pp. 320–332, 2018.
https://doi.org/10.1007/978-3-319-99978-4_25

lines or under the road surface and collect information about the road conditions including temperature, pressure, humidity, etc., see details in Sect. 3.1. Some of them even allow making video recordings. This information can be used for road monitoring and helps in organizing deicing activities and snow removal. It allows to increase safety of roads and decrease maintenance cost. Certain Road Weather monitoring systems are equipped with intelligent modeling capabilities and even allow predicting adversarial conditions. For that usually either statistical [1] or physical models [2] are used. METRo is a physical model allowing to forecast local road surface conditions based on information, collected from sensors located under the road, and global weather forecasts [3], see Sect. 3.2 for details. Figures 1 and 2 present examples of RWIS sensors and location of the RWIS station on a sideway.

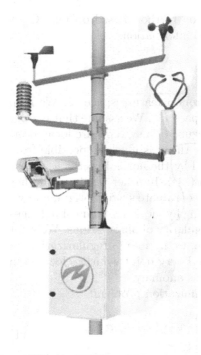

Fig. 1. RWIS and its sensors   **Fig. 2.** Installed RWIS

Forecasting accuracy depends on quality of collected data. If for some reason a sensor is broken we will not be able to make accurate predictions. This issue explains why the problem of detecting malfunctioning sensors is so crucial.

A viable approach to detect broken sensors is to recognize anomalous patterns in transmitted data using anomaly detection techniques based on One Class Classification [4], for example One Class Support Vector Machine (One Class SVM) [5] or Support Vector Data Description (SVDD) [6]. These techniques allow to build a model of a system regular operating state using historical data and then detect broken sensors based on a distance to the regular data patterns.

Usually the most significant symptoms of a broken sensor can be found in data patterns only after some period of time after the failure. To incorporate this post-failure information, we develop an anomaly detection technique based on Learning Using Privileged Information framework (LUPI) [7,8]. As a post-failure information we use forecasts of the METRo model for a local region around the considered RWIS. Such information is not available during the test phase, when we use the trained model for anomaly detection, but can be used during the training phase.

2   Anomaly Detection

Below we briefly describe the anomaly detection method based on One Class SVM and its modification with using privileged information.

2.1   One Class SVM

We have a data set consisting of l points x_i from a feature space X. We map them by a function $\phi(\cdot)$ to some kernel Hilbert space H_ϕ. We assume that regular points (corresponding to a regular operating regime of a system) are located far away from the origin of H_ϕ and try to separate them as much as possible from the origin by a hyperplane $w \cdot \phi(x) = \rho$, defined by the normal vector w. Every point produces the loss $\ell(x) = \max(0, \rho - (w \cdot \phi(x)))$. In other words, we do not want the points to lie in one subspace with the coordinate system origin, and we do not want them to be too close to the separating hyperplane. Value of ρ defines at what distance we become confident in the regularity of observations. We want to have the confidence ρ as big as possible. Also, we impose ℓ_2 regularization on the normal vector of the separating hyperplane. Every point, which is located in the same half-space with the origin, is labeled as anomaly.

This intuition can be summed up in an optimization problem:

$$\nu l \cdot \|w\|^2/2 - \nu l \rho + \sum_{i=1}^{l} \xi_i \to \min_{w,\rho,\boldsymbol{\xi}} \quad s.t. \begin{cases} (w \cdot \phi(x_i)) \geq \rho - \xi_i, \\ \xi_i \geq 0. \end{cases} \tag{1}$$

The problem (1) is convex. However, in this case it is easier to solve the dual problem because it is a quadratic optimization problem.

To formulate the dual problem, we write down the Lagrangian for (1):

$$L_{w,\boldsymbol{\xi},\rho,\boldsymbol{\alpha},\boldsymbol{\beta}} = \nu l \frac{\|w\|^2}{2} + \sum_{i=1}^{l} \xi_i - \nu l \rho - \sum_{i=1}^{l} \alpha_i \left[(w \cdot \phi(x_i)) - \rho + \xi_i \right] - \sum_{i=1}^{l} \beta_i \xi_i.$$

From Karush Kuhn Tucker conditions we get that

$$\partial L/\partial w = \nu l w - \sum_{i=1}^{l} \alpha_i \phi(x_i) \Rightarrow w = \frac{1}{\nu l} \sum_{i=1}^{l} \alpha_i \phi(x_i),$$

$$\partial L/\partial \rho = -\nu l + \sum_{i=1}^{l} \alpha_i \Rightarrow \sum_{i=1}^{l} \alpha_i = \nu l, \ \partial L/\partial \xi_i = 1 - \alpha_i - \beta_i = 0.$$

We regroup the summations in the Lagrangian and apply Karush Kuhn Tucker conditions:

$$L_{w,\xi,\rho,\alpha,\beta} = \nu l \frac{\|w\|^2}{2} + \rho \underbrace{\left(\sum_{i=1}^{l} \alpha_i - \nu l\right)}_{0} + \sum_{i=1}^{l} \xi_i \underbrace{[1 - \alpha_i - \beta_i]}_{0} - \sum_{i=1}^{l} \alpha_i (w \cdot \phi(x_i)).$$

Replacing w with $\frac{1}{\nu l} \sum_{i=1}^{l} \alpha_i \phi(x_i)$ we get that:

$$L_{w,\xi,\rho,\alpha,\beta} = -\frac{1}{2\nu l} \sum_{i=1}^{l} \sum_{j=1}^{l} \alpha_i \alpha_j (\phi(x_i) \cdot \phi(x_j)).$$

In matrix form the dual problem can be written down as follows:

$$- \alpha^T K \alpha \rightarrow \max_{\alpha} \ s.t. \begin{cases} 0 \le \alpha \le 1, \\ e^T \alpha = \nu l. \end{cases} \qquad (2)$$

Here $K(x,z) = (\phi(x) \cdot \phi(z))$ is a kernel function representing a dot product in H_ϕ, and $K = \{K(x_i, x_j)\}$ is a kernel matrix.

The solution of the dual problem allows to write down the solution of the primal one:

$$\hat{f}(x) = (w \cdot \phi(x)) = \frac{1}{\nu l} \sum_{i=1}^{l} \alpha_i K(x, x_i).$$

Based on Karush Kuhn Tucker conditions we can show that for every x_i, such that $0 < \alpha_i < 1$, the value of ρ can be calculated as follows: $\rho = \hat{f}(x_i)$.

In [5] the authors demonstrated that ν is an upper bound of support vectors fraction and a lower bound of anomaly points fraction; moreover, if $l \rightarrow \infty$, the fraction of points marked as anomalies converges to ν [5].

2.2 One Class SVM+

Let assume that for every element from the train set x_i there exists an element x_i^* from the privileged feature space X^*. This additional (privileged) information is not available on the test phase, i.e. we are going to train the decision rule on pairs of patterns $\{(x_i, x_i^*)\}_{i=1}^l \in X \times X^*$, but when making predictions during the test phase we can use only features $x \in X$. Let us describe how to perform Learning Using Privileged Information (LUPI) for One Class SVM.

In order to incorporate privileged information into the anomaly detection framework, we construct the approximation $\xi_i \approx (w^* \cdot \phi^*(x_i^*)) + b^*$. Here $\phi^*(\cdot)$ is a mapping into some privileged kernel Hilbert space H_{ϕ^*}. In fact using such approximation we assume that thanks to the privileged patterns $(x_1^*, \ldots, x_l^*)$ we can refine the location of the separating boundary w.r.t. the sample of training objects. Also we impose a regularization on w^*.

As a result the optimization problem takes the form:

$$\nu l \cdot \frac{\|w\|^2}{2} + \gamma \cdot \frac{\|w^*\|^2}{2} - \nu l \rho + \sum_{i=1}^l (w^* \cdot \phi^*(x_i^*)) + b^* + \zeta_i \to \min_{w, w^*, \rho, \xi, \zeta}, \quad (3)$$

$$s.t. \begin{cases} (w \cdot \phi(x_i)) \geq \rho - (w^* \cdot \phi^*(x_i^*)) - b^*, \\ (w^* \cdot \phi^*(x_i^*)) + b^* + \zeta_i \geq 0, \ \zeta_i \geq 0. \end{cases}$$

In this formulation, γ is a regularization parameter for the slack variables approximation. We add ζ_i to prevent those approximations becoming negative. Note that if γ is small and the space H_{ϕ^*} is rich enough, the solution of (3) will be close to the solution of (1).

As before solving the dual of (3) is easier than solving the initial problem. In order to formulate the dual problem we write down the Lagrangian for (3):

$$L_{w,w^*,b^*,\rho,\zeta,\alpha,\beta,\nu} = \frac{\nu l}{2}\|w\|^2 - \nu l \rho + \frac{\gamma}{2}\|w^*\|^2 + \sum_{i=1}^l [(w^* \cdot \phi^*(x_i^*)) + b^* + \zeta_i] - \sum_{i=1}^l \kappa_i \zeta_i$$

$$- \sum_{i=1}^l \alpha_i [(w \cdot \phi(x_i)) - \rho + (w^* \cdot \phi^*(x_i^*)) + b^*] - \sum_{i=1}^l \beta_i [(w^* \cdot \phi^*(x_i^*)) + b^* + \zeta_i].$$

From Karush Kuhn Tucker conditions:

$$\partial L/\partial w = \nu l w - \sum_{i=1}^l \alpha_i \phi(x_i) \Rightarrow w = \frac{1}{\nu l} \sum_{i=1}^l \alpha_i \phi(x_i),$$

$$\partial L/\partial \zeta_i = \underbrace{1 - \beta_i}_{\delta_i} - \kappa_i = 0 \Rightarrow \delta_i = \kappa_i \geq 0, \ \partial L/\partial w^* = \gamma w^* - \sum_{i=1}^l (\alpha_i - \delta_i)\phi^*(x_i^*),$$

$$\partial L/\partial \rho = -\nu l + \sum_{i=1}^l \alpha_i \Rightarrow \sum_{i=1}^l \alpha_i = \nu l, \ \partial L/\partial b^* = l - \sum_{i=1}^l \alpha_i - \sum_{i=1}^l \beta_i = 0.$$

Using these results we get:

$$L_{w,w^*,b^*,\rho,\zeta,\alpha,\beta,\nu} = -\frac{1}{2\nu l} \sum_{i=1}^l \sum_{j=1}^l \alpha_i \alpha_j (\phi(x_i) \cdot \phi(x_j))$$

$$- \frac{1}{2\gamma} \sum_{j=1}^l (\delta_i - \alpha_i)(\delta_j - \alpha_j)(\phi^*(x_i^*) \cdot \phi^*(x_j^*)). \quad (4)$$

In matrix form the dual problem can be written down as follows:

$$-\frac{1}{\nu l}\alpha^T K \alpha - \frac{1}{\gamma}(\alpha - \delta)^T K^*(\alpha - \delta) \to \max_{\alpha,\delta} \ s.t. \begin{cases} e^T \alpha = \nu l, \ e^T \delta = \nu l, \\ 0 \leq \delta \leq 1, \ \alpha \geq 0. \end{cases} \quad (5)$$

Here $K^*(x^*, z^*) = (\phi^*(x^*) \cdot \phi^*(z^*))$ is a kernel function representing a dot product in some kernel Hilbert space H_{ϕ^*}, and $K^* = \{K^*(x_i^*, x_j^*)\}$ is a kernel matrix, $e = (1, \ldots, 1) \in \mathbb{R}^l$. By comparing (2) and (5) we get that the second summand in (5), containing privileged information, acts as a regularizer with the regularization parameter γ.

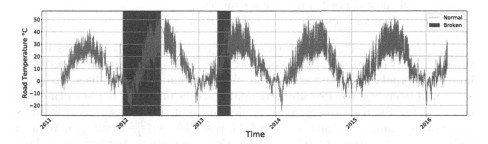

Fig. 3. Road temperature data

The decision function can be represented as follows:

$$\hat{f}(x) - (w \cdot \phi(x)) = \frac{1}{\nu l} \sum_{i-1}^{l} \alpha_i K(x, x_i).$$

The approximation of ξ can be represented as follows:

$$\hat{f}^*(x^*) = (w^* \cdot \phi^*(x^*)) = \frac{1}{\gamma} \sum_{i=1}^{l} (\alpha_i - \delta_i) K^*(x^*, x_i^*) + b^*.$$

Based on Karush Kuhn Tucker conditions we get that for every pair (x_i, x_i^*), such that $\alpha_i > 0$, $0 < \delta_i < 1$, the values of ρ and b^* can be calculated as follows:

$$\rho = \hat{f}(x_i) + \hat{f}^*(x_i^*), \ b^* = -\hat{f}^*(x_i^*).$$

3 Sensor Fault Detection in Road Weather Information Systems

Let us demonstrate the proposed approach by solving the problem of sensor fault detection in a Road Weather Information System.

3.1 Sensor Data

We collected data from eight RWIS stations, located in different regions of Russian Federation. Every station collects information about

- air temperature and humidity,
- wind direction, speed and gustiness,
- atmospheric pressure,
- dew point temperature,
- type and intensity of precipitation,
- road surface temperature,
- temperature in the depth of the pavement,
- amount of snow deposits on the road surface,
- presence of reagents on the surface of the road.

In Fig. 3 we provide an example of road temperature data. Sensors record these values at irregular moments of time. Thus we pre-processed time-series by linearly interpolating observations and obtained temperature values separated by 30 min intervals. We also used calendar features including day d of the year and time t of the day. To represent cyclic nature of this data we used features based on trigonometric functions: $\sin(\frac{2\pi d}{365})$, $\cos(\frac{2\pi d}{365})$, $\sin(\frac{2\pi t}{24})$, $\cos(\frac{2\pi t}{24})$. We also included lagged features to take into account information about the last three hours. Finally we used all these features to define the input space X.

3.2 METRo Model

The data obtained from the sensors can be used to build models and systems for road surface condition monitoring and prediction. One of the most common "physical" model is the so-called METRo model [3]. This model calculates the heat flow through the road surface, representing it as a sum:

$$R = (1 - \alpha)S + \varepsilon I - \varepsilon\sigma T^4 - H - L_a E \pm L_f P + A,$$

where

- $(1 - \alpha)S$ is a solar radiation,
- εI is an absorbed infrared radiation,
- $\varepsilon\sigma T^4$ is a radiated energy,
- H is a turbulent heat flow,
- $L_a E$ is a latent heat flux,
- $\pm L_f P$ is a heat change due to phase transition,
- A represents anthropomorphic sources.

The METRo model allows predicting the road surface conditions including the probability of icing. At the moment, its effectiveness has been tested in the Czech Republic [9], Canada [3] and USA [10]. In all of these cases they obtained sufficient forecasting accuracy using the METRo model.

It is important to note that in practice to get the most accurate predictions the METRo model requires to know the flow of heat, associated not only with weather conditions, but also with anthropomorphic sources; unfortunately, it is rather difficult to simulate this quantity [11]. Another problem is that the METRo forecast is deterministic, i.e. it is not possible to estimate forecast uncertainty. Although, the Monte Carlo approach can be used, it requires solution of the METRo equations with different initial conditions multiple times [12,13].

The described problems can be partially solved using machine learning methods. Machine learning methods allow to directly build forecasts based on the available sensor measurements [14,15], and can be used to aggregate forecasts obtained using physical models [16–20].

To build road condition forecasts using the METRo physical model, we also need the global weather forecasts including information about temperature, pressure, precipitation, etc. (see the METRo documentation for more details [3]). Unfortunately, we were not able to obtain the historical information about the global weather forecasts, and so we used real (observed) values of these parameters. As a result using the METRo model we built local forecasts of such parameters as pressure, temperature, humidity, different road conditions, type and amount of precipitation.

3.3 Learning Sample

The dataset was labeled manually based on sensor measurements. The fraction of records which were measured by broken sensors varied from 2% to 15% for different RWIS stations and types of sensors.

We use the output of the METRo model as the privileged information because we want to detect measurements from broken sensors before the forecasting system utilizes these measurements. Both privileged and original data was normalized by subtracting its mean value and dividing by the standard deviation.

As a training data we randomly selected 20000 measurements from correctly working sensors. As a test data we used all broken sensors' values and other correct measurements.

3.4 Anomaly Detection Accuracy Metric

Malfunction sensor detection is a highly imbalanced problem, and the cost of missing a broken sensor is higher than a false alarm. In this case the classification accuracy is not the best option as a measure of the effectiveness of anomaly detection. Precision and recall better reflect requirements of the considered application: recall is a fraction of all broken sensors that we are able to detect; precision is the number of correctly detected broken sensors divided by the number of all detections. We want to get both high precision and high recall on the test sample.

One Class SVM predicts not only binary labels but also a distance to the separating hyperplane which characterizes confidence of the prediction. This

328   D. Smolyakov et al.

property allows us to select different thresholds which can lead to different trade-offs between precision and recall. Thus as an aggregated accuracy metric we used the Area Under the Precision-Recall Curve (AUCPR): for every possible threshold value we calculate precision and recall and plot these values by using recall as the abscissa axis and precision as the ordinate axis. The bigger area under this curve is the better classifier is.

3.5 One Class SVM Parameters

In the considered case on average the actual fraction of anomalies, observed in data, is around 0.1. Since ν is a lower bound on the fraction of anomalous observations (see [21,22]), and our previous results showed that varying ν does not affect AUCPR significantly (see [23,24]), we set the value of ν to 0.1.

We used the Gaussian kernel both for the original and for the privileged features: $K_\sigma(x, x') = \exp\left\{-\frac{\|x-x'\|^2}{2\sigma^2}\right\}$. This kernel function corresponds to the feature transformation, which maps all points to the surface of a hypersphere, since $\|\phi(x)\|^2 = K_\sigma(x,x) = 1$. This property is crucial for One Class SVM since it builds linear decision rules.

The smaller the value of σ is, the sharper the decision rule we get, but this also increases the risk of overfitting. Large values of σ allow to get more stable decision rules which although can suffer from underfitting.

We fixed the width of the privileged space kernel to $(2\sigma^2) = 10^3$, and set the privileged space regularization parameter to $\gamma = 10^{-2}$. We calculated decision rules and accuracy metrics for various values of the original space kernel width from the interval $[10^{-4}, \ldots, 10^3]$. We selected this range because it contains both small values, producing overfitted decision rules, and big values, producing underfitted decision rules; therefore, we were able to demonstrate how the LUPI principle can work as a regularization technique.

3.6 Results

Typical results for eight RWIS stations are presented in Table 1. AUCPR depends on the proportion of anomalous observations, i.e. for random guessing AUCPR does not equal to 0.5. Thus in order to estimate it correctly to get some reference value we emulated an anomaly detector: for each observation from the test sample we generated anomaly detection score uniformly randomly in some interval and calculated AUCPR for such predictions. Results are provided in the column "Random guessing". By bold font we indicated values of AUCPR for One Class SVM+, which are different from those values for One Class SVM at the 5% statistical significance level. We used the bootstrap procedure to test the statistical significance.

Obtained results demonstrate that although on average the performance is not very impressive privileged information provides statistically significant increase in accuracy of predictions.

Table 1. AUCPR

Station	One Class SVM	One Class SVM+	Random guessing
RWIS Station 1	0.12	**0.14**	0.04
RWIS Station 2	0.19	0.2	0.13
RWIS Station 3	0.06	0.06	0.2
RWIS Station 4	0.04	0.04	0.02
RWIS Station 5	0.17	**0.19**	0.04
RWIS Station 6	0.18	**0.21**	0.07
RWIS Station 7	0.23	**0.25**	0.15
RWIS Station 8	0.08	**0.1**	0.06

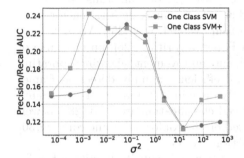

Fig. 4. Results on the validation set

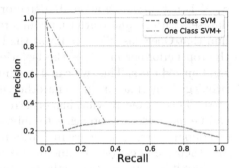

Fig. 5. Precision-recall on the test set

Table 2. Precision values for different recall values

Recall	**0.1**	**0.3**	0.5	0.7	0.9
One Class SVM	0.2	0.24	0.25	0.23	0.17
One Class SVM+	**0.8**	**0.3**	0.25	0.23	0.17
Random guessing	0.14	0.14	0.14	0.14	0.14

Figure 4 demonstrates some intuition about how the LUPI principle works. In this picture, using validation data from some RWIS station, we plotted AUCPR for different values of the kernel width. We can see that for One Class SVM+ the optimal kernel width is smaller than for original One Class SVM. In other words for One Class SVM+, the optimal solution came from a richer set of decision functions. The observed behavior illustrates the ability of privileged information to work as a regularizer, see the discussion after the formula (5) in Sect. 2.2.

Figure 5 shows the precision-recall curve for the test set. In order to obtain this figure, we used the best models, selected according to Fig. 4. We also provided a slice of the precision-recall curve in Table 2. The table demonstrates all available trade-offs between coverage of failures and detection accuracy. By bold font we indicated precision-recall values for One Class SVM+, which are different from those values for One Class SVM at the 5% statistical significance level. We used the bootstrap procedure to test the statistical significance.

4   Conclusions

We demonstrated practical applicability of the LUPI principle for anomaly detection: by incorporating privileged information into the One Class SVM approach we can improve accuracy of anomaly patterns recognition. The dependence of AUCPR on the kernel width demonstrated that with privileged information we can use richer hypothesis sets without the risk of overfitting. The reason that we obtained only marginal improvement is in the fact that original One Class SVM is not a reliable anomaly detector—experiments on numerous benchmark datasets with ground truth that compared popular anomaly detection algorithms find that One Class SVM ranks at the bottom (Table 1, p. 4 [25]). We note that the top performer in [25] is the Isolation Forest algorithm [26], an ensemble of randomized trees. Thus we can get significant improvement when incorporating privileged information in some other more efficient anomaly detection framework. This is exactly what has been recently done in [27]. Further, these results can be improved if approaches to imbalanced classification are used [28], as well as nonparametric confidence measures based on the conformal framework [29,30] and their modifications for anomaly detection [31–33].

Also it is clear that the applied problem at hand presents itself quite naturally as a time-series modeling task, since all involved quantities depend on time (see e.g. Fig. 3). In fact, in order to capture the time-related phenomena we used some naive and heuristic features (day, time, the trigonometric functions), see Sect. 3.1. Thus another direction of improvement would be to consider a proper time-series model (e.g. time-series change-point detection [34,35]), and/or a sequence processing learning machine (e.g. recurrent neural net, hybrid SVM/hidden Markov model) and to develop methods for incorporating privileged information in these models.

References

1. Hertl, S., Schaffar, G.: An autonomous approach to road temperature prediction. Meteorol. Appl. **5**(3), 227–238 (1998)
2. Sass, B.H.: A numerical model for prediction of road temperature and ice. J. Appl. Meteorol. **31**(12), 1499–1506 (1992)
3. Crevier, L.P., Delage, Y.: Metro: a new model for road-condition forecasting in Canada. J. Appl. Meteorol. **40**(11), 2026–2037 (2001)
4. Garcia-Font, V., Garrigues, C., et al.: A comparative study of anomaly detection techniques for smart city wireless sensor networks. Sensors **16**(6), 868 (2016)
5. Schölkopf, B., Williamson, R.C., et al.: Support vector method for novelty detection. In: Advances in Neural Information Processing Systems, pp. 582–588 (2000)
6. Tax, D.M., Duin, R.P.: Data domain description using support vectors. In: ESANN, vol. 99, pp. 251–256 (1999)

7. Vapnik, V., Vashist, A.: A new learning paradigm: learning using privileged information. Neural Netw. **22**(5–6), 544–557 (2009)
8. Pechyony, D., Vapnik, V.: On the theory of learning with privileged information. In: Advances in Neural Information Processing Systems, pp. 1894–1902 (2010)
9. Sokol, Z., Zacharov, P.: First experience with the application of the metro model in the Czech Republic. Atmos. Res. **143**, 1–16 (2014)
10. Rutz, J.J., Gibson, C.V.: Integration of a road surface model into NWS operations. Bull. Am. Meteorol. Soc. **94**(10), 1495–1500 (2013)
11. Khalifa, A., Marchetti, M., Bouilloud, L., Martin, E., Bues, M., Chancibaut, K.: Accounting for anthropic energy flux of traffic in winter urban road surface temperature simulations with the TEB model. Geosci. Model Dev. **9**, 547–565 (2016). https://doi.org/10.5194/gmd-9-547-2016
12. Chapman, L.: Probabilistic road weather forecasting. In: Proceedings of the Standing International Road Weather Conference (SIRWEC 2012) (2012)
13. Berrocal, V., Raftery, A., et al.: Probabilistic weather forecasting for winter road maintenance. J. Am. Stat. Assoc. **105**(490), 522–537 (2010)
14. Zaytar, M.A., El Amrani, C.: Sequence to sequence weather forecasting with long short-term memory recurrent neural networks. Int. J. Comput. Appl. **143**(11), 7–11 (2016)
15. Gensler, A., Henze, J., et al.: Deep learning for solar power forecastingan approach using autoencoder and LSTM neural networks. In: 2016 IEEE International Conference on Systems, Man, and Cybernetics (SMC), pp. 002858–002865. IEEE (2016)
16. Martin, R., Aler, R., Valls, J.M., Galván, I.M.: Machine learning techniques for daily solar energy prediction and interpolation using numerical weather models. Concurr. Comput.: Pract. Exp. **28**(4), 1261–1274 (2016)
17. Lu, S., Hwang, Y., et al.: Machine learning based multi-physical-model blending for enhancing renewable energy forecast-improvement via situation dependent error correction. In: 2015 European Control Conference (ECC), pp. 283–290. IEEE (2015)
18. Zaytsev, A., Burnaev, A.: Minimax approach to variable fidelity data interpolation. In: Proceedings of the 20th International Conference on AISTATS, PRML, vol. 54, pp. 652–661 (2017)
19. Zaytsev, A., Burnaev, E.: Large scale variable fidelity surrogate modeling. Ann. Math. Artif. Intell. **81**(1), 167–186 (2017)
20. Burnaev, E., Zaytsev, A.: Surrogate modeling of multifidelity data for large samples. J. Commun. Technol. Electron. **60**(12), 1348–1355 (2015)
21. Wang, S., Liu, Q., et al.: Hyperparameter selection of one-class support vector machine by self-adaptive data shifting. Pattern Recognit. **74**, 198–211 (2018)
22. Thomas, A., Feuillard, V., et al.: Calibration of one-class SVM for MV set estimation. CoRR abs/1508.07535 (2015)
23. Burnaev, E., Smolyakov, D.: One-class SVM with privileged information and its application to malware detection. In: 2016 IEEE 16th International Conference on Data Mining Workshops (ICDMW), pp. 273–280. IEEE (2016)
24. Smolyakov, D., Erofeev, P., Burnaev, E.: Model selection for anomaly detection. In: Proceedings of SPIE 9875, Eighth International Conference on Machine Vision, vol. 9875 (2015)
25. Emmott, A., Das, S., et al.: Systematic construction of anomaly detection benchmarks from real data. In: Proceedings of KDD ODD (2013)
26. Liu, F., Ting, K., Zhou, Z.H.: Isolation forest. In: Proceedings of ICDM (2008)

27. Shekhar, S., Akoglu, L.: Incorporating privileged information to unsupervised anomaly detection. CoRR abs/1805.02269 (2018)
28. Papanov, A., Erofeev, P., Burnaev, E.: Influence of resampling on accuracy of imbalanced classification. In: Proceedings of SPIE 9875, Eighth International Conference on Machine Vision, vol. 9875 (2015)
29. Vovk, V., Burnaev, E.: Efficiency of conformalized ridge regression. Proceedings of COLT abs/1404.2083 (2014)
30. Nazarov, I., Burnaev, E.: Conformalized kernel ridge regression. In: 15th IEEE International Conference on Machine Learning and Applications, ICMLA 2016, pp. 45–52 (2016)
31. Ishimtsev, V., Bernstein, A., Burnaev, E., Nazarov, I.: Conformal k-NN anomaly detector for univariate data streams. In: Proceedings of the 6th Workshop on Conformal and Probabilistic Prediction and Applications, PRML, vol. 60, pp. 213–227 (2017)
32. Safin, A., Burnaev, E.: Conformal kernel expected similarity for anomaly detection in time-series data. Adv. Syst. Sci. Appl. **17**(3), 22–33 (2017)
33. Volkhonskiy, D., Burnaev, E., Nouretdinov, I., Gammerman, A., Vovk, V.: Inductive conformal martingales for change-point detection. In: Proceedings of the 6th Workshop on Conformal and Probabilistic Prediction and Applications, PRML, vol. 60, pp. 132–153 (2017)
34. Artemov, A., Burnaev, E., Lokot, A.: Nonparametric decomposition of quasi-periodic time series for change-point detection. In: Proceedings of SPIE 9875, Eighth International Conference on Machine Vision, vol. 987520 (2015)
35. Artemov, A., Burnaev, E.: Ensembles of detectors for online detection of transient changes. In: Proceedings of SPIE 9875, Eighth International Conference on Machine Vision, vol. 98751Z (2015)

Capturing Suprasegmental Features of a Voice with RNNs for Improved Speaker Clustering

Thilo Stadelmann[1(✉)], Sebastian Glinski-Haefeli[1], Patrick Gerber[1], and Oliver Dürr[1,2]

[1] ZHAW Datalab, Zurich University of Applied Sciences, Winterthur, Switzerland
stdm@zhaw.ch, sebastian.glinski@gmail.com, gerber.pat@gmail.com,
oliver.duerr@gmail.com
[2] Institute for Optical Systems, Konstanz University of Applied Sciences,
Konstanz, Germany

Abstract. Deep neural networks have become a veritable alternative to classic speaker recognition and clustering methods in recent years. However, while the speech signal clearly is a time series, and despite the body of literature on the benefits of prosodic (suprasegmental) features, identifying voices has usually not been approached with sequence learning methods. Only recently has a recurrent neural network (RNN) been successfully applied to this task, while the use of convolutional neural networks (CNNs) (that are not able to capture arbitrary time dependencies, unlike RNNs) still prevails. In this paper, we show the effectiveness of RNNs for speaker recognition by improving state of the art speaker clustering performance and robustness on the classic TIMIT benchmark. We provide arguments why RNNs are superior by experimentally showing a "sweet spot" of the segment length for successfully capturing prosodic information that has been theoretically predicted in previous work.

Keywords: Speaker clustering · Speaker recognition
Recurrent neural network

1 Introduction

Automatic speaker recognition comes in many flavors, of which speaker clustering is the most unconstrained and hence the most difficult one [3,41]. It can be defined as the task of judging if two short utterances come from the same (previously unknown) speaker, and thus forms a suitable benchmark for the general ability of a system to capture what makes up a voice: speaker clustering can only be solved satisfactory by regarding all available cues in the utterances themselves. This distinguishes speaker clustering from a more complex experimental setup like e.g. speaker diarization, where engineering a complex system of many components has a not negligible influence on the final result besides the pure voice modeling [2]; and for example from speaker identification, where

© Springer Nature Switzerland AG 2018
L. Pancioni et al. (Eds.): ANNPR 2018, LNAI 11081, pp. 333–345, 2018.
https://doi.org/10.1007/978-3-319-99978-4_26

more available data enables the creation of models that work well just because of the sheer amount of collected training statistics [34]. Previous work [41] hence suggests that the bottleneck for speaker clustering performance lies in exploiting the supra-frame information present in the audio signal's evolution in time. This information of how single audio frames depend on each other in a speaker-characteristic way can be identified with the prosodic features of a voice—with its "sound".

Recently, deep neural networks (DNN) have been successfully applied to various speaker recognition tasks [7,22,24,28,37,45], reaching and exceeding state of the art results of classic GMM- [35] or i-vector-based [8] systems. With few exceptions, systems based on convolutional neural network (CNN) architectures have been used on spectrograms for their unprecedented performance on visual recognition tasks [21]. While spectrograms encode time information of an audio signal explicitly on the image's horizontal axis, and CNNs can in principle learn temporal patterns by having filters with an extent in the horizontal direction, CNNs are per se not sequence learning models. Recurrent neural networks (RNN) [36,38] instead are explicitly built with temporal modeling in mind [12] and have shown exceptional performance on other audio recognition tasks [4,13,17,19]. However, RNNs have only very recently been successfully applied to the task of speaker recognition for the first time (see Sect. 2.1). This discrepancy may be related to the reported difficulty to train recurrent models successfully [32].

In this paper, we present a quantitative and qualitative analysis of RNNs for speaker clustering on the TIMIT database. We demonstrate results that slightly improve state of the art on the evaluation set of 40 speakers, while being more robust to hyperparameter choice and model initializations than previously used CNN models. More importantly, extensive experiments allow the conclusion that the RNN model achieves this performance through an ability to model voice prosody effectively[1]. This contributes a strong rationale with empirical evidence to the recently published first experiments with RNNs for speaker recognition, and provides valuable insights into the workings of deep learning approaches on audio data. Specifically, we empirically confirm the "sweet spot" of the segment length to capture time-dependent (prosodic) information that has been predicted in earlier work [41]. Section 2 provides an introduction and the background to our approach, including related work. Section 3 reports on our experimental setup and results, before findings are discussed in Sect. 4. Section 5 finally provides conclusions and suggestions for future work.

---

[1] Prosody can be defined on the application level as the "use of suprasegmental features to convey sentence-level pragmatic meanings" [20]. While we do not study the mediation of meaning, we do claim to capture *"those elements of speech that are not [elements of] individual phonetic segments (vowels and consonants) but [...] of syllables and larger units of speech"* https://en.wikipedia.org/wiki/Prosody_(linguistics).

2   Speaker Clustering with RNNs

2.1   Related Work

Speaker embeddings are fixed-size vectorial representations of a voice, formed by the activations of neurons at higher-level layers of a neural network during the forward pass of a respective speech utterance of that voice. Learning speaker embeddings has been approached with different neural network architectures such as Siamese [6], fully connected [43], and CNN [10,24,25]. Only very recently have RNNs been used successfully [23,33,44].

While for classification problems, where the network learns to distinguish known classes, the cross entropy is the natural choice for the loss function [11], there is no natural choice for learning embeddings and hence there exist many different approaches to choose a suitable loss. *Garcia-Romera et al.* [10] learn embeddings through the training of a deep neural network. They consider a learnable distance function defined in the spirit of probabilistic linear discriminant analysis (PLDA) [5]. This distance function is then used as the input to a binary cross entropy to classify if two segments come from the same speaker or not. To account for the fact that there are more pairs of segments between different speakers than between identical speakers, they introduce a weighting in the loss function. Using the distance function in an agglomerative hierarchical clustering and further refining the result with variational Bayes re-segmentation [40], diarization error rates (DER) between 11.2% and 9.9% on the CALLHOME corpus are achieved. While the network training and in particular the loss function of this approach is comparable to ours, the temporal information is still extracted by a classical feed-forward neural net instead of a sequence-learning RNN.

Cyrta et al. [33] suggest to learn the embeddings by training a recurrent convolutional neural network for the task of speaker classification. Although they utilize recurrent layers to retrieve temporal information, the feature extraction is still done by convolutional layers. Furthermore, the training is done using a surrogate classification task with the standard cross-entropy as a loss function, and no clustering specific loss is defined. They reportedly outperform the DER of a GMM-based baseline [27] with 30% relative improvement, and a CNN [24] with 12% relative improvement on a novel dataset.

Li et al. [23] experiment with two architectures: a residual CNN [14], and a RNN using gated recurrent units on top of a conv layer. Two successive training stages are performed, initially with cross entropy loss, followed by triplet loss [39] to minimize intra-speaker distances while maximizing inter-speaker distances. Best results are achieved by the residual CNN with both losses. They report an equal error rate (EER) of 1.13% and accuracy (ACC) of 96.83% for text-independent speaker identification on the Mandarin and English UIDs dataset, a relative improvement of EER and ACC by 50% and 60%, respectively, over a DNN-based i-vector approach. Additionally, they show good transfer learning capabilities from Mandarin to English speaker recognition. However, in our work, we achieve state of the art speaker clustering performance without using an

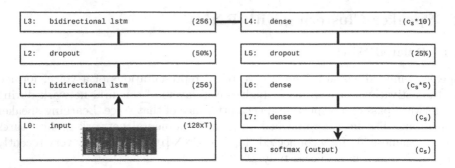

Fig. 1. Neural network architecture used for all experiments, including important hyperparameters of each layer like input size, number of neurons and dropout factor. The width T (in pixels) of each input segment is varied in our experiments.

additional convolutional network in front of the RNN or dual training, showing the sufficiency of a recurrent architecture alone for speaker modeling.

Finally, *Wang et al.* [44] use long short-term memory (LSTM) cells indirectly to convert MFCC feature vectors to embeddings that are fed into a subsequent off-line spectral clustering [31] process. They reach an absolute DER of 12.0% on the CALLHOME dataset, where other DNN- and i-vector based studies achieved between 14.9% and 12.1% DER (all numbers without additional resegmentation). The approach is similar to ours in using only a RNN to extract speaker embeddings, but differs in training (network architecture, loss function) as well as front-end (features) and post-processing (clustering approach).

2.2 Overview of Our Approach

To learn speaker embeddings, our neural network (cp. Figure 1) uses a combination of bidirectional LSTM (BLSTM) layers (L1 and L3) with additional fully connected layers (L4, L6 and L7) right in front of the output layer L8 with c_s neurons. We chose c_s to equal the number of speakers in the training set and input the audio in form of an image (spectrogram). We chose BLSTM layers because their awareness of previous and upcoming sequence steps, i.e., past and future time steps in the signal. They are thus able to relate current information to its relevant temporal context. It is shown in [24] that it is not advisable to use the final layer trained for a surrogate task for extracting the embeddings; we thus add three dense layers between the last BLSTM and the output to later experiment with from where to actually extract our embeddings.

Training. We treat the c_s-dimensional output vector from layer L8 as a distribution that should be similar for all spectrograms from the same speaker, and dissimilar for spectrograms from different speakers. We train the network to give this output by using a loss function based on the pairwise Kullback-Leibler divergence (PKLD), as described in [15]: it enforces said within-speaker similarity/between-speaker dissimilarity between all possible pairs of spectrograms in each mini batch (see below). This helps the neural network to be fit for clustering previously unseen speakers, as it is not specifically forcing the network to learn a one-hot encoding of the speaker identity as would be the case when using cross entropy. Rather, it ensures a voice-specific arbitrary discrete distribution.

For each mini batch, the loss is computed as follows: first, the mini batch is created by randomly selecting c_m segments of length $T \cdot 10\,\text{ms}$ from the training set (a mini batch thus does not contain a fixed number of speakers). Then, each segment within a mini batch is converted to a mel spectrogram with T time steps (columns) according to [24] and passed forward through the network, resulting in an output distribution at layer L8. Finally, the loss function is calculated for all possible pairs of output distributions (p, q) within a mini batch as follows.

If the two outputs p and q are from the same speaker, the Kullback-Leibler (KL) divergence is calculated:

$$KL(\mathbf{p} \parallel \mathbf{q}) = \sum_i^{c_s} p_i \log \frac{p_i}{q_i}. \tag{1}$$

Otherwise, if a paired output is from different speakers, the hinge loss is calculated:

$$HL(\mathbf{p} \parallel \mathbf{q}) = \max(0, \text{margin} - KL(\mathbf{p} \parallel \mathbf{q})) \tag{2}$$

where the margin hyperparameter defines the maximum distance between two elements of a pair. Both loss terms are combined as follows:

$$\text{loss}(\mathbf{p} \parallel \mathbf{q}) = I_s \cdot KL(\mathbf{p} \parallel \mathbf{q}) + I_{ds} \cdot HL(\mathbf{p} \parallel \mathbf{q}) \tag{3}$$

where I_s equals 1 for pairs from the same speaker and 0 for pairs from different speakers. Inversely, I_{ds} equals 1 for pairs from different speakers and 0 for pairs from the same speaker. Finally, we symmetrize the loss function via:

$$L(\mathbf{p}, \mathbf{q}) = \text{loss}(\mathbf{p} \parallel \mathbf{q}) + \text{loss}(\mathbf{q} \parallel \mathbf{p}) \tag{4}$$

Clustering. To perform speaker clustering on a completely disjunct set of test speakers, the trained neural network is applied to their utterances (chopped into $T \cdot 10\,\text{ms}$ long segments), and the respective embeddings are extracted during a forward pass. We experiment with different layers as potential sources to extract the embedding vectors from, ranging from L3 to L8. Then, hierarchical agglomerative clustering [29] is used off-line to perform the actual clustering of these vectors (see Fig. 2).

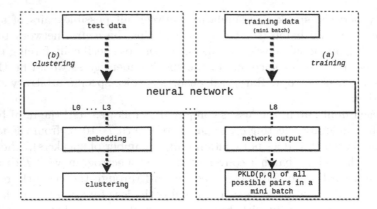

Fig. 2. Schematic of the training and clustering process with our neural network: (a) training is executed by computing the loss between all possible pairs of outputs from L8 within a mini batch; (b) AHC clustering is fed with embeddings extracted from a lower layer during a forward pass on the trained neural network.

3   Experimental Evaluation

The goal of these experiments is to provide support for the assumption that RNNs capture prosodic features of a voice. The experiments are evaluated on the TIMIT [9] corpus, which has been chosen for two reasons: (a) its cleanness, to prevent detraction from the pure voice modeling aspect; and (b) to compare directly with related work on speaker clustering [24,25,41]. Despite speaker recognition progress in more challenging environments like meeting diarization [1] or less constrained speaker identification [30], speaker clustering of even just 40 speakers from TIMIT only recently gave reasonable results [24], while attempting to cluster all 630 speakers failed altogether [41]. Studying pure voice modeling capability in isolation thus still seems appropriate, despite the clean studio recordings of sufficient length in TIMIT. The code for our experiments can be found online[2].

3.1   Experimental Setup

In accordance with [25], we perform all training of neural networks on the `speaker_100_50w_50m_not_reynolds` subset of 100 TIMIT speakers[3]. Of these speakers, randomly selected 80% are used as training data and 20% for validation during the training procedure. All audio is converted to mel spectrograms of 128 pixels height (mel-scaled frequency bins). Unlike [24,25], who used a fixed input width of 100 pixels (1,000 ms) for the network, we experiment with different segment length below. We chose a batch size of $c_m = 100$ in conjunction with the Adam optimizer [16] (and unchanged standard parameters learningrate = 0.001, $\beta_1 = 0.9$, $\beta_2 = 0.999$, $\epsilon = 1e^{-8}$, and decay = 0.0). All trainings run for 10,000

---

[2] See https://github.com/stdm/ZHAW_deep_voice.

[3] See https://github.com/stdm/ZHAW_deep_voice/tree/master/common/data/speaker_lists on GitHub.

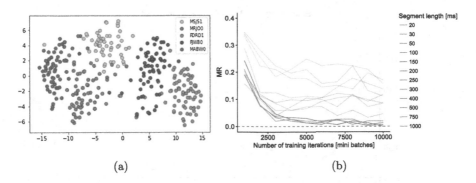

(a)   (b)

Fig. 3. (a) A t-SNE visualization [26] of all single embeddings extracted from all available data of 5 TIMIT speakers, colored by speaker identity. (b) MR as a function of training iterations for all evaluated segment lengths, evaluated every 1,000 mini batches on the 38 speaker evaluation set. For visual clarity, segment lengths of 300 ms and below are faded.

mini batches. For the margin parameter of the PKLD, Hsu et al. [15] suggested a value of 2; we determined margin $= 3$ to work best after grid search within $\{1.5, 2, 2.5, 3\}$ for speaker clustering tasks with $N = \{40, 60, 80\}$ speakers.

Evaluation is based on the `speakers_40_clustering_vs_reynolds` list from [41] in two stages: intermediate experiments are performed on a 38 speaker subset of this list, where the 10 sentences per speaker are randomly split into 2 utterances, 8 and 2 sentences long, respectively[4]. Final evaluations are done in accordance with [41] on the complete list of 40 speakers, using the first 8 sentences (lexicographically ordered by filename) of each speaker to form utterance 1 and the remaining 2 sentences for the second utterance[5]. As in [24,25], we finally use agglomerative hierarchical clustering (AHC) with complete-linkage and the cosine distance between embeddings, and average multiple embeddings per utterance prior to entering AHC for utterances longer than the segment length. Figure 3a visually confirms this practice of averaging.

We evaluate each clustering result using the misclassification rate (MR) as introduced by Kotti et al. [18]: $MR = \frac{1}{N} \sum_{j=1}^{N_{cl}} e_j$, where N is the overall number of embeddings to cluster, N_{cl} the number of found clusters, and e_j is defined as the number of embeddings in cluster j that are not assigned to the *correct* cluster. The unique correct cluster for any speaker is arguably the one that fulfills the following two conditions: (a) it is the cluster with the largest number of embeddings from this speaker; and (b) if there are also embeddings from other speakers in this cluster, their number is smaller. However, previous work [24,25] used a slightly more conservative definition, adding two more necessary conditions: (c) clusters holding only one embedding cannot be correct; and (d)

---

[4] Both changes to the setup of [41] are due to unintentional anomalies in the data loading process that got corrected later. The missing speakers are the well-clustering FPKTO and FAKSO (see Fig. 5), thus results aren't expected to change much.

[5] Evidence in the source code suggests that [24,25] used random allocation here, too.

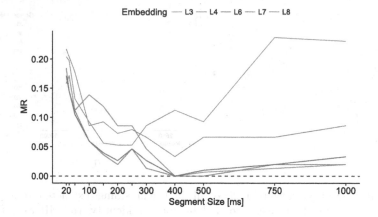

Fig. 4. MR on the 38 speaker evaluation set as a function of the input segment length used for different embedding layers (L3/4/6/7/8), averaged over 2 runs. All layers show a "sweet spot" for the segment length; the best performing layers L3–L6 have this "sweet spot" at around 400 ms with MR = 0.0.

all embeddings in a cluster with mixed speakers are incorrect. In this paper, we additionally report MR using the more reasonable (and likely more popular) interpretation of correct clusters with only conditions (a)–(b).

3.2   Results

Our goal in the following experiments is to vary the input segment length available for the RNN to learn the temporal dependencies, in order to verify if a "sweet spot" and thus evidence for prosodic feature modeling as shown by [41] is found. To use an optimal number of training steps, we first evaluate the MR against the number of training iterations as the first intermediate experiment.

Figure 3b shows how different networks for the varying segment lengths perform on the clustering task, depending on how much they have been trained in terms of number of mini batches. Two things can be seen: first, the networks that perform well/average/poorly after full training do so as well after shorter training. Second, training (at least for models trained with 400 ms segments or longer) seems to stabilize somewhat after ca. 5,000 mini batches; it does not appear as if longer training would have significantly altered the results presented above nor do we observe significant overfitting. In the following, we therefore set the number of training steps to "full" 10,000 mini batches for all experiments, also to avoid cherry picking of particularly well-behaving training snapshots.

Next, we study the effect of varying segment sizes for the different embedding layers (see Fig. 4) as the second intermediate experiment. The results are averages of two runs with independently trained networks of different random weight initializations to account for potential instabilities in the training. With the layers L3, L4, and L6, we achieve a perfect clustering, while the layers L7 and L8 perform much worse. For all layers, we observe the following universal behavior:

Table 1. Comparison of model performances on the TIMIT 40 speaker evaluation set, given in terms of MR and the more strict legacy variant.

Method	MR	MR (legacy)
RNN/w PKLD	$\mathbf{2.19\%} \left(\frac{1.25\% + 2.5\% + 1.25\% + 3.75\%}{4} \right)$	$\mathbf{4.38\%}$ (average of 4 runs)
CNN/w PKLD [25]	-	5%
CNN/w cross entropy [24]	-	5%
ν-SVM [41]	6.25%	-
GMM/MFCC [41]	12.5%	-

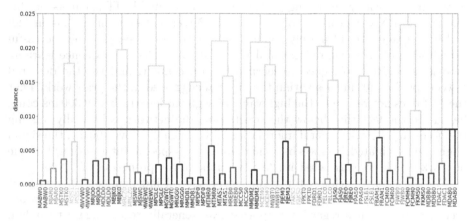

Fig. 5. The (lower part of a) dendrogram of all 40 speakers from the evaluation set, created with one of our best performing models (segment length 400 ms, embeddings extracted from L3). Misclassifications arise only for speaker MCCS0.

with increasing segment size, the MR first decreases and reaches a minimum before it rises again. The minimum or "sweet spot" is at about 400 ms for the lower layers L3–L6 and at about 150–250 ms for the final layer L8.

Finally, we evaluate one of our best-performing network configurations (segment length 400 ms, embedding extraction from L3) on the full 40 speakers test set to compare with previous results. Table 1 shows that the pure clustering performance is slightly better in terms of the more strict "legacy" MR than the work of Lukic et al. [24, 25]. It is computed as the average over 4 independently trained networks with identical parameter settings, showing the independence of the result from random fluctuations in the optimization. This has to be seen in contrast to the reported dependence of the results in [25] from optimizer hyperparameters and large number of training rounds. Figure 5 shows a dendrogram of one of the best performing clusterings (MR = 0.0125) for reference.

4   Discussion

Stadelmann and Freisleben [41] suggested that using prosodic features of a voice is likely to improve speaker clustering performance by an order of magnitude. They considered a temporal context lengths of 32–496 ms length as potentially reasonable and demonstrated that an implementation using MFCCs and one-class SVMs was able to realize parts of this potential using a context length of 130 ms, reaching MR = 0.0625 on the 40 speaker subset of TIMIT.

Our experimental results above suggest that for our RNN implementation, a temporal context length around 400 ms is optimal, slightly increasing the state of the art in speaker clustering performance on the 40 TIMIT speakers. Is this due to the RNN capturing voice prosody? Perceptual experiments as in [42] will be necessary to ultimately confirm it, but two facts suggest the successful exploitation of these suprasegmental features: (a) RNNs are sequence learning models, specifically designed to learn temporal context; and (b) the detected "sweet spot" of 400 ms lies within the interval suggested in the prior work. That the MR does not return to the high value of segment lengths below 300 ms could be due to the fact that below the sweet spot, the network is basically missing the necessary suprasegmental information in the signal; above 400 ms, the signal might be dominated by speech- rather than voice-specific information, but the network is able to learn what it shall by virtue of the properties of RNNs.

We consistently achieve top results over several independent runs without specifically tuning the hyperparameters of the Adam optimizer or the network architecture; we thus regard our approach as generally robust against random influences (weight initialization, mini batch constitution, architectural choices). We observe however a dependency on the data: some particularly well clustering speakers like e.g. MREB0 (according to low distances in Fig. 5) where already mentioned by [41] to be easily clustered by humans, while on the other hand the few misclassifications consistently involve speaker MCSS0. Finally, results fluctuate depending on which sentences are grouped to utterances—thus, for comparability it is important to have returned to the original experimental setup of [41] in this paper with respect to MR definition and utterance assembly.

The results with respect to which layer we use for extracting embeddings are interesting: we originally chose the PKLD loss function to train on a task as close as possible to our actual goal of clustering. Related work [24] had shown that a network learned purely for speaker identification (a surrogate task) is not ideal for later determining the similarity of unknown voices. This could be mitigated if an earlier than the final layer is used. The intuition behind this is that at the final layer is too well adapted to the specific voices seen during training, whereas the lower embeddings are more abstract. A continuation of said work [25] suggested to use the PKLD to improve this issue. It is thus interesting that we still need to extract embeddings from a lower layer—a layer farther away from the trained task—in order to achieve optimal results for speaker clustering. This suggests that even training for a pairwise similarity or dissimilarity of distributions is still a surrogate task far away from the actual task of speaker clustering;

and that learning speaker clustering end-to-end with the actual clustering task incorporated in the loss function and network output could improve this.

5 Conclusions and Future Work

In this work, executed simultaneously to first published results on RNNs for speaker embeddings, we demonstrated that recurrent neural networks can model prosodic, i.e., suprasegmental, aspects of a voice. We were able to show that specifically bidirectional LSTMs in combination with the PKLD loss function perform better than any other machine learning approach tested so far for speaker clustering on the TIMIT corpus and seem fit for voice modeling per se. Furthermore, our results show a "sweet spot" for extracting temporal context information with this kind of RNN at around 400 ms for a range of embedding extraction layers without extensive tuning of the optimizer and other hyperparameters.

Future work will include more challenging data as VoxCeleb [30]. Additionally, the following aspects in the presented approach offer room for further analysis: how to better inform the clustering stage about the common bond of multiple embeddings from the same utterance, beyond averaging; how to use deeper RNN architectures to exploit more speaker data during training; how to scale to considerable more than 40 speakers/80 utterances; how to formulate a more suitable surrogate task for training, towards end-to-end neural clustering.

Acknowledgements. We thank Timon Gygax and Jörg Egli for their detective work on the MR, Benjamin Heusser and Savin Niederer for unifying the code base, and Niclas Simmler and Amin Trabi for helpful discussions. We are thankful to the anonymous reviewers for helpful comments.

References

1. Anguera, X., Bozonnet, S., Evans, N., Fredouille, C., Friedland, G., Vinyals, O.: Speaker diarization: a review of recent research. TASLP **20**, 356–370 (2012)
2. Barras, C., Zhu, X., Meignier, S., Gauvain, J.L.: Multistage speaker diarization of broadcast news. TASLP **14**, 1505–1512 (2006)
3. Beigi, H.: Fundamentals of Speaker Recognition. Springer, Boston (2011). https://doi.org/10.1007/978-0-387-77592-0
4. Boulanger-Lewandowski, N., Bengio, Y., Vincent, P.: Audio chord recognition with recurrent neural networks. In: ISMIR (2013)
5. Burget, L., Plchot, O., Cumani, S., Glembek, O., Matějka, P., Brümmer, N.: Discriminatively trained probabilistic linear discriminant analysis for speaker verification. In: ICASSP (2011)
6. Chen, K., Salman, A.: Extracting speaker-specific information with a regularized siamese deep network. In: NIPS (2011)
7. Chen, K., Salman, A.: Learning speaker-specific characteristics with a deep neural architecture. Trans. Neural Netw. **22**, 1744–1756 (2011)
8. Dehak, N., Kenny, P.J., Dehak, R., Dumouchel, P., Ouellet, P.: Front-end factor analysis for speaker verification. TASLP **19**, 788–798 (2011)

9. Fisher, W., Doddington, G., Marshall, G.K.: The DARPA speech recognition research database: specification and status. In: Proceedings of the DARPA Speech Recognition Workshop, pp. 93–100 (1986)
10. Garcia-Romero, D., Snyder, D., Sell, G., Povey, D., McCree, A.: Speaker diarization using deep neural network embeddings. In: ICASSP (2017)
11. Goodfellow, I., Bengio, Y., Courville, A.: Deep Learning. MIT Press, Cambridge (2016)
12. Graves, A., Fernández, S., Gomez, F., Schmidhuber, J.: Connectionist temporal classification: labelling unsegmented sequence data with recurrent neural networks. In: ICML (2006)
13. Graves, A., Mohamed, A.R., Hinton, G.: Speech recognition with deep recurrent neural networks. In: ICASSP. IEEE (2013)
14. He, K., Zhang, X., Ren, S., Sun, J.: Deep residual learning for image recognition. In: CVPR (2016)
15. Hsu, Y.C., Kira, Z.: Neural network-based clustering using pairwise constraints. arXiv:1511.06321 (2015)
16. Kingma, D.P., Ba, J.: Adam: a method for stochastic optimization. arXiv:1412.6980 (2014)
17. Kolbæk, M., Yu, D., Tan, Z.H., Jensen, J.: Multi-talker speech separation and tracing with permutation invariant training of deep recurrent neural networks. arXiv:1703.06284 (2017)
18. Kotti, M., Moschou, V., Kotropoulos, C.: Speaker segmentation and clustering. Signal Process. **88**, 1091–1124 (2008)
19. Koutnik, J., Greff, K., Gomez, F., Schmidhuber, J.: A clockwork RNN. arXiv:1402.3511 (2014)
20. Ladd, D.R.: Intonational Phonology. Cambridge University Press, Cambridge (2008)
21. LeCun, Y., Bengio, Y., Hinton, G.: Deep learning. Nature **521**, 436 (2015)
22. Lee, H., Pham, P., Largman, Y., Ng, A.Y.: Unsupervised feature learning for audio classification using convolutional deep belief networks. In: NIPS (2009)
23. Li, C., et al.: Deep speaker: an end-to-end neural speaker embedding system. arXiv:1705.02304 (2017)
24. Lukic, Y., Vogt, C., Dürr, O., Stadelmann, T.: Speaker identification and clustering using convolutional neural networks. In: MLSP (2016)
25. Lukic, Y., Vogt, C., Dürr, O., Stadelmann, T.: Learning embeddings for speaker clustering based on voice equality. In: MLSP (2017)
26. van der Maaten, L., Hinton, G.: Visualizing data using t-SNE. JMLR **9**, 2579–2605 (2008)
27. Meignier, S., Merlin, T.: LIUM SpkDiarization: an open source toolkit for diarization. In: CMU SPUD Workshop (2010)
28. Milner, R., Hain, T.: DNN-based speaker clustering for speaker diarisation. In: Interspeech (2016)
29. Murtagh, F.: A survey of recent advances in hierarchical clustering algorithms. Comput. J. **26**, 354–359 (1983)
30. Nagrani, A., Chung, J.S., Zisserman, A.: Voxceleb: a large-scale speaker identification dataset. arXiv:1706.08612 (2017)
31. Ng, A.Y., Jordan, M.I., Weiss, Y.: On spectral clustering: analysis and an algorithm. In: NIPS (2002)
32. Pascanu, R., Mikolov, T., Bengio, Y.: On the difficulty of training recurrent neural networks. In: ICML (2013)

33. Cyrta, P., Trzciúski, T., Stokowiec, W.: Speaker diarization using deep recurrent convolutional neural networks for speaker embeddings. arXiv:1708.02840 (2017)
34. Reynolds, D.A.: Speaker identification and verification using Gaussian mixture speaker models. Speech Commun. **17**, 91–108 (1995)
35. Reynolds, D.A., Quatieri, T.F., Dunn, R.B.: Speaker verification using adapted Gaussian mixture models. Digit. Signal Process. **10**, 19–41 (2000)
36. Robinson, A., Fallside, F.: The utility driven dynamic error propagation network. University of Cambridge Department of Engineering (1987)
37. Rouvier, M., Bousquet, P.M., Favre, B.: Speaker diarization through speaker embeddings. In: EUSIPCO. IEEE (2015)
38. Schmidhuber, J.: Deep learning in neural networks: an overview. Neural Netw. **61**, 85–117 (2015)
39. Schroff, F., Kalenichenko, D., Philbin, J.: Facenet: a unified embedding for face recognition and clustering. In: CVPR (2015)
40. Sell, G., Garcia-Romero, D.: Diarization resegmentation in the factor analysis subspace. In: ICASSP (2015)
41. Stadelmann, T., Freisleben, B.: Unfolding speaker clustering potential: a biomimetic approach. In: ACM Multimedia (2009)
42. Stadelmann, T., Wang, Y., Smith, M., Ewerth, R., Freisleben, B.: Rethinking algorithm design and development in speech processing. In: ICPR (2010)
43. Variani, E., Lei, X., McDermott, E., Moreno, I.L., Gonzalez-Dominguez, J.: Deep neural networks for small footprint text-dependent speaker verification. In: ICASSP (2014)
44. Wang, Q., Downey, C., Wan, L., Mansfield, P.A., Moreno, I.L.: Speaker diarization with LSTM. arXiv:1710.10468 (2017)
45. Yella, S.H., Stolcke, A., Slaney, M.: Artificial neural network features for speaker diarization. In: SLT Workshop. IEEE (2014)

Trace and Detect Adversarial Attacks on CNNs Using Feature Response Maps

Mohammadreza Amirian[1,2(✉)], Friedhelm Schwenker[2], and Thilo Stadelmann[1]

[1] ZHAW Datalab & School of Engineering, Winterthur, Switzerland
{amir,stdm}@zhaw.ch
[2] Institute of Neural Information Processing, Ulm University, Ulm, Germany
friedhelm.schwenker@uni-ulm.de

Abstract. The existence of adversarial attacks on convolutional neural networks (CNN) questions the fitness of such models for serious applications. The attacks manipulate an input image such that misclassification is evoked while still looking normal to a human observer—they are thus not easily detectable. In a different context, backpropagated activations of CNN hidden layers—"feature responses" to a given input—have been helpful to visualize for a human "debugger" what the CNN "looks at" while computing its output. In this work, we propose a novel detection method for adversarial examples to prevent attacks. We do so by tracking adversarial perturbations in feature responses, allowing for automatic detection using average local spatial entropy. The method does not alter the original network architecture and is fully human-interpretable. Experiments confirm the validity of our approach for state-of-the-art attacks on large-scale models trained on ImageNet.

Keywords: Model interpretability · Feature visualization · Diagnostic

1 Introduction

The success of deep neural nets for pattern recognition [35] has been a main driver behind the recent surge of interest in AI. A substantial part of this success is due to the Convolutional Neural Net (CNN) [5, 20] and its descendants, applied to image recognition tasks. Respective methods have reached the application level in business and industry [38] and lead to a wide variety of deployed models for critical applications like automated driving [2] or biometrics [46].

However, concerns regarding the reliability of deep neural networks have been raised through the discovery of so-called adversarial examples [41]. These inputs are specifically generated to "fool" [28] a classifier by visually appearing as some class (to humans), but being misclassified by the network with high confidence through the addition of barely visible perturbations (see Fig. 1). The perturbations are achieved by an optimization process on the input: the network weights are fixed, and the input pixels are optimized for the dual criterion of (a) classifying the input differently than the true class, and (b) minimizing the

© Springer Nature Switzerland AG 2018
L. Pancioni et al. (Eds.): ANNPR 2018, LNAI 11081, pp. 346–358, 2018.
https://doi.org/10.1007/978-3-319-99978-4_27

changes to the input. A growing body of literature confirms the impact of this discovery on practical applications of neural nets [1]. It raises questions on how— and in what respect different from humans—they achieve their performance, and threatens serious deployments with the possibility of tailor-made adversarial attacks.

For instance, Su et al. [40] report on successfully attacking neural networks by modifying a single pixel. The attack works without having access to the internal structure nor the gradients in the network under attack. Moosavi-Dezfooli et al. [27] furthermore show the existence of universal adversarial perturbations that can be added to any image to fool a specific model, whereas transferability of perturbations from one model to another is for example shown by Xu et al. [44]. The impact of similar attacks extends beyond classification [26], is transferable to other modalities than images [6], and also works on models distinct from neural networks [31]. Finally, adversarial attacks have been shown to work reliably even after perturbed images have been printed and captured again via a mobile phone camera [18]. Apparently, such research touches a weak spot.

On the other hand, there is a recent interest in the interpretability of AI agents and in particular machine learning models [30,42]. It goes hand in hand with societal developments like the new European legislation on data protection that is impacting any organization using algorithms on personal data [13]. While neural networks are publicly perceived as "black boxes" with respect to how they arrive at their conclusions [15], several methods have been developed recently to allow insight into the representation and decision surface of a trained model, improving interpretability. Prime candidates amongst these methods are feature visualization approaches that make the operations in hidden layers of a CNN visible [29,37,45]. They can thus serve a human engineer as a diagnostic tool in support of reasoning over success and failure of a model on the task at hand.

In this paper, we propose to use a specific form of CNN feature visualization, namely feature response maps, to not only *trace* the effect of adversarial inputs on algorithmic decisions throughout the CNN; we subsequently also use it as input to a novel automated *detection* approach, based on statistical analysis of the feature responses using average of image local spatial entropy. The goal is to decide if a model is currently under attack by the given input. Our approach has the advantage over existing methods of not changing the network architecture, i.e., not affecting classification accuracy; and of being interpretable both to humans and machines, an intriguing property also for future work on the method. Experiments on the validation set of ImageNet [34] with VGG19 networks [36] shows the validity of our approach for detecting various state-of-the-art attacks.

Below, Sect. 2 reviews related work in contrast to our approach. Section 3 presents the background on adversarial attacks and feature response estimation before Sect. 4 introduces our approach in detail. Section 5 reports on experimental evaluations, and Sect. 6 concludes with an outlook to future work.

2   Related Work

Work on adversarial examples for neural networks is a very active research field. Potential attacks and defenses are published at a high rate and have been surveyed recently by Akhtar and Mian [1]. Amongst potential defenses, directly comparable to our approach are those that focus on the sole detection of a possible attack and not on additionally recovering correct classification.

On one hand, several detection approaches exist that exploit specific abnormal behavioral traces that adversarial examples leave while passing through a neural network: Liang et al. [22] consider the artificial perturbations as noise in the *input* and attempt to detect it by quantizing and smoothing image filters. A similar concept underlies the SqueezeNet approach by Xu et al. [43], that compares the network's *output* on the raw and filtered input, and raises a flag if detecting a large difference between both. Feinman et al. [9] observe the network's output confidence as estimated by dropout in the forward pass [11], and Lu et al's SafetyNet [23] looks for abnormal patterns in the ReLU activations of *higher layers*. In contrast, our method performs detection based on statistics of activation patterns in the complete *representation learning* part of the network as observed in feature response maps, whereas Li and Li [21] directly observe convolutional filter statistics there.

On the other hand, a second class of detection approaches trains sophisticated classifiers for directly sorting out malformed inputs: Meng and Chen's MagNet [24] learns the manifold of friendly images, rejects far away ones as hostile and modifies close outliers to be attracted to the manifold before feeding them back to the network under attack. Grosse et al. [14] enhance the output of an attacked classifier by an additional class and retrain the model to directly classify adversarial examples as such. Metzen et al. [25] have a similar goal but target it via an additional subnetwork. In contrast, our method uses a simple threshold-based detector and pushes all decision power to the human-interpretable feature extraction via the feature response maps.

Finally, as shown in [1], different and mutually exclusive explanations for the existence of adversarial examples and the nature of neural network decision boundaries exist in the literature. Because our method enables a human investigator to trace attacks visually, it can be helpful in this debate in the future.

3   Background

We briefly present adversarial attacks and feature response estimation in general before assembling both parts into our detection approach in the next Section.

3.1   Adversarial Attacks

The main idea of adversarial attacks is to find a small perturbation for a given image that changes the decision of the Convolutional Neural Network. Pioneering work [41] demonstrated that negligible and visually insignificant perturbations

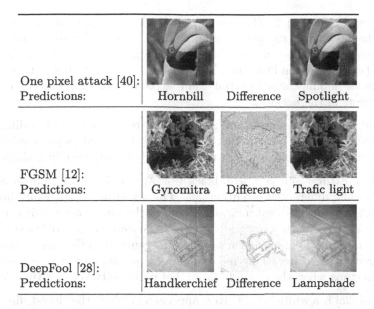

One pixel attack [40]:			
Predictions:	Hornbill	Difference	Spotlight
FGSM [12]:			
Predictions:	Gyromitra	Difference	Trafic light
DeepFool [28]:			
Predictions:	Handkerchief	Difference	Lampshade

Fig. 1. Examples of different state-of-the-art adversarial attacks on a VGG19 model: original image and label (left), perturbation (middle) and mislabeled adversarial example (right). In the middle column difference of zero is encoded white and maximum difference is black because of visual enhancement.

could lead to considerable deviations in the networks' output. The problem of finding a perturbation η for a normalized clean image $I \in \mathbb{R}^m$, where m is the image width × height, is stated as follows [41]:

$$\min_{\eta} \| \eta \|_2 \quad \text{s.t.} \quad \mathscr{C}(I + \eta) \neq \ell \; ; \quad I + \eta \in [0,1]^m \tag{1}$$

where $\mathscr{C}(.)$ presents the classifier and ℓ is the ground truth label. Szegedy et al. [41] proposed to solve the optimization problem in Eq. 1 for an arbitrary label ℓ' that differs from the ground truth to find the perturbation. However, box-constrained Limmited-memory Broyden-Fletcher-Goldfarb-Shanno (L-BFGS) [10] is alternatively used to find perturbations satisfying Eq. 1 to improve computational efficiency. Optimization based on the L-BFGS algorithm for finding adversarial attacks are computational inefficient compared with gradient-based methods. Therefore, we use a couple of gradient-based attacks, a one-pixel attack, and boundary attack to compute adversarial examples (see Fig. 1).

Fast Gradient Sign Method (FGSM) [12] is a method suggested to compute adversarial perturbations based on the gradient $\nabla_I J(\theta, I, \ell)$ of the cost function with respect to the original image pixel values:

$$\eta = \epsilon \, \text{sign}(\nabla_I J(\theta, I, \ell)) \tag{2}$$

where θ represents the network parameters and ϵ is a constant factor that constrains the max-norm l_∞ of the additive perturbation η. The ground truth label is presented by ℓ in Eq. 2. The sign function is Eq. 2 computes the elementwise sign of the gradient of the loss function with respect to the input image. Optimizing the perturbation in Eq. 2 in a single step is called Fast Gradient Sign Method (FGSM) in the literature. This method is a white box attack, i.e. the algorithm for finding the adversarial example requires the information of weights and gradients of the network.

Gradient attack is a simple and straightforward realization of finding adversarial perturbations in the FoolBox toolbox [33]. It optimizes pixel values of an ori ginal image to minimize the ground truth label confidence in a single step.

One pixel attack [40] is a semi-black box approach to compute adversarial examples using differential evolution [39]. The algorithm is not white box since it does not need the gradient information of the classifier; however, it is not fully black box as it needs the class probabilities. The iterative algorithm starts with randomly initialized parent perturbations. The generated offspring compete with their parent at each iteration, and the winners advance to the next step. The algorithm stops when the ground truth label probability is lower than 5%.

DeepFool [28] is a white box iterative approach in which the closest direction to the decision boundary is computed in every step. It is equivalent to finding the corresponding path to the orthogonal projection of the data point onto the affine hyperplane which separates the binary classes. The initial method for binary classifiers can be extended to a multi-class task by considering it as multiple one-versus-all binary classifications. After finding the optimal updates toward the decision boundary, the perturbation is added to the given image. The iterations continue with estimating the optimal perturbation and apply it to the perturbed image from the last step until the network decision changes.

Boundary attack is a reliable black-box attack proposed by Brendel et al. in [3]. The iterative algorithm already starts with an adversarial image and iteratively optimize the distance between this image and the original image. It searches for an adversarial example with minimum distance from the original image.

3.2   Feature Response Estimation

The idea of visualizing CNNs through feature responses is to find out which region of the image leads to the final decision of the network. Computing feature responses enhances the interpretability of the classifier. In this paper, we use this visualization tool to track the effect of the adversarial attacks on a CNN's decision as well as to detect perturbed examples automatically.

Erhan et al. [8] used backpropagation for visualizing feature responses of CNNs. This is implemented by evaluating an arbitrary image in the forward pass, thereby retaining the values of activated neurons at the final convolutional layer, and backpropagating these activations to the original image. The feature response has higher intensities in the regions that cause larger values of activation in the network (see Fig. 2). The information of max-pooling layers in the forward pass can further improve the quality of visualizations. Zeiler et al. [45] proposed

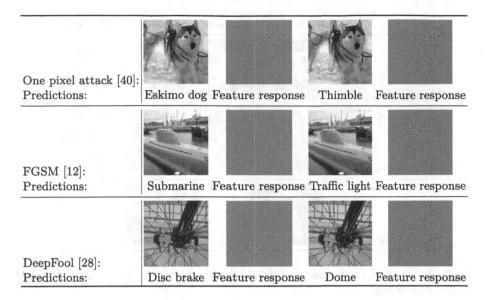

One pixel attack [40]: Predictions:	Eskimo dog	Feature response	Thimble	Feature response
FGSM [12]: Predictions:	Submarine	Feature response	Traffic light	Feature response
DeepFool [28]: Predictions:	Disc brake	Feature response	Dome	Feature response

Fig. 2. Effect of adversarial attacks on feature responses: original image and feature response (left), perturbed versions (right).

to compute "switches", the position of maxima in all pooling regions, and then construct the feature response using transposed convolutional [7] layers.

Ultimately, Springenberg et al. [37] proposed a combination of both methods called guided backpropagation. In this approach, the information of "switches" (max-pooling spatial information) is kept, and the activations are propagated backwards with the guidance of the "switch" information. This method leads to the best performance in network innards visualization, therefore we use guided backpropagation for computing feature response maps in this paper.

4 Human-Interpretable Detection of Adversarial Attacks

After reviewing the necessary background in the last Section, we will now present our work on tracing adversarial examples in feature response maps, which inspired a novel approach to automatic detection of adversarial perturbations in images. Using visual representations of the inner workings of neural network in this manner additionally provides a human expert guidance in developing deep convolutional networks with increased reliability and interpretability.

4.1 Tracing Adversarial Attacks in Feature Responses

The research question followed in this work is to obtain insight into the reasons behind misclassification of adversarial examples. Their effect in the feature response of a CNN is for example traced in Fig. 2. The general phenomenon

| | Original | Adversarial | Original | Adversarial |

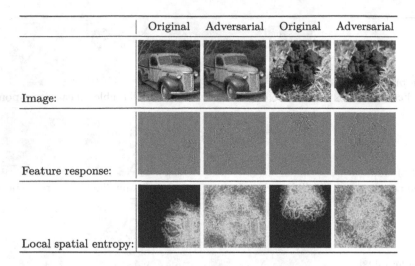

Fig. 3. Input, feature response and local spatial entropy for clean and perturbed images, respectively.

observed in all experiments is the broader feature response of adversarial examples. In contrast, Fig. 2 demonstrates that the network looks at a smaller region of the image—is more focused—in case of not manipulated samples.

The adversarial images are visually very similar to the original ones. However, they are not correctly recognizable by deep CNNs. The original idea which triggered this study is that the focus of CNNs changes during an adversarial attack and lead to the incorrect decision. Conversely, the network makes the correct decision once it focuses on the right region of the image. Visualizing the feature response provides this and other interesting information regarding the decision making in neural networks: for instance, the image of the submarine in Fig. 2 can be considered a good candidate for an adversarial attack since the CNN is making the decision based on an object in the background (see the feature response of the original submarine in Fig. 2).

4.2 Detecting Adversarial Attacks Using Spatial Entropy

Experiments for tracing the effect of adversarial attacks on feature responses thus suggested that a CNN classifier focuses on a broader region of the input if it has been maliciously perturbed. Figure 2 demonstrates this connection for decision making in case of clean inputs compared with manipulated ones. The effect of adversarial manipulation is visible in the local spatial entropy of the grayscale feature responses as well (see Fig. 3). The feature responses are initially converted to gray scale images, and local spatial entropies are computed based on transformed feature responses as follows [4]:

$$S_k = -\sum_i \sum_j \boldsymbol{h}_k(i,j) \log_2(h_k(i,j)) \qquad (3)$$

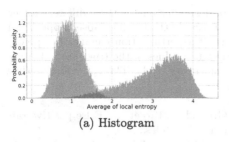

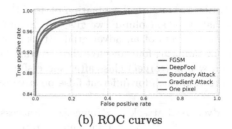

(a) Histogram   (b) ROC curves

Fig. 4. (a) Distribution of average local spatial entropy in clean images (green) versus adversarial examples (red) as computed on the ImageNet validation set [34]. (b) Receiver operating characteristic (ROC) curve of the performance of our detection algorithm on different attacks. (Color figure online)

where S_k is the local spatial entropy of a small part (patch) of the input image and h_k represents the normalized 2D histogram value of the kth patch. The indices i and j scan through the height and width of the image patches. The patch size is 3×3 and the same as the filter size of the first layer of the used CNN (VGG19 [36]). The local spatial entropies of corresponding feature responses are presented in Fig. 3, and their difference for clean and adversarial examples suggests a likely chance to detect perturbed images based on this feature.

Accordingly, we propose to use the average local spatial entropy of an image as the final single measure to decide whether an attack has occurred or not. The average local spatial entropy $\bar{S}$ is defined as:

$$\bar{S} = \frac{1}{K} \sum_k S_k \qquad (4)$$

where K is the number of patches on the complete feature response and S_k shows the local spatial entropy as defined in Eq. 3 and depicted in the last row of Fig. 3. Our detector makes the final decision by comparing the average local spatial entropy from Eq. 4 with a selected threshold, i.e., we use this feature to measure the spatial complexity of an input image (feature response).

5   Experimental Results

To confirm the value of our final metric in Eq. 4, we first perform experiments to visually compare the approximated distribution of the averaged local spatial entropy of feature responses in clean and perturbed images. We use the validation set of ImageNet [34] with more than $50,000$ images from $1,000$ classes and again the VGG19 CNN [36]. Perturbations for this experiment are computed only via the Fast Gradient Sign Attack (FGSM) method for computational reasons. Figure 4(a) shows that the clean images are separable from perturbed examples although there is some overlap between the distributions.

Computing adversarial perturbations using evolutionary and iterative algorithms is demanding regarding time and computational resources. However, we

Table 1. Numerical evaluation of detection performance on the three different adversarial attacks. Column two gives the amount of tested attacks and elapsed approx. run time. Success of an adversarial attack is given if a perturbation changes the prediction. Columns four and five show average confidence values of the true (ground truth) and wrong (target) class after successful attack, respectively. The last columns show detection rates for different false positive rates.

Adversarial attack	#Images (run time [days])	Success rate	Ground truth confidence	Target class confidence	False positive rate		
					1%	5%	10%
FGSM [12]	50,014 (3)	0.925	0.022	0.588	0.954	0.974	0.983
Gradient attack [33]	50,014 (15)	0.499	0.052	0.371	0.922	0.954	0.969
One pixel attack [40]	50,014 (32)	0.620	0.037	0.463	0.917	0.951	0.966
DeepFool [28]	47,858 (42)	0.606	0.041	0.446	0.936	0.963	0.976
Boundary attack [3]	4,013 (17)	0.940	0.023	0.583	0.934	0.960	0.972

Table 2. Performance of similar adversarial attack detection methods. The Area Under Curve (AUC) is the average value of all attacks in the third and last row.

Method	Dataset	Network	Attack	Performance		
				Recall	Precision	AUC
Uncertainty density estimation [9]	SVHN [17]	LeNet [19]	FGSM	-	-	0.890
Adaptive noise reduction [22]	ImageNet (4 classes)	CaffeNet	DeepFool	0.956	0.911	-
Feature squeezing [43]	ImageNet-1000	VGG19	Several attacks	0.859	0.917	0.942
Statistical analysis [14]	MNIST	Self-designed	FGSM ($\epsilon = 0.3$)	0.999	0.940	-
Feature response (our approach)	ImageNet validation	VGG19	Several attacks	0.979	0.920	0.990

would like to apply the proposed detector to a wide range of adversarial attacks. Therefore, we have drawn a number of images from the validation set of ImageNet for each attack and present the detection performance of our method in Fig. 4. The selection of images is done sequentially by class and file name up to a total number of images per method that could be processed in a reasonable amount of time (see Table 1). We base our experiments on the FoolBox benchmarking implementation[1], running on a Pascal-based TitanX GPU.

---

[1] https://github.com/bethgelab/foolbox.

Original	Adversarial	Original	Adversarial

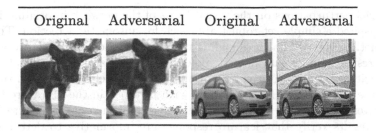

Fig. 5. Successful adversarial examples created by DeepFool [28] for binary and ternary classification tasks are only possible with notable visible perturbations.

Figure 4b presents the Receiver Operating Characteristics (ROC) of the proposed detector, and numerical evaluations are provided in Table 1. Our detection method performs better for gradient-based perturbations compared to the single pixel attack. Furthermore, Table 1 suggests that the best adversarial attack detection performance is achieved for FGSM and boundary attack perturbations, where the network confidences are changed the most. This observation suggests that the proposed detector is more sensitive to attacks which are stronger in fooling the network (i.e., change the ground truth label and target class confidence more drastically). By using feature responses, we detect more than 91% of the perturbed samples with a low false positive rate (1%).

In general, it is difficult to directly compare different studies on attack detectors since they use a vast variety of neural network models, datasets, attacks and experimental setups. We present a short overview of the performances of current detection approaches in Table 2. Our approach is most similar to the methods of Liang et al. [22] and Xu et al. [43]. The proposed detector in this paper outperforms both based on the presented results in their work; however, we cannot guarantee identical implementations and parameterizations of the used attacks (e.g., subset of used images, learning rates for optimization of perturbations). Similarly, adaptive noise reduction in the original publication [22] is applied to only four classes of the ImageNet dataset and defended a model based on CaffeNet, which differs from our experimental setup.

6   Discussion and Conclusion

The presented results demonstrate that the reality of adversarial attacks: improving the robustness of CNNs is necessary. However, we conducted further preliminary experiments on binary (cat versus dog [32]) and ternary (among three classes of cars [16]) classification tasks as proxies for the kind of few-class classifications settings frequently arising in practice. They suggest that it is more challenging to find adversarial examples in such a setting without plenty of "other classes" to pick from for misclassification. Figure 5 illustrates these results.

In this paper, we have presented an approach to detect adversarial attacks based on human-interpretable feature response maps. We traced the effect of adversarial perturbations on the visual focus of the network in original images, which inspired a simple yet robust approach for automatic detection. This proposed method is based on thresholding the averaged local spatial entropy of the feature response maps and detects at least 91% of state-of-the-art adversarial attacks with a low false positive rate on the validation set of ImageNet. However, the results are not directly comparable with methods in the literature because of the diversity in the experimental setups and implementations of attacks.

Our results verify that feature response are informative to detect specific cases of failure in deep CNNs. The proposed detector applies to increase the interpretability of neural network decisions, which is an increasingly important topic towards robust and reliable AI. Future work, therefore, will concentrate on developing reliable and interpretable image classification methods for practical use cases based on our preliminary results for binary and ternary classification.

Acknowledgements. We are grateful for the support by Innosuisse grant 26025.1 PFES-ES "QualitAI".

References

1. Akhtar, N., Mian, A.: Threat of adversarial attacks on deep learning in computer vision: A survey. arXiv preprint arXiv:1801.00553 (2018)
2. Bojarski, M., et al.: End to end learning for self-driving cars. arXiv preprint arXiv:1604.07316 (2016)
3. Brendel, W., Rauber, J., Bethge, M.: Decision-based adversarial attacks: reliable attacks against black-box machine learning models. arXiv preprint arXiv:1712.04248 (2017)
4. Chanwimaluang, T., Fan, G.: An efficient blood vessel detection algorithm for retinal images using local entropy thresholding. In: International Symposium on Circuits and Systems (ISCAS), vol. 5 (2003)
5. Cireşan, D., Meier, U., Masci, J., Schmidhuber, J.: A committee of neural networks for traffic sign classification. In: IJCNN, pp. 1918–1921. IEEE (2011)
6. Cisse, M., Adi, Y., Neverova, N., Keshet, J.: Houdini: fooling deep structured prediction models. arXiv preprint arXiv:1707.05373 (2017)
7. Dumoulin, V., Visin, F.: A guide to convolution arithmetic for deep learning. arXiv preprint arXiv:1603.07285 (2016)
8. Erhan, D., Bengio, Y., Courville, A., Vincent, P.: Visualizing higher-layer features of a deep network. Univ. Montr. **1341**(3), 1 (2009)
9. Feinman, R., Curtin, R.R., Shintre, S., Gardner, A.B.: Detecting adversarial samples from artifacts. arXiv preprint arXiv:1703.00410 (2017)
10. Fletcher, R.: Practical Methods of Optimization. Wiley, Hoboken (2013)
11. Gal, Y., Ghahramani, Z.: Dropout as a Bayesian approximation: representing model uncertainty in deep learning. In: ICML (2016)
12. Goodfellow, I.J., Shlens, J., Szegedy, C.: Explaining and harnessing adversarial examples. In: ICLR (2015)
13. Goodman, B., Flaxman, S.: EU regulations on algorithmic decision-making and a "right to explanation". In: ICML Workshop on Human Interpretability in Machine Learning (WHI) (2016)

14. Grosse, K., Manoharan, P., Papernot, N., Backes, M., McDaniel, P.: On the (statistical) detection of adversarial examples. arXiv preprint arXiv:1702.06280 (2017)
15. Gunning, D.: Explainable Artificial Intelligence (XAI). Defense Advanced Research Projects Agency (DARPA) (2017)
16. Krause, J., Stark, M., Deng, J., Fei-Fei, L.: 3D object representations for fine-grained categorization. In: ICCV Workshops (2013)
17. Krizhevsky, A., Hinton, G.: Learning multiple layers of features from tiny images (2009)
18. Kurakin, A., Goodfellow, I., Bengio, S.: Adversarial examples in the physical world. In: ICRL Workshop Track (2016)
19. LeCun, Y., et al.: Backpropagation applied to handwritten zip code recognition. Neural Comput. $\mathbf{1}$(4), 541–551 (1989)
20. LeCun, Y., Bottou, L., Bengio, Y., Haffner, P.: Gradient-based learning applied to document recognition. Proc. IEEE $\mathbf{86}$(11), 2278–2324 (1998)
21. Li, X., Li, F.: Adversarial examples detection in deep networks with convolutional filter statistics. arXiv preprint arXiv:1612.07767 (2016)
22. Liang, B., Li, H., Su, M., Li, X., Shi, W., Wang, X.: Detecting adversarial examples in deep networks with adaptive noise reduction. arXiv preprint arXiv:1705.08378 (2017)
23. Lu, J., Issaranon, T., Forsyth, D.: Safetynet: Detecting and rejecting adversarial examples robustly. arXiv preprint arXiv:1704.00103 (2017)
24. Meng, D., Chen, H.: Magnet: a two-pronged defense against adversarial examples. In: ACM SIGSAC Conference on Computer and Communications Security (2017)
25. Metzen, J.H., Genewein, T., Fischer, V., Bischoff, B.: On detecting adversarial perturbations. In: ICLR (2017)
26. Metzen, J.H., Kumar, M.C., Brox, T., Fischer, V.: Universal adversarial perturbations against semantic image segmentation. arXiv preprint arXiv:1704.05712 (2017)
27. Moosavi-Dezfooli, S.M., Fawzi, A., Fawzi, O., Frossard, P.: Universal adversarial perturbations. arXiv preprint arXiv:1610.08401 (2017)
28. Moosavi Dezfooli, S.M., Fawzi, A., Frossard, P.: Deepfool: a simple and accurate method to fool deep neural networks. In: CVPR (2016)
29. Olah, C., Mordvintsev, A., Schubert, L.: Feature visualization. Distill (2017). https://doi.org/10.23915/distill.00007
30. Olah, C., et al.: The Building Blocks of Interpretability. Distill (2018). https://doi.org/10.23915/distill.00010
31. Papernot, N., McDaniel, P., Goodfellow, I.: Transferability in machine learning: from phenomena to black-box attacks using adversarial samples. arXiv preprint arXiv:1605.07277 (2016)
32. Parkhi, O.M., Vedaldi, A., Zisserman, A., Jawahar, C.: Cats and dogs. In: CVPR (2012)
33. Rauber, J., Brendel, W., Bethge, M.: Foolbox v0.8.0: A python toolbox to benchmark the robustness of machine learning models. arXiv preprint arXiv:1707.04131 (2017)
34. Russakovsky, O., et al.: Imagenet large scale visual recognition challenge. Int. J. Comput. Vis. (IJCV) $\mathbf{115}$(3), 211–252 (2015). https://doi.org/10.1007/s11263-015-0816-y
35. Schmidhuber, J.: Deep learning in neural networks: an overview. Neural Netw. $\mathbf{61}$, 85–117 (2015)
36. Simonyan, K., Zisserman, A.: Very deep convolutional networks for large-scale image recognition. In: ICLR (2015)

37. Springenberg, J.T., Dosovitskiy, A., Brox, T., Riedmiller, M.: Striving for simplicity: The all convolutional net. arXiv preprint arXiv:1412.6806 (2014)
38. Stadelmann, T., Tolkachev, V., Sick, B., Stampfli, J., Dürr, O.: Beyond imagenet - deep learning in industrial practice. In: Braschler, M., Stadelmann, T., Stockinger, K. (eds.) Applied Data Science - Lessons Learned for the Data-Driven Business. Springer (2018, to appear). https://stdm.github.io/data-science-book/
39. Storn, R., Price, K.: Differential evolution-a simple and efficient heuristic for global optimization over continuous spaces. J. Glob. Optim. **11**(4), 341–359 (1997)
40. Su, J., Vargas, D.V., Kouichi, S.: One pixel attack for fooling deep neural networks. arXiv preprint arXiv:1710.08864 (2017)
41. Szegedy, C., et al.: Intriguing properties of neural networks. In: ICLR (2014)
42. Vellido, A., Martín-Guerrero, J.D., Lisboa, P.J.: Making machine learning models interpretable. In: ESANN. vol. 12, pp. 163–172 (2012)
43. Xu, W., Evans, D., Qi, Y.: Feature squeezing: detecting adversarial examples in deep neural networks (2018)
44. Xu, X., Chen, X., Liu, C., Rohrbach, A., Darell, T., Song, D.: Can you fool AI with adversarial examples on a visual turing test? arXiv preprint arXiv:1709.08693 (2017)
45. Zeiler, M.D., Fergus, R.: Visualizing and understanding convolutional networks. In: Fleet, D., Pajdla, T., Schiele, B., Tuytelaars, T. (eds.) ECCV 2014. LNCS, vol. 8689, pp. 818–833. Springer, Cham (2014). https://doi.org/10.1007/978-3-319-10590-1_53
46. Zhu, J., Liao, S., Yi, D., Lei, Z., Li, S.Z.: Multi-label CNN based pedestrian attribute learning for soft biometrics. In: International Conference on Biometrics (ICB). IEEE (2015)

Video and Audio Data Extraction for Retrieval, Ranking and Recapitulation (VADER³)

Volkmar Frinken[1](✉), Satish Ravindran[2], Shriphani Palakodety[1],
Guha Jayachandran[1], and Nilesh Powar[2]

[1] Onu Technology, San Jose, CA 95129, USA
{volkmar,spalakod,guha}@onai.com
[2] University of Dayton Research Institute, Dayton, OH 45469, USA
satishr@g.clemson.edu, Nilesh.Powar@udri.udayton.edu

Abstract. With advances in neural network architectures for computer vision and language processing, multiple modalities of a video can be used for complex content analysis. Here, we propose an architecture that combines visual, audio, and text data for video analytics. The model leverages six different modules: action recognition, voiceover detection, speech transcription, scene captioning, optical character recognition (OCR) and object recognition. The proposed integration mechanism combines the output of all the modules into a text-based data structure. We demonstrate our model's performance in two applications: a clustering module which groups a corpus of videos into labelled clusters based on their semantic similarity, and a ranking module which returns a ranked list of videos based on a keyword. Our analysis of the precision-recall graphs show that using a multi-modal approach offers an overall performance boost over any single modality.

Keywords: Multi modal video analytics · LSTM · CNN

1 Introduction

Recently, there has been considerable focus on trying to extract relevant information from video content, rather than just the metadata [2,8]. Understanding semantic content greatly improves access to video corpora through improved searching and ranking. Trying to extract relevant information using a single modality like the image or audio is prone to errors, either because of lack of accuracy of the processing algorithm or because of lack of underlying information in the modality under consideration. Fusing information from multiple modalities helps in providing more relevant results for video analytics. In this paper, we propose a novel way to integrate the information from a wide spectrum of information sources in a video. We will demonstrate our approach in two applications: ranking of videos in response to a search query, and clustering a corpus of videos based on semantic similarity.

© Springer Nature Switzerland AG 2018
L. Pancioni et al. (Eds.): ANNPR 2018, LNAI 11081, pp. 359–370, 2018.
https://doi.org/10.1007/978-3-319-99978-4_28

Even recent state-of-the-art techniques for video analytics either focus on extracting key frames from a video [7,15] or provide a textual summary of the video [10]. Since these approaches rely on visual information only and also focus on key subjects in the frame, they miss out on much of the contextual information that could be provided by the audio and background text.

2   Approach

Our approach addresses the shortcomings of the current state of the art by utilizing the information available in all the modalities in a video, i.e. the individual frames, audio and text. To our knowledge, this is the first time that a technique has been proposed which combines such a wide spectrum of information sources. Each of the independent modules operates on the input video after which the outputs from each module is combined into a text-based data structure.

We developed and tested the independent modules and will describe each of them in detail in this section.

2.1   Action Recognition

To recognize actions in videos, we combined deep learning based semantic segmentation approaches with recurrent neural networks. A high level overview of the bounding box detection network is given in Fig. 1. The first layers fulfill the function of semantic image segmentation. For this, we use the *DeepLap-MSc-COCO-LargeFOV* network provided by the University of California, Los Angeles [1]. Output activations from intermediate layers (as low level representation) as well as the pixel-wise output probabilities are fed into a long short-term memory (LSTM) layer [5]. The LSTM layer forms the recurrent part of the network and binds several frames together. The output of the network given a frame at time t therefore not only depends on the current frame, but also on previously read frames. At the top, a softmax output layer is used with cross-entropy training to recognize an action happening in the video frames.

For the textual representation, we divide the image into a pyramid: not only is the entire frame classified, but also the top-left, top-right, bottom-left, and bottom-right zoomed-in sum-frame, as seen in Fig. 2. Thus, each frame has five potential outputs, which are simply written in a line. If no action can be detected (the output activation of the *no-action* node is the largest activation), the output from that sub-frame is simply the empty string ε.

2.2   Voiceover Detection

The voiceover detection system is a neural network which evaluates whether the sound (such as voiceover text or music) in the video is added in a clean postprocessing step or part of the original recording, captured at the same time (and with the same device) that recorded the video.

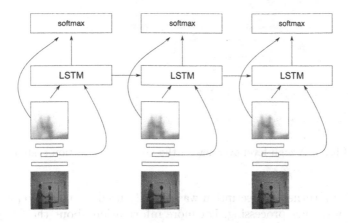

Fig. 1. High level overview of the action recognition neural network

Fig. 2. The action recognition module is executed five times in parallel on each frame to cover actions at different scales.

The neural network designed to detect voiceover text is outlined in Fig. 3. Given the video's audio track, we extract a sequence of 13 Mel-frequency cepstral coefficients (MFCC) [14] with a frame rate of 25 ms frame width and a step size of 20ms. A larger step size than normally found in the literature allows for a faster processing and simpler model. Each group of 10 consecutive MFCC samples is recognized in a feed-forward neural network (with hidden layers of size 128, 64, 32, 16, 8, and 2) in a binary classification. The averaged binarized value is returned as the voiceover score, i.e., the fraction of time steps in which the *yes* output node has a larger activation than the *no* output node.

2.3   Speech Recognition

After comparison and research into the state of the art within the field, we settled upon using the Google Cloud Speech API[1]. From an input video, our module extracts the audio track and performs an API call to the Google cloud server. The length of the audio file accepted by Google is limited, so for longer audio transcriptions, we split the audio track into smaller segments with one second

---

[1] https://cloud.google.com/speech/.

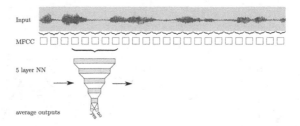

Input

MFCC

5 layer NN

average outputs

Fig. 3. A high-level overview of the voiceover detection system.

overlap. The returned transcription was directly used as textual representation, without any further processing. For more information about the Google Cloud Speech API, we refer to its documentation[2].

2.4   Automated Scene Captioning

Image and video captioning have seen much work in the past decade. As is typical in computer vision, an early emphasis was on still images rather than video. We used an encoder-decoder model where the image or video is first encoded into a semantically-rich embedding space and then the embedded representation is decoded using LSTM architectures. We leveraged the open source code associated with [16] for our application.

2.5   Text Detection and Recognition (OCR) and Object Recognition

For text detection, we initially tried a text specific object proposal algorithm described in [3]. Ultimately, we settled on using the OCR module in the Google Vision API[3] since it gave superior results.

For object recognition, we leverage the current state of the art CNNs to detect objects of interest in our database. We also evaluated other architectures including YOLO [12], DenseNet [6] and Resnet [4], but the Inception V3 architecture [13], released by Google performed much better in our tests.

2.6   Language Model Based Video Similarity

As explained in the Introduction, the previously introduced modules are run in parallel on an input video. Each of the modules returns a textual description of the different aspects of the video, such as speech, actions, objects, etc. The textual outputs are concatenated, cleaned, and normalized in the following manner: The URLs are first extracted and saved as words in the dictionary. The remaining text is transformed to lowercase. The Python NLTK word stemmer[4] is then

[2] https://cloud.google.com/docs/.
[3] https://cloud.google.com/vision/docs/ocr.
[4] http://www.nltk.org/api/nltk.stem.html.

applied to each word. We then save in a hash table, all word stemming transformation for a reverse lookup that is used later. Stop words from the NLTK stop word list[5] are removed. All the resulting words are then added to the dictionary. Finally, a token <UKN> symbolizing an unknown out-of-vocabulary word is added to the dictionary.

2.7   Video Ranking and Retrieval

All text documents created from the video database are represented as a bag-of-words. Similarities are computed using vector similarities between two frequency-inverted document frequency (tf-idf) [11] vectors of those bag-of-words. This provides a unified view for videos (which results in a matrix of pairwise distances) for arbitrary text queries. A query is transformed into a bag-of-words through the same steps outlined above. Words not occurring in the videos in the database are mapped onto the <UKN> word. Afterwards the vector similarities to all vectors in the database are computed and ranked. This provides a fast and robust method to retrieve videos that correspond to any arbitrary query. A sample demonstration is shown in Fig. 4.

2.8   Clustering

The pairwise video distances derived from the NLP-based text dissimilarities lend themselves well to hierarchical clustering, in our case agglomerative bottom-up clustering with single-linkage cluster distances. Starting from each video as a cluster of its own, a threshold is gradually increased (x-axis). As soon as that threshold is larger than the distance between two clusters, they merge into a new cluster, until finally all elements are part of one cluster.

The quality of the clustering is not easily measured by its own because it is not clear what a good cluster is without extensive ground truth. For the two main clusters, graduation speeches and TV commercials, we have an implicit ground truth given, but not at a finer level. Furthermore, there are ambiguous outliers. For example, consider a TV commercial with text in Spanish and a questions such as, "Are English language graduation speeches closer to English TV commercials than Spanish TV commercials to English TV commercials?" Since there is no clear answer to that, we jointly evaluate the clustering accuracy combined with the semantic cluster labels introduced next.

2.9   Semantic Labeling

After creating the clusters, we want to automatically generate cluster labels using the semantic information extracted from the individual modules. This is done using mutual information [9]. In a nutshell, considering the textual description of a video, we identify those words, whose occurrence (or lack thereof) serves best

---

[5] https://raw.githubusercontent.com/nltk/nltk_data/gh-pages/packages/corpora/
   stopwords.zip.

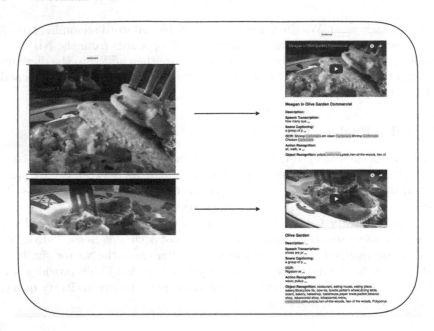

Fig. 4. The output for the query word "Carbonara". In the top video on the right the query word was detected by multiple modules (OCR and Object Recognition) resulting in a higher score. In the second video, the word was only detected by the Object recognition module.

to predict whether or not a video is part of a cluster. Mathematically speaking, consider the mutual information of two random variables X, and Y:

$$I(X;Y) = \sum_{x \in X} \sum_{y \in Y} p(x,y) \log \frac{p_{X,Y}(x,y)}{p_X(x)p_Y(y)} \tag{1}$$

where $p_{X,Y}$ is the joint probability distribution of X and Y, and p_X and p_Y are the marginal probability distribution of X and Y. In our case, given a cluster C and a document D, X_C is a random variable to indicate membership in a cluster ($X_C = 1$) or not ($X_C = 0$), and Y_W indicates the occurrence of a word W ($Y_W = 1$) or the lack of it ($Y_W = 0$). Hence, $p(X_C)$ is the probability of a document being part of the cluster C, $p(Y_W = w)$ is the probability of a document containing the word W, and $p_{X_C,Y_W}(X_C,Y_W)$ is the probability of a document being a member (or not) of cluster C while containing the word W (or not). The mutual information becomes

$$I(X_C, Y_W) = \sum_{c=0,1} \sum_{d=0,1} p(X_C = c, Y_W = w) \log \frac{p_{X_C,Y_W}(X_C = c, Y_W = w)}{p_{X_C}(X_C = c)p(Y_W = w)} \tag{2}$$

The values $p_{X_C,Y_W}(X_C = 0, Y_W = 0)$, $p_{X_C,Y_W}(X_C = 0, Y_W = 1)$, $p_{X_C,Y_W}(X_C = 1, Y_W = 0)$, and $p_{X_C,Y_W}(X_C = 1, Y_W = 1)$ as well as $p(X_C = 0)$, $p(X_C = 1)$,

```
ceremony/ceremonial/ceremonies;  graduates/graduated/graduate/graduation/graduating;
colleg/college; nations/national/nation; sciences/science; next; commencement; ramp;
school/schools; NOT commercials/commercial
```

Fig. 5. A cluster of videos and a ranked list of labels with the largest mutual information. Labels are word stems extracted during the pre-processing phase, which can result in multiple word instances as a label, e.g., ceremony, ceremonial, ceremonies.

$p(Y_W = 0)$, and $p(Y_W = 1)$ can be efficiently estimated for a given cluster and word either by counting the entire set or a randomly sampled set of document.

For each cluster, we consider the ten words with the highest mutual information. The mutual information is a measure by how much a cluster becomes predictable upon knowledge of the occurrence of a word. This is symmetric in both directions, where the existence or non existence of a word can provide information about the cluster. Therefore, we compute the mutual information between each cluster and word. Any word whose occurrence is negatively correlated with a cluster is appended with the prefix "NOT". Figure 5 shows a cluster of videos and the labels, and Fig. 6 shows all videos in a forced-directed graph, segmented into four clusters of at least two videos each, as well as a few singular videos.

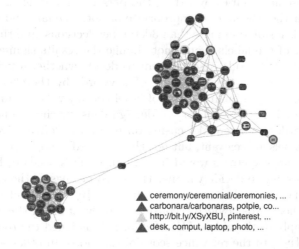

▲ ceremony/ceremonial/ceremonies, ...
▲ carbonara/carbonaras, potpie, co...
▲ http://bit.ly/XSyXBU, pinterest, ...
▲ desk, comput, laptop, photo, ...

Fig. 6. An example where the threshold is set so that clusters arise with more than one video each.

3   Experimental Evaluation

3.1   Dataset

We manually annotated videos from YouTube belonging to two categories, *Commercials* and *Graduation Ceremonies*. The former consists of advertisements for phones, shoes, restaurants, and various other products or services. The ground truth included the object category, brand, a brief description of the activity and the text in the video. This category consists of 44 videos. The Graduation Ceremonies category consists of commencement ceremonies of various schools and colleges around the country. The labels includes the school name, the grade, date, and whether it is indoor or outdoor. We had 22 videos in this category. We selected them due to the readily available data and the diversity of content within these categories. The modules developed were tested on these datasets but can be applied to any video corpus.

3.2   Clustering and Labeling

We evaluated the cluster and the semantic labeling jointly using the following protocol. First, we took the list of the most important semantic keywords of all clusters, i.e., the list of all words that have the highest mutual information for at least 10 clusters. Those are the 53 labels shown in Fig. 7. For all videos in the database, we manually decided for each label whether it is an appropriate label or not. At times the labeling was ambiguous—for example, the label "room" can be seen in nearly all videos or "clock" is an object that may appear in the background in many videos. Also, for example, negative labels, such as "NOT loaf," are not easy to assign if a frame of an Olive Garden commercial shows a loaf of bread somewhere, yet the focus of the commercial is not the loaf. We handled all these ambiguities by letting the person annotating the video decide subjectively whether the label is appropriate or not. The large number of labels and commercials resulted in more than 3000 label decisions, and thus some inaccuracy in a few of the labels should not change the results significantly. In the next step, we created a rule-based system to decide whether a video should be part of a cluster or not. Given the list of keywords by the cluster labels, we consider for each video a binary vector of label relevancy. For example, a cluster might have the labels "olive/oliver, garden/gardens, mashed, consomme" then for an Olive Garden commercial focusing on pasta, the "olive/oliver" and "garden/gardens" labels are relevant, but not the "mashed" and "consomme" labels. Hence, the relevancy vector v would be $(1, 1, 0, 0)$. This needs to be reduced to a single yes/no-value to decide whether the video belongs to the cluster or not.

The *Min* rule assigned the minimal value $\min_i\{v_i\}$ to the relevance score of the video. In other words, a video is considered relevant under the *Min* rule, if all of the labels apply to the video. The *Median* rule assigned the rounded median value $\lfloor \text{median}\{v_i\} \rfloor$ to the relevance score of the video. In other words, a video is considered relevant under the *Median* rule, if at least half of the labels apply to the video. The *Max* rule assigned the value $\max_i\{v_i\}$ to the relevance score

carbonara/carbonaras	salad	room/rooms
olive/oliver	walk	riding/ride
garden/gardens	pour	NOT mashed
mashed	pinterest	NOT loaf
consomme	http://bit.ly/XSyXBU	NOT consomme
potpie	shoot	NOT burrito
potato	NOT potpie	live/living/lives
ceremony/ceremonial	colleg/college	http://bit.ly/XjG32m
cauliflower	breadsticks	history
pot	sciences/science	cradle
hotpot	NOT potato	channels
bench	lamp	cake
commencement	ramp	NOT pot
next	man	machine
nations/national/nation	kroc	clock
tennis	http://bit.ly/WUOGUw	universitys/universities/university
peopl	http://bit.ly/VFrhsx	/universe/universal
group	check	graduates/graduated/graduate/
frisbe	degree/degrees	graduation/graduating
president/preside	wrote	

Fig. 7. The labels that occur in the semantic labelling of at least 10 cluster. Note that the stemming joins different to the same stem, such as "graduat". Since "graduat" is not an English word, thus the returned label is the combination of all words mapped to it. The labels also include proper names, URLs, etc.

of the video. In other words, a video is considered relevant under the *Max* rule, if at least one of the labels applies to the video.

Figure 8 shows three recall-precision plots, for the three different rules. Each disk in the plot is one cluster. The *Min, Median*, and *Max* rule determine which videos should be part of the cluster. This is compared to the actual members of the cluster. From that we can compute the number of True Positives (TP), False Positives (FP), True Negatives (TN), and False Negatives (TN), which in turn is used to compute the precision of a cluster P and its recall R. Precision is a measure of a cluster's purity, the higher the precision, the less irrelevant videos are in the cluster. Recall gives the fraction of relevant videos being found. The larger the recall, the more videos that should be part of the cluster, are actually part of it.

The stricter the rule, the fewer videos should be member of a cluster. Sometimes, no video in the database should be part of a cluster, hence True Positive and False Negative must be 0, and the recall is undefined. In those cases, we do not plot any disk at all.

3.3   Individual Modules

In this subsection, we compare the performances of the individual modules with the combined analysis that take all modules into account.

Figure 9 shows three separate recall-precision plots for different cluster evaluation rules. A setting where half of the labels of a cluster must apply to video for it to be relevant appears similar to how a human user would evaluate correctness, but we include the extremes below for comparison.

A more detailed picture of the *Median* evaluation rule is shown in Fig. 10. Each circle represents a given cluster threshold. The size of the circle represents

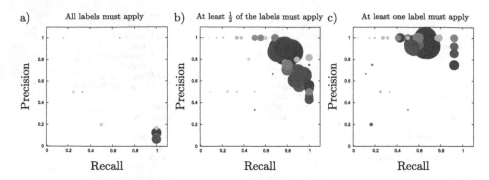

Fig. 8. Recall-Precision plots for all clusters given the three rules that determine whether a video should be part of a cluster. Each disk indicates one cluster, with the diameter of the disk indicating the size of the cluster while the color indicates the threshold.

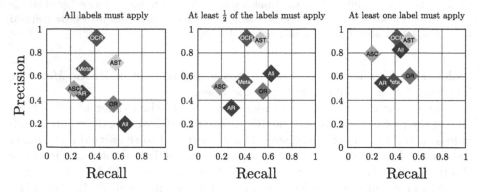

Fig. 9. Average precision and average recall values for the clusters generated using the individual modules (in color) and the combined system (grey). Evaluation of the cluster is done using the strict *Min* rule in the left plot, the more realistic *Median* rule in the central plot and using the relaxed *Max* rule in the right plot. (OCR = Optical Character Recognition, Meta = Video title and description, AST = Automatic Speech Transcription, ASC = Automatic Scene Captioning, AR = Action Recognition, OR = Object Recognition, All = Combined analysis) (Color figure online)

the number of clusters at that threshold. The black circles are from integrating together all the analyses. Note that all the black circles are towards the upper right corner, as desired (high precision and high recall). Certain individual analyses have high precision but fail to consistently accomplish both precision and recall.

For example, we can see that highly informative modules such as OCR return results with outstanding precision, yet they lack the power to find all videos, as can be seen by the comparatively low average recall value. Combining the modules gives a clear advantage as it finds more relevant videos, even at the cost of introducing some noise to the clusters.

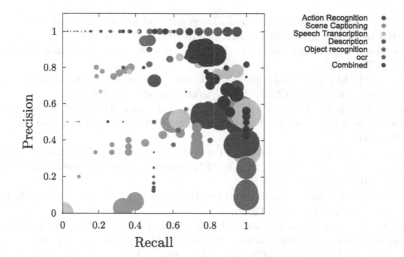

Fig. 10. Recall-Precision for all cluster created when considering only individual modules (in color) compared to a combined analysis (grey). Cluster ground truth is given by the *Median* decision rule. (Color figure online)

4   Conclusion

In this paper, we presented a mechanism for combining information from different modalities for video analytics. The visual, audio and textual information present in the video was converted into a combined text document. Latent Semantic Analysis was then used to compute a similarity metric between a corpus of documents, each document representing a video. We demonstrated two applications of our video analytics platform in this paper: (1) Video retrieval and ranking based on a keyword search and (2) Clustering of a corpus of videos based on semantic similarity of video contents. Our analysis show that combining the different modalities improves the overall robustness and performance of the system.

Acknowledgements. Supported by US Office of Naval Research

References

1. Chen, L., Papandreou, G., Kokkinos, I., Murphy, K., Yuille, A.: Semantic image segmentation with deep convolutional nets and fully connected CRFs. http://arxiv.org/abs/1412.7062 (2015)
2. Dong, J., Li, X., Lan, W., Huo, Y., Snoek, C.G.: Early embedding and late reranking for video captioning. In: Proceedings of the 2016 ACM on Multimedia Conference, pp. 1082–1086. ACM (2016)
3. Gómez, L., Karatzas, D.: Textproposals: a text-specific selective search algorithm for word spotting in the wild. Pattern Recognit. **70**, 60–74 (2017)

4. He, K., Zhang, X., Ren, S., Sun, J.: Deep residual learning for image recognition. CoRR abs/1512.03385 (2015). http://arxiv.org/abs/1512.03385
5. Hochreiter, S., Schmidhuber, J.: Long short-term memory. Neural Comput. **9**, 1735–1780 (1997)
6. Huang, G., Liu, Z., Weinberger, K.Q.: Densely connected convolutional networks. CoRR abs/1608.06993 (2016). http://arxiv.org/abs/1608.06993
7. Ji, Z., Xiong, K., Pang, Y., Li, X.: Video summarization with attention-based encoder-decoder networks. CoRR abs/1708.09545 (2017). http://arxiv.org/abs/1708.09545
8. Kaufman, D., Levi, G., Hassner, T., Wolf, L.: Temporal tessellation: a unified approach for video analysis. In: The IEEE International Conference on Computer Vision (ICCV), vol. 8 (2017)
9. Manning, C.D., Raghavan, P., Schütze, H.: Introduction to Information Retrieval. Cambridge UP, Cambridge (2008)
10. Otani, M., Nakashima, Y., Rahtu, E., Heikkilä, J., Yokoya, N.: Video summarization using deep semantic features. In: Lai, S.-H., Lepetit, V., Nishino, K., Sato, Y. (eds.) ACCV 2016. LNCS, vol. 10115, pp. 361–377. Springer, Cham (2017). https://doi.org/10.1007/978-3-319-54193-8_23
11. Ramos, J., et al.: Using TF-IDF to determine word relevance in document queries. In: Proceedings of the First Instructional Conference on Machine Learning, vol. 242, pp. 133–142 (2003)
12. Redmon, J., Farhadi, A.: Yolo9000: Better, faster, stronger. arXiv preprint arXiv:1612.08242 (2016)
13. Szegedy, C., Vanhoucke, V., Ioffe, S., Shlens, J., Wojna, Z.: Rethinking the inception architecture for computer vision. In: Proceedings of the IEEE Conference on Computer Vision and Pattern Recognition, pp. 2818–2826 (2016)
14. Xu, M., Duan, L.-Y., Cai, J., Chia, L.-T., Xu, C., Tian, Q.: HMM-based audio keyword generation. In: Aizawa, K., Nakamura, Y., Satoh, S. (eds.) PCM 2004. LNCS, vol. 3333, pp. 566–574. Springer, Heidelberg (2004). https://doi.org/10.1007/978-3-540-30543-9_71
15. Zhang, K., Chao, W.-L., Sha, F., Grauman, K.: Video summarization with long short-term memory. In: Leibe, B., Matas, J., Sebe, N., Welling, M. (eds.) ECCV 2016. LNCS, vol. 9911, pp. 766–782. Springer, Cham (2016). https://doi.org/10.1007/978-3-319-46478-7_47
16. Zhou, L., Xu, C., Koch, P., Corso, J.J.: Image caption generation with text-conditional semantic attention. arXiv preprint arXiv:1606.04621 (2016)

ATM Protection Using Embedded Deep Learning Solutions

Alessandro Rossi$^{(\boxtimes)}$, Antonio Rizzo, and Francesco Montefoschi

University of Siena, Siena, Italy
{alessandro.rossi2,antonio.rizzo,francesco.montefoschi}@unisi.it

Abstract. Last decade advances in Deep Learning methods lead to sensible improvements in state of the art results in many real world applications, thanks to the exploitation of particular Artificial Neural Networks architectures. In this paper we present an investigation of the application of such kind of structures to a Video Surveillance case of study, in which the special nature and the small amount of available data increases the difficulties during the training phase.

The analyzed scenario involves the protection of Automatic Teller Machines (ATM), representing a sensitive problem in the world of both banking and public security. Because of the critical issues related to this environment, even apparently small improvements in either accuracy or responsiveness of surveillance systems can produce a fundamental contribution. Even if the experimentation has been reproduced in an artificial scenario, the results show that the implemented architecture is able to classify depth data in real-time on an embedded system, detecting all the test attacks in a few seconds.

Keywords: Deep Learning · Convolutional Neural Networks
Recurrent Neural Networks · Computer Vision · Bank security
Embedded systems · Depth images

1 Introduction

In recent years the global digitalization and the consolidation of information technologies sensibly changed our daily life and the way we interact together, both at local and global level. This digital revolution is also changing how users access banks and financial services, turning a relationship based on the peer-to-peer trust into a mainly online service, with sporadic human interactions. Such mutation and the resulting change in the bank branch structure obviously affect the criminal behavior related to this environment. In this scenario, ATMs are an easy target for fraud attacks, like card skimming/trapping, cash trapping, malware and physical attacks. Assaults based on explosives are a rising problem in Europe and many other parts of the world. A report from the EAST association

---

Supported by Banca Monte dei Paschi di Siena.

L. Pancioni et al. (Eds.): ANNPR 2018, LNAI 11081, pp. 371–382, 2018.
https://doi.org/10.1007/978-3-319-99978-4_29

shows a rise of 80% of such attacks between the first six months of 2015 and 2016. This trend is particularly worrying, not only for the stolen cash, but also for the significant collateral damages to buildings and equipment [1]. Sectorial international studies [2] show that despite the use of explosives and other physical assaults continue to spread, in the long term the attacks will focus on the cyber and logical approaches. In fact, ATM malware and logical security attacks were reported by seven countries in Europe during the year 2017. Moreover, statistics from ABI (Italian Banking Association) show a sensible increase of attacks to the ATMs in opposition to a reduction to bank branches robberies. This is due both to the juridical categorization of the committed crime and to the lower amount of money that can be stolen in a robbery. Indeed, security systems are in general concentrated on the branch rather than on the ATM area, which is usually located outside of the buildings. This also allows perpetrators to perform the assaults during nightly hours. An important issue to consider about these gestures is related to the collateral effects. In fact, the violence necessary in such attacks often lead to serious physical damages to buildings and objects in the neighborhood of the targeted area; this is when considering the best scenario, where no human is involved.

After these premises it is clear how can be fundamental to develop technologies capable of preventing in some way this kind of situation. Crucial features of such a system are the low rate of false alarms and effective promptness in detecting the potential risk, both to alarm the interested control systems and, in the first place, to try to automatically discourage the underway criminal action with some deterrents. In this paper we propose ATMSENSE, an automatic surveillance system, based on video stream analysis of depth frames, that can run on Low Power Single Board Computer. This approach allows a feasible installation of the system in order to analyze the actions performed in front of the ATM, while preserving the privacy of customers. Even if the tests are carried out on data recorded in our laboratory, the goodness of the obtained results lays the groundwork for an in-depth experimentation on the field. The results show that the implemented architecture is able to classify depth data in real-time on an embedded system, detecting all the test attacks in a few seconds.

Since the acquired Depth images are processed by an Artificial Neural Networks (ANNs) in order to predict the nature of the running situation, the scientific contributions of the paper regards mainly the application of ANNs to the specific task under the framework of Action Recognition. However, we think that even the investigation of the behavior of different kind of architectures can be of interested. Indeed, we will show that, even if static Convolutional Neural Networks (CNNs) can achieve good results even basing the prediction on a single frame, the global responsiveness of the system can be improved by adding a LSTM modules or 3-D convolutions. In the composition of such architectures, even some issues related to the training process could be of interest, involving the mixing of different type of techniques like data augmentation and multi-stage learning.

2 Related Works

2.1 Video Surveillance and Action Recognition

Recent advances in Deep Learning techniques and, in particular, in those approaches dedicated to Computer Vision [4,5] lead to a cutting-edge improvement in Image and Video Analysis algorithms. Even if methodologies for Video Surveillance and, more in general, for Action Recognition [6] based on different approaches had been investigated in the past, allowing us to reach good results in restricted scenarios, Deep Learning methods can provide state-of-the-art achievements, at least in the short term. As remarkable examples, different methods have been proposed in order to modify the basic CNNs structure when dealing with sequential input frames coming from video streams, such as 3-Dimensional convolutional filters [7] or temporal pooling [8]. Other interesting application involve the combination of Convolutional and Long-Short Term Modules (LSTM) [9] both for supervised classification tasks [10] and for unsupervised next-frame prediction [11].

Taking in account these results and the possibility of fast and portable prototyping of such algorithms, it seems reasonable to follow this direction and to go towards technologies that should be even more widespread and consolidated in the future. Moreover, such approaches should also allow us a direct scalability when facing new kind of specific situation and typologies of attacks.

2.2 ATM

As ATMs started to play a central role in the customers services [12], many works have been developed trying to improve the security of these interactions. Several systems designed to deal with identity thefts [13–15], interactions with forged documents and certificates [16] and the detection of specific dangerous situations [17,18] had been developed through the investigation and the integration of various hardware devices. However, the most common approach is the analysis by surveillance cameras trying to recognize those actions characterizing a potential critical scenario [19]. In other cases, more specific systems had been oriented towards face detection and tracking [20] or to the recognition of partially occluded faces and bodies [21,22].

In our approach, we head towards a quite new technology like the images analysis throughout depth cameras, which is, at the best of our knowledge, unexplored. This should allow us to join the representation capabilities of videos processing and the need for customer privacy protection, both for ethical and juridical reasons.

3 ATMSense

ATMSENSE is intended to discriminate people's behaviour exhibited in front of an ATM, in order to detect risky situations at an early stage. The sensor used

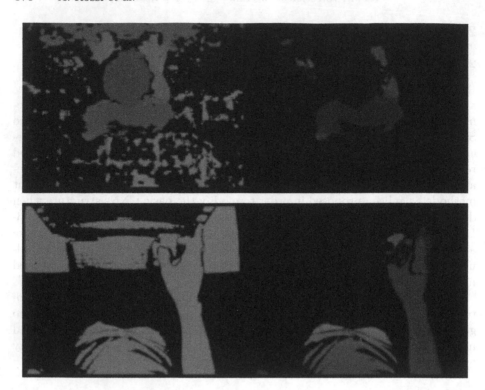

Fig. 1. Images acquired from the employed depth cameras. At the top, images from the long-range camera, while at the bottom frames from the short-range ones. In the left column we shown raw images as they are acquired, in the right one the output after noise removal and background subtraction.

to analyze the scene is the Intel RealSense depth camera. Using depth images instead of the RGB ones provides great advantages: we can avoid dealing with personal data and privacy issues; frames are unaffected by lighting conditions; from a computational point of view, we can rely on a slight improvement by reducing the input channels from three to one. Depth images are processed on a Low Power Single Board Computer with image processing techniques and Convolutional/Recurrent Neural Networks. We found that an SBC with an Intel N3710 processor (1.6 GHz) with 4 Gb RAM is capable of running the system on CPU at the provided frame rate of 6 fps. Again, we would like to specify that ATMSENSE is designed as a supporting surveillance device. Indeed, ATMs are usually monitored from a Surveillance Control Room, but it is not possible to keep under control all the installed machines, so that the assault detection is often late enough to not allow to prevent at least an initial damage. Hence, the predictions provided by our system can be exploited to focus the attention on potential dangerous situations and allow to security agents to promptly take appropriate actions.

3.1 Depth Cameras

Intel RealSense is a family of depth cameras proving several video streams: RGB, depth and infrared. Our experimentation focused on two camera models: a short-range (SR) and a long-range (LR) cameras. LR cameras are intended to be placed above the ATM, focusing the whole interested scene in a portion of machine neighborhood. SR cameras can be placed in the ATM chassis, so that the images are restricted to a small area around the ATM keyboard area. Examples of acquired frames are shown in Fig. 1. As we can observe, SR raw images are less affected by external noise. However, final performances are similar for both cameras. The short-range camera should be embedded in new ATMs, the long-range fits better as an external plugin for already installed ATMs. Whichever camera is employed, the depth video stream is used to classify what is going on in the ATM area. For debugging purposes RGB streams can be collected, but they are not used, neither for the system training, nor for the runtime.

In fact, relying on the RGB stream would create a dependency on factors that we do not want to depend on, like light conditions. Moreover, dealing with faces and other personal images can be an issue for the privacy laws. Having only a low-resolution shape of the person does not allow the personal identification.

3.2 Image Pre-processing

Depth images collected from the cameras are preprocessed before the classification. In this phase we want to remove both the noise and the background objects. The noise is intrinsic in the camera sensor and is reduced using a cascade,of standard image processing filters (i.e. median filtering, erosion, depth clipping and so on). This technique leads to the generation of one video frame starting from 5 frames read from the depth camera. Although the dynamics of the system scales down from 30 fps to 6 fps, the information necessary to classify the images is preserved. The background suppression is related to the environment in which the ATM is located, and includes the device itself. The background is subtracted (using kNN based techniques) making the solution independent from the ATM machines and environments.

The difference between the original image read from the camera and the cleaned version is visible in Fig. 1 for both the camera versions.

3.3 Deep Neural Networks

Starting from what reported in Sect. 2.1, we decided to drive our experimentation on different Deep Learning architectures. However, in order to characterize the main features and to outline an initial baseline for the analyzed scenario, we avoid to exploit hand-crafted techniques too specific for the task. At the moment, the problem is faced as a supervised 2-class classification task, even if we would like to explore different approaches in the future (see Sect. 6).

The tested architectures are a classic Multi-Layer Perceptron (MLP), a standard Convolutional Neural Network (CNN), a CNN with 3-D convolutional filters (CNN3D), a Multi-Layer Perceptron combined with an LSTM module (LSTM) and network assembling Convolutional layers with an LSTM module (CNN+LSTM). All the models are described in Table 1. The *Rectifier Linear Unit (ReLU)* is maintained as activation function for all the layers, except from the LSTM one, which exploits the *Hyperbolic Tangent*. The output layers of all the architectures exploit the *Softmax* function to force a probabilistic scoring within the classes, paired with the *Cross-Entropy* loss function. All the networks are optimized through the *AdaDelta* learning methods [23].

All the technical choices, such as the number of units, convolutional filters, type of pooling and optimization algorithms have been selected after an initial experimental analysis, which however does not shown essential improvements in order to prefer one particular direction. Hence, we decide to assemble the proposed architectures by relying on general common practices. The number of units in the different layers have been selected in order to be large enough to produce good results, while keeping the number of parameters as small as possible (for example, we tested MLP with more parameters and layers without obtaining remarkable improvements).

Table 1. General layer description of tested architectures. FC notation stands for Fully-connected layers.

MLP	LSTM	CNN	CNN+LSTM	CNN3D
FC (2048 HU)	FC (1024 HU)	Conv ($32 \times 5 \times 5$)	Conv ($32 \times 5 \times 5$)	Conv ($32 \times 3 \times 5 \times 5$)
FC output	Lstm (512 HU)	Conv ($32 \times 5 \times 5$)	Conv ($32 \times 5 \times 5$)	Conv ($32 \times 3 \times 5 \times 5$)
Softmax	Drop-out ($p = 0.1$)	Max-pooling	Max-pooling	Max-pooling
	FC (1024 HU)	Conv ($64 \times 5 \times 5$)	Conv ($64 \times 5 \times 5$)	Conv ($64 \times 3 \times 5 \times 5$)
	FC output	Conv ($64 \times 5 \times 5$)	Conv ($64 \times 5 \times 5$)	Conv ($64 \times 3 \times 5 \times 5$)
	Softmax	Max-pooling	Max-pooling	Max-pooling
		Conv ($128 \times 5 \times 5$)	Lstm (512 HU)	Conv ($128 \times 3 \times 5 \times 5$)
		Max-pooling	Drop-out ($p = 0.1$)	Max-pooling
		FC (1024 HU)	FC (1024 HU)	FC (1024 HU)
		FC output	FC output	FC output
		Softmax	Softmax	Softmax

4 Experiments

The real working environment has been reproduced in our laboratory by installing ATMSENSE on a dismissed ATM provided by Monte dei Paschi di Siena Bank. The dataset has been built by recording and tagging depth videos in which different people stage withdrawals and attacks, replicating the actions the thieves do, thanks to the knowledge of the Security Department of the bank. As a prototype, we taped the SR camera to the ATM frame, and we installed the LR camera on the top of a support above the ATM. With both the cameras connected, we recorded 132 depth videos simulating both the *Withdrawal* and the *Attack* scenarios, representing the two classes to be discriminated by the system. However, we reported only the results on the LR camera data, since the final performances are similar. To improve variability and generalization, these videos have been staged by several actors in different sessions, using different light conditions (which only slightly affect the acquired images). Videos have been manually labelled frame by frame. Background profiling has been carried out by recording 25 videos without any kind of interaction with the ATM. In the training phase, pre-processed videos (as stated in Sect. 3.2) are split, as reported in Table 2, among train and test sets. To reduce the computational cost, the images, acquired at 640×480 pixels resolution, are down-sampled to 80×60 pixels. Hence, the dataset is generated by separating and shuffling sequences of consecutive frames, together with the correspondent labels. In this way we obtained about 250,000 and 30,000 labeled samples for training and test respectively. A part of the training data has been used to built validation set to implement Early Stopping during the optimization process. All the experimentation has been carried out within the Keras [25] framework with Tensorflow [26] backend, while for the Image Processing part we exploited the OpenCV [27] package.

Since we would like to exploit the temporal order of data, we tested our architectures in three configurations. In the static one, a single frame is provided as input to the networks. Hence, we tested two dynamic configurations by composing the input as a sequence of 5 or 10 video frames. Of course, temporal architectures (LSTM, CNN+LSTM, CNN3D) are not tested in the static configuration, while frames are concatenated in the dynamic settings in order to deal with static networks (MLP, CNN). Because of the relatively small amount of training data, we also tested all the configurations on an augmented dataset, generating 5 input samples from each original one. Apart from standard transformations as rotation, horizontal flip, blurring and translation, we also apply z-axis translation. Indeed, each pixel express, in fact, the vertical distance from the camera and these translations should help to lose dependencies on camera distance and actors height. This techniques is not only useful in the classic sense to improve generalization on our data, but we hope that it could be also useful to make the system less dependent on the specific ATM environment and to increase the flexibility when installing it in different machines.

Results of the F1 score are reported in Table 3. Even if, all the tested architectures obtain high scores, DL methods reach best performances and show to obtain more benefits from data augmentation. Since CNNobtains good perfor-

Table 2. Number of videos recorded to assemble the dataset

	Videos		Total samples
	Withdrawal	*Attack*	
Train	54	42	250,000
Test	18	18	30,000
Total	72	60	280,000

mances even when dealing with a single input frame, while LSTM modules could have problems with gradient back-propagation, we exploited the fact that the first part of the architectures is the same and tried to help the CNN+LSTM optimization by initializing the weights from the ones learned by CNN in its best configuration. Nevertheless, this procedure does not lead to any improvements (p-CNN+LSTM).

Apart from the predictions quality of each model, we are also interested in the capability of a system to actually detect an attack with a reasonable promptness and, at the same time, in its stability in order not to provide too much false alarms. Hence, we set an evaluation in a more practical configuration, considering the predictions over a window of frames and raising an alarm only when all the predictions inside the window are positive. Even if we could have carried out a more systematic analysis on the window length and the alarm threshold (AUC, ROC, Precision-Recall curve etc.), we would like to find a way to reduce the False Positive Rate and to compare the architectures promptness in an homogeneous configuration. Hence, we fixed the length of the window to 20 frames, allowing to all the architectures to detect, sooner or later, every attack in the test videos. In Table 4 and Fig. 2 we reported the number of False Positive Rates (FPR) for each architecture, plus the average (Avg.) and the standard deviation (Std.) among all the measures of times employed by each model to detect an attack since its beginning. From these statistics, we can see that the effect of data augmentation is useful not only to improve F1-score but also the promptness and the stability of the predictions, even if the measures are related in some way.

In addition to a better F1-score, DL approach and temporal models shown in general a better promptness and, moreover, a lower variability in the reaction times. This could be interpreted as an index of stability and, thus, reliability of the system itself. In general we could point out CNN3Das the best model in all the reported measures. The fact that to consider the sequential order of data did not lead to remarkable performance enhancements could be addressed mainly to two issues. The first one is, as already said, related to the size of the training data. Indeed, the optimization of recurrent modules requires in general a wider statistics on the real phenomenon, as we can see from the general performance decay when employing sequences of 10 frames as input. The second one is the fact that each input frame is generated by averaging 5 original frames. Even if we mainly exploit this technique in order to clean data, the produced static samples contain an intrinsic dynamic statistics which is often employed in Action Recognition [6].

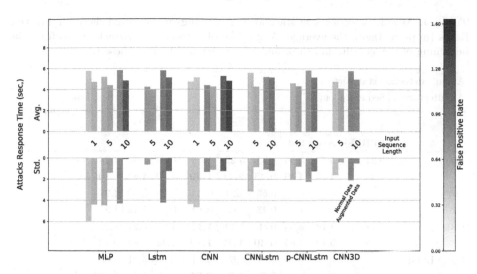

Fig. 2. Comparison on False Positive Rate (bars color) with average (Avg.) and standard deviation (Std.) of attack detection times among all the tested architectures. Avg. and Std. are reported on top and bottom bars respectively, for each model in each training configuration: varying the length of the input sequences in {1, 5, 10} and with original (Normal) data or by performing data augmentation (Augmented). (Color figure online)

Table 3. Evaluation of F1-score on each model in each training configuration.

F1-score (%)				
Architecture	Sequences length			Data augmentation
	1	5	10	
MLP	92.93	92.02	91.40	None
	94.51	93.63	92.94	5×
LSTM	-	93.40	92.91	
	-	93.58	93.72	
CNN	93.77	90.91	89.83	
	95.04	**95.18**	92.63	
CNN+LSTM	-	92.26	89.81	
	-	**95.52**	93.78	
p-CNN+LSTM	-	90.39	91.50	
	-	**95.48**	93.48	
CNN3D	-	92.38	86.56	
	-	**95.55**	94.46	

Table 4. Assault detection statistics: for each settings we reported the False Positive Rates (in percentage), the average (Avg.) time of detection on *Attack* videos from the beginning of the assault, and the standard deviation (Std.) on these times.

Assault detection statistics										
Architecture	Sequence length									Data augmentation
	1			5			10			
	FPR(%)	Avg.	Std.							
Mlp	0.21	5.77	5.96	0.35	5.22	4.46	0.63	5.83	4.28	None
	0.33	4.73	4.39	**0.43**	**4.41**	**1.39**	1.10	4.83	0.10	5×
Lstm	-	-	-	0.53	4.24	0.63	0.78	5.80	4.21	
	-	-	-	0.48	4.00	0.10	0.98	5.12	1.24	
Cnn	0.18	4.73	4.35	0.66	4.39	1.32	1.03	5.27	1.22	
	0.18	5.13	4.65	**0.40**	**4.27**	**1.12**	1.44	4.82	0.10	
Cnn+Lstm	-	-	-	0.26	5.59	3.14	0.58	5.21	1.06	
	-	-	-	**0.35**	**4.28**	**0.85**	0.79	5.17	1.18	
p-Cnn+Lstm	-	-	-	0.31	4.62	2.03	0.52	5.83	2.23	
	-	-	-	**0.48**	**4.33**	**0.80**	0.83	5.17	1.24	
Cnn3DCnn3D	-	-	-	0.18	4.74	1.62	0.64	5.76	2.09	
	-	-	-	**0.37**	**4.09**	**0.40**	0.60	4.95	0.51	

5   Conclusions

In this work we propose an application of Automatic Video Analysis to improve the surveillance and the security on ATMs. From laboratory tests, the system can detect attacks very quickly, both when the depth camera is integrated into the ATM itself, and when it is installed nearby. Moreover, the approach employs off-the-shelf technologies of a total cost which is quite inexpensive when compared with an ATM cost or with the potential financial and general damages. The software solution is general for the approach, even if an additional data collection and a re-training phase will be necessary, depending on particular needs of specific situations. Although the current solution is customized for a single mode of assault, the obtained results allowed us a short terms scheduling of a more real experimentation phase on the field. Indeed, the very fast attack detection time will allow to the Surveillance Control Room to promptly intervene, raising a reasonable amount of false alarms. In this work, we also verified that standard and specific data augmentation techniques could be exploited to improve the models performances within all the proposed measures.

6   Future Works

Detection accuracy in a real-world scenario could be improved by collecting further data, statistically enlarging the events analyzed by the system and, hence, helping the DL methods to better generalize, instead of over-fit on training examples. The depth footage recorded for the training is focused on explosive-based attacks. New videos could be recorded with the perspective to detect additional kind of ATM assaults, providing a more complete surveillance equipment. The downside of having more depth videos is the need of manually tagging the frames. A complementary approach could be to introduce by Novelty Detection [28] algorithms. As an example, the solution we proposed in [29] to bank branch Audio-Surveillance can be redesigned in this scenario. This algorithm would be totally unsupervised, and capable of detecting any kind of anomaly which comes from unexpected users behavior. An arbiter would take as input the outputs of both the algorithms, and rule a final decision.

Acknowledgments. This work was supported by Monte dei Paschi Bank grant DIS-POC017/6. We thank Antonio J. Peña and the BSC/UPC NVIDIA GPU Center of Excellence for many useful advices regarding computational issues and to provide us the computational capability required for Deep Networks training through their multi-GPU cluster.

References

1. European Association for Secure Transactions. ATM explosive attacks surge in Europe (2016). https://www.association-secure-transactions.eu/atm-explosive-attacks-surge-in-europe/
2. European Association for Secure Transactions. EAST publishes European fraud update 3-2017 (2017). https://www.association-secure-transactions.eu/east-publishes-european-fraud-update-3-2017/
3. Author, F., Author, S.: Title of a proceedings paper. In: Editor, F., Editor, S. (eds.) CONFERENCE 2016, LNCS, vol. 9999, pp. 1–13. Springer, Heidelberg (2016). https://doi.org/10.10007/1234567890
4. Deng, J., et al.: A large-scale hierarchical image database. In: IEEE Conference on Computer Vision and Pattern Recognition (CVPR), pp. 248–255 (2009)
5. Krizhevsky, A., et al.: Imagenet classification with deep convolutional neural networks. In: Advances in Neural Information Processing Systems (NIPS), pp. 1097–1105 (2012)
6. Herath, S., et al.: Going deeper into action recognition: a survey. Image Vis. Comput. **60**, 4–21 (2017)
7. Ji, S., et al.: 3D convolutional neural networks for human action recognition. IEEE Trans. Pattern Anal. Mach. Intell. **35**(1), 221–231 (2013)
8. Ng, J.Y.-H., et al.: Beyond short snippets: deep networks for video classification. In: Proceedings of IEEE Conference on Computer Vision and Pattern Recognition (CVPR), pp. 4694–4702 (2015)
9. Hochreiter, S., Schmidhuber, J.: Unsupervised learning of video representations using LSTMs, pp. 473–479 (1997)
10. Donahue, J.L.A., et al.: Long-term recurrent convolutional networks for visual recognition and description. In: Proceedings of the IEEE Conference on Computer Vision and Pattern Recognition (CVPR), pp. 2625–2634 (2015)
11. Srivastava, N., et al.: Long-term recurrent convolutional networks for visual recognition and description. In: International Conference on Machine Learning (ICML), pp. 843–852 (2015)

12. De Luca, A., et al.: Towards understanding ATM security: a field study of real world ATM use. In: Proceedings of the Sixth Symposium on Usable Privacy and Security (2010)
13. Puente, F., et al.: Improving online banking security with hardware devices. In: Annual International Carnahan Conference on Security Technology (CCST), pp. 174–177 (2005)
14. Lasisi, H., Ajisafe, A.A.: Development of stripe biometric based fingerprint authentications systems in Automated Teller Machines. In: International Conference on Advances in Computational Tools for Engineering Applications (ACTEA), pp. 172–175 (2012)
15. AshokaRajan, R., et al.: A novel approach for secure ATM transactions using fingerprint watermarking. In: International Conference on Advanced Computing (ICoAC), pp. 547–552 (2013)
16. Sako, H., et al.: Self-defense-technologies for automated teller machines. In: International Machine Vision and Image Processing Conference (IMVIP), pp. 177–184 (2007)
17. Raj, M.M.E., Julian, A.: Design and implementation of anti-theft ATM machine using embedded systems. In: International Conference on Circuit, Power and Computing Technologies (ICCPCT), pp. 1–5 (2015)
18. Shriram, S., et al.: Smart ATM surveillance system. In: International Conference on Circuit, Power and Computing Technologies (ICCPCT), pp. 1–6 (2016)
19. Ding, N., et al.: Energy-based surveillance systems for ATM machines. In: World Congress on Intelligent Control and Automation (WCICA), pp. 2880–2887 (2010)
20. Tang, Y., et al.: ATM intelligent surveillance based on omni-directional vision. In: World Congress on Computer Science and Information Engineering (WRI), pp. 660–664 (2009)
21. Chen, I.-P., et al.: Image processing based burglarproof system using silhouette image. In: International Conference on Multimedia Technology (ICMT), pp. 6394–6397 (2011)
22. Zhang, X.: A novel efficient method for abnormal face detection in ATM. In: International Conference on Audio, Language and Image Processing (ICALIP), pp. 695–700 (2014)
23. Zeiler, M.D.: ADADELTA: an adaptive learning rate method. arXiv preprint arXiv:1212.5701 (2012)
24. Simonyan, K., Zisserman, A.: Very deep convolutional networks for large-scale image recognition (2014). arXiv:1409.1556
25. Chollet, F., et al.: Keras (2015). https://keras.io
26. Abadi, M., et al.: TensorFlow: large-scale machine learning on heterogeneous systems (2015). https://www.tensorflow.org/
27. Bradski, G.: The OpenCV library. Dr. Dobb's J. Softw. Tools (2000)
28. Pimentel, M.A.F., et al.: A review of novelty detection. Signal Process. **99**, 215–249 (2014)
29. Rossi, A., et al.: Auto-associative recurrent neural networks and long term dependencies in novelty detection for audio surveillance applications. In: IOP Conference Series: Materials Science and Engineering (2017)

Object Detection in Floor Plan Images

Zahra Ziran$^{(\boxtimes)}$ and Simone Marinai

Dipartimento di Ingegneria dell'Informazione (DINFO),
Università degli Studi di Firenze, Florence, Italy
{zahra.ziran,simone.marinai}@unifi.it

Abstract. In this work we investigate the use of deep neural networks for object detection in floor plan images. Object detection is important for understanding floor plans and is a preliminary step for their conversion into other representations.

In particular, we evaluate the use of object detection architectures, originally designed and trained to recognize objects in images, for recognizing furniture objects as well as doors and windows in floor plans. Even if the problem is somehow easier than the original one in the case of this research the datasets available are extremely small and therefore the training of deep architectures can be problematic. In addition to the use of object detection architectures for floor plan images, another contribution of this paper is the creation of two datasets that have been used for performing the experiments covering different types of floor plans with different peculiarities.

Keywords: Floor plan analysis · Object detection
Convolutional neural networks · Transfer learning

1 Introduction

Detecting and recognizing objects in floor plans is an essential task for the understanding of these graphical documents. Our research on this topic is part of the overall task of understanding of graphical documents for generating accessible graphical documents for visually impaired people [4,13]. A comprehensive perception of a floorplan is crucially important for blind people, allowing them to find their path as they face a new building. It is important to clarify that floorplans available in real estate websites or other floorplans in other websites are nearly always in image format even if they have been produced with CAD tools. CAD files are in general only available to their authors and not distributed. Object detection in natural images is basically defined as finding the location of objects in one image and labeling them. In many cases, the object location is based on the identification of the bounding box surrounding it. Also in this application the identification of the object bounding box is sufficient for our purposes. Starting from widely studied architectures based on convolutional neural networks, a few object detectors have been recently proposed, such as: Faster R-CNN [17], R-FCN, Multibox, SSD [11] and YOLO [16].

© Springer Nature Switzerland AG 2018
L. Pancioni et al. (Eds.): ANNPR 2018, LNAI 11081, pp. 383–394, 2018.
https://doi.org/10.1007/978-3-319-99978-4_30

1.1 Previous Work

As in several domains also the document analysis community faced a growing use of deep learning in recent research. When looking to the use of deep learning in the area of graphics recognition there are a limited, but interesting research works. Among various techniques object detectors have been used to address various problems in document analysis. Symbol detection in on-line graphical documents is proposed in [9] where the authors use Faster R-CNN do address the task. In particular, the work addresses the recognition of mathematical expressions and flowcharts in handwritten documents by using the Tensorflow Object Detection API [8]. Another application of the latter API is related to handwritten music object detection [14] where the Faster R-CNN is used to recognize musical symbols. In both papers the number of training items is relatively high and the results are evaluated only considering the accuracy of the model without taking into account the recall. Other authors used Faster R-CNN for page layout identification [18], for comic character face detection [15], and for arrow localization on handwritten industrial inspection sheets [5].

One recent effort to extract structural information from floor plan images is described in [2] where the authors parse floor plan images to estimate the size of the rooms for interactive furniture fitting. They first perform wall segmentation by using a fully convolutional neural network, subsequently they detect objects using a Faster R-CNN, and finally, they do optical character recognition to obtain the rooms dimensions. One interesting feature of this work is the combination of three methods to achieve the overall floor plan understanding. Unfortunately, very few details are provided in the paper about the use of Faster R-CNN for object location. Moreover, the floor plan dataset created by the authors only contains the ground-truth about the wall position.

In the work described in [6] the authors address the floor plan understanding by segmenting walls, windows, and doors. One of the main focuses of the paper is to address images with different notations (e.g. for walls or for furniture objects). The proposed techniques are tested on four floor plan datasets (named CVC-FP) which are freely accessible to the public. As discussed also in Sect. 3 the CVC-FP dataset only contains objects of 6 classes: sink, toilet, shower, bath, door, and window, without include furniture objects.

From the point of view of the neural architecture one important paper for this work is [7] where the authors evaluate and compare different object detection architectures. The goal of [7] is to identify the most successful architectures and support users when choosing one architecture on the basis of various perspectives: speed, memory, and accuracy. To this end, the authors in [7] implement some modern convolutional detectors: Faster R-CNN, R-FCN and SSD in a unified framework, as a part of the Tensorflow Object Detection API [8]. The authors pre-trained the architectures on several datasets, but the best performance were achieved by pre-training with the COCO dataset [10].

In this work we explore the use and adaptation of the Tensorflow Object Detection API [8] to identify floor plan objects in two datasets that have been built to address this task.

The rest of the paper is organized as follows. In Sect. 2 we describe the architecture of the Faster R-CNN model considered and provide some information about how we modified it in order to obtain information about the recall of the system. In Sect. 3 we analyze the peculiarities of the floor plan datasets that we built and used in the experiments discussed in Sect. 4. Our conclusions and pointers for future work are in Sect. 5.

2 The Architecture

In this research we work with the widely used Tensorflow Object Detection API [8] for an easy comparison of alternative architectures. We initially evaluated one COCO-pre-trained Single Shot Detector with MobileNets that we fine-tuned with floor plan images. We selected this architecture because it is a small and flexible model that has the benefit of fast training times compared to larger models, while it does not sacrifice much in terms of accuracy. In these preliminary tests we also compared the SSD with Faster R-CNN with ResNet 50 and with ResNet 101. After these preliminary experiments it turned out that Faster R-CNN performs significantly better than SSD. Moreover, comparing the performance of ResNet 50 with ResNet 101 on the floor plan datasets, there was no real difference. We therefore used Faster R-CNN with ResNet 50 as a basis model for our work.

Faster R-CNN is one of the most accurate and fast neural object detectors proposed so far. The internal structure of the network is as follows (see Fig. 1): first, the image is passed through some convolutional layers to produce several feature maps. Then, the main component of Faster R-CNN, the region proposal network (RPN), uses a 3 × 3 sliding window and takes the previous feature maps as input. The output of the RPN is a tensor in a lower dimension. At this stage, each window location generates some bounding boxes, based on fixed-ratio anchor boxes (e.g. 2.0, 1.0, 0.3) and an "objectness" score for each box. These are the region proposals for the input image which provide approximate coordinates of the objects in the image. The "objectness" scores, if above a given threshold, determine which region proposal can move forward in the network. Subsequently, the good regions pass through a pooling layer, then a few fully-connected layers, and finally a softmax layer for classification and a regressor for bounding box refinement.

As previously mentioned to perform our experiments we use the Tensorflow Object Detection API. This is an open source framework, built on top of the widely used Tensorflow library, that takes care of the training and evaluation of the different architectures implemented. One interesting feature of the API is that it makes it easy to train different models on the same dataset and compare their performance. In addition to the average precision performance per category, we extended the API to calculate the number of false negatives as well as the average recall per class.

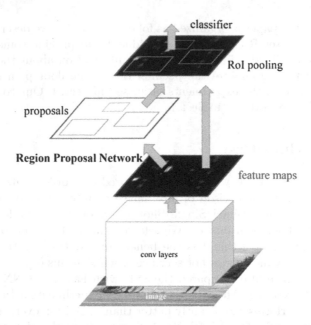

Fig. 1. The internal architecture of the Faster R-CNN as a single, unified network for object detection (Image from [17]).

2.1 False Negative Calculation

By default, the Tensorflow Object Detection API supports the PASCAL Visual Object Classes (VOC) 2007 [3] detection metric. This metric is designed to evaluate visual object detection and recognition models, which helps machine learning researchers have standard evaluation procedures.

In the detection metric, for a detection bounding box to be a true positive, three conditions must be true:

- The area of the intersection of the detected bounding box B_d and the ground truth bounding box B_{gt} over the union area of the two bounding boxes must be greater than 0.5, according to the following equation:

$$r_i = \frac{area(B_d \cap B_{gt})}{area(B_d \cup B_{gt})} > 0.5 \qquad (1)$$

- The class label of the detection bounding box and the ground truth bounding box must be the same.
- The probability of the object's recognition must be greater than some specific thresholds. In most cases, and also in this work, we consider the object as found if the probability is higher than 0.50.

To find false negative detections we first matched all the detections to objects in the ground truth. True/false positives are determined and detections matched to difficult boxes are ignored. In the next stage the ground truth objects that have not been detected are determined as false negatives.

After computing the true positives and false negatives number for each category it is easy to calculate the average recall in addition to the average precision computed by the API: $Recall = \frac{TP}{TP+FN}$.

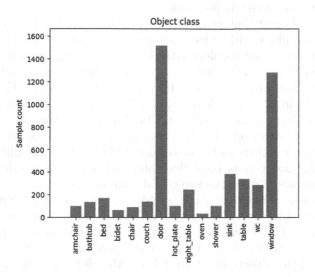

Fig. 2. Object class distribution of the *d1* dataset.

3   The Floor Plan Datasets

In order to evaluate the object detection in floor plans we obviously need one or more labeled datasets. In the past decade, some datasets have been proposed for evaluating research on floor plan analysis. The SESYD dataset [1] contains synthetic floor plans where furniture objects are randomly placed on a few fixed floor plan layouts. Even if this approach for dataset generation is very interesting, the actual dataset contains only ten floor plan layouts and the objects come from a limited number of categories (for instance, there is only one model for the bed). Moreover, the generated floor plans are somehow unrealistic with very small beds or similar mistakes. Another dataset widely used for this research has been proposed in [12]. This dataset contains 90 actual floor plans provided by one architectural firm. While more realistic than the others, these floor plans contain only few objects and therefore are not suitable for the research carried out in this work.

In order to work with realistic images we first created one small dataset (referred to as *d1*) using the images that show up in Google's image search. This dataset consists of 135 images of variable size containing objects in 15 classes and a total of 4973 objects (in the experiments we considered 2697 objects in the training set, 1165 objects in the validation set, and 1111 objects in the test set). In Fig. 4 we show one example of floor plan in this collection, while Fig. 2 shows the distribution of objects in the different classes. Of course some types of objects are not present in all the images, for instance the floor plan of an office might not have any bed in it. Among all the classes, the oven is the rarest one and the door is the most frequent one.

The second dataset that we gathered is called *d2*. This dataset contains Middle Eastern floor plans, with object shapes different from the ones in *d1*. Another important feature is that the floor plans in *d2* come from one architectural firm and are therefore more homogeneous in their content. The *d2* dataset consists of 300 images, but only 160 images have been labeled so far. The 160 images contain objects in 12 classes and a total of 7788 (in the experiments we considered 4535 objects in the training set, 1457 objects in the validation set, and 1796 objects in the test set). In Fig. 5 we show one example of floor plan in this collection, while Fig. 3 shows the distribution of objects in the different classes. As a particular property of these floor plan datasets, it is worth to note that the images are mostly grayscale and contain simple shapes. As we will see in the experimental part this property has a positive effect on the performance of the model, compared to datasets that contain images with more complex features and more noise. The dataset *d1* has the greatest imbalance in the number of objects in each class. Moreover, images in *d1* have more diversity. For example, almost none of the objects in *d2* are filled with color, while in *d1* all the floor plans are painted, for presentation purposes.

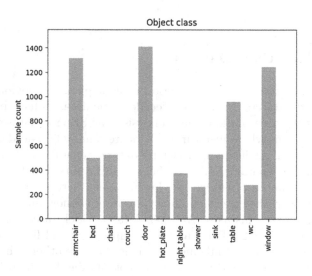

Fig. 3. Object class distribution of the *d2* dataset (Color figure online)

4   Experiments

The Faster R-CNN is first trained on the *d1* dataset, which contains 135 images. As mentioned earlier, this dataset is substantially diverse and contains random non-standard images collected from the Internet.

Table 1. Final evaluation results

Dataset	Objects	False negatives		Mean average precision		Mean average recall	
		Val	Test	Val	Test	Val	Test
d1	1111	411	445	0.32	0.31	0.60	0.56
d2	1796	87	102	0.83	0.86	0.92	0.92

Final evaluation results of the d1 *dataset after 46916 training steps, and the* d2 *dataset after 18550 training steps.*

In the first experiments performed on a smaller dataset with the default configuration of the API the results on the validation set were not satisfying with a maximum mean average precision of about 0.26. To aid generalization, we threw in a few of data augmentation options. In particular we considered **random horizontal flip**, **random vertical flip**, **random rotation 90**, and **random RGB to gray**. These options are provided by the Tensorflow Object Detection API. In addition to data augmentation we also changed the scales and aspect ratios of anchor generator in order to take into account the peculiarities of the floor plan objects.

With the above mentioned modified configuration we also ran our experiments on the *d2* dataset. After stopping the training considering the validation set, the mean average precision and recall on the test set for both datasets are shown in Table 1. Details about the performance achieved for each class are reported in Table 2.

Taking into account the features of the two datasets it is not surprising that the best results are achieved on the *d2* dataset with a mean average precision of 0.86, and a mean average recall of 0.92. Part of the difference in performance is probably related to the special nature of the dataset: compared to *d1*, the objects are cleaner, less diverse and not different across images other than rotation and scale. As it turns out, the performance of the model is not too much affected by an imbalanced dataset (*d2*). For instance the model achieves 0.80 average precision for the **couch** class that is the less frequent one. At the same time the model achieved 0.93 average precision for the **door** class which has at least 10 times more samples than the **couch** class. Concerning the average recall it is very interesting to notice that the test images had zero false negatives for the **hot-plate** and **bed** classes. On the basis of the superior performance of the network on the *d2* dataset we wanted to explore more in details the possibility of doing some transfer learning of this network to improve the performance on the *d1* dataset. As we can see in Table 3 by finetuning on the *d1* dataset one network previously trained on the *d2* dataset we achieve a 0.39 mean average precision and 0.69 mean average recall. This is better than the previously mentioned results obtained by finetuning on the *d1* dataset one network previously trained on the COCO dataset.

Table 2. Performance by category

Class	Average recall(Val)		Average recall(Test)		Average precision(Val)		Average precision(Test)	
	d1	d2	d1	d2	d1	d2	d1	d2
armchair	0.92	0.97	0.50	0.95	0.21	0.92	0.21	0.92
bathtub	0.72	N/A	0.80	N/A	0.57	N/A	0.57	N/A
bed	0.56	1.00	0.70	1.00	0.47	0.98	0.47	0.98
bidet	0.57	N/A	0.54	N/A	0.11	N/A	0.11	N/A
chair	0.58	0.68	0.44	0.76	0.12	0.50	0.12	0.62
couch	0.75	0.80	0.52	0.88	0.46	0.52	0.48	0.80
door	0.63	0.97	0.63	0.95	0.60	0.94	0.60	0.93
hot_plate	0.67	0.97	0.52	1.00	0.40	0.92	0.39	0.96
night_table	0.65	0.93	0.34	0.81	0.37	0.86	0.37	0.79
oven	0.12	N/A	0.55	N/A	0.01	N/A	0.01	N/A
shower	0.55	0.97	0.71	0.93	0.19	0.90	0.16	0.90
sink	0.57	0.95	0.47	0.94	0.28	0.80	0.28	0.85
table	0.58	0.94	0.46	0.98	0.33	0.86	0.33	0.93
wc	0.45	1.00	0.65	0.95	0.13	0.79	0.13	0.78
window	0.58	0.91	0.54	0.88	0.49	0.87	0.48	0.85

Average precision and recall calculated by category, for the d1 *and* d2 *dataset after 46916 and 18550 training steps, respectively.*

Table 3. More transfer learning

Dataset	Objects	False negatives	Mean average precision	Mean average recall
COCO - *d1*	1111	445	0.31	0.60
d2 - d1	1111	368	0.39	0.69

The final evaluated results of the pre-trained models that is fine tuned on dataset *d1*.

Fig. 4. An inference result of the model trained on the *d1* dataset. False negatives have black bounding boxes and detection bounding boxes are colorful. Note how new shapes and colors in this test image damage the performance of the model. (Color figure online)

4.1 Discussion

From the experiments performed on the two datasets we can notice that by using convolutional object detectors, the recognition performance are not too much influenced from class imbalance in the training set.

The only exception is related to the `oven` class from dataset *d1* whose extremely low performance are probably due to the very low sample size and the variability of the appearance of this object in the dataset. On the other hand, the `door` and `window` classes are responsible for 52% of the false negatives in d1 validation set, while they make up 58% of the object samples. In these two classes the model performs relatively well in terms of average precision, but it

Fig. 5. Example of results of the model trained on the *d2* dataset.

is not capable of detecting many objects. At first this result might contradict the intuition that more samples led to better performance. However it is important to recall that the performance of the model in one class heavily depends on the diversity of object samples. It is useful to remark that in the case of doors and walls the objects are connected to the walls while other objects are usually more isolated in the rooms. The diversity of walls and doors is therefore higher with respect to other classes because of the variable context. Another source of errors for windows is the higher variability of the aspect-ratio with respect to other objects that in most cases are simply scaled and rotated, in particular in dataset *d2* where reasonable performance on the dataset is obtained. Regarding the three most frequent classes in *d2* `armchair`, `door`, and `window`, it can be seen from Table 2 that the model is nearly perfect in terms of average precision and the class bed (whose items are more regular) achieves an average precision of 98%.

5   Conclusions

In this work two different floor plan datasets have been created to cover different architectures and drawing conventions of floor plans from all over the world. The performance of an object detector which is originally designed for detecting objects in natural images was tested to identify objects in the floor plans in these datasets. The floor plan has an essential misrepresentation issue in terms of the sample size of objects. To better analyze the performance, false negative objects of each class have individually been counted in order to find out whether the detection results suffer from differences in the number of samples for each class. We noticed that the performance of the model in a class heavily depends on the diversity of object samples, object rotation and scale somehow outweighing the role of sample size. It is interesting also to notice how a network that pre-trained from another domain (COCO pre-trained Faster-RCNN with Res Net 50) can perform well on the floor plan datasets using just a one hundred images. To further improve the results on this task, it is recommended to either collect larger datasets to cover different graphical conventions, or implement data augmentation techniques more suitable for the object detection in floor plans.

References

1. Delalandre, M., Valveny, E., Pridmore, T., Karatzas, D.: Generation of synthetic documents for performance evaluation of symbol recognition & spotting systems. Int. J. Doc. Anal. Recognit. (IJDAR) **13**(3), 187–207 (2010)
2. Dodge, S., Xu, J., Stenger, B.: Parsing floor plan images. In: 2017 Fifteenth IAPR International Conference on Machine Vision Applications (MVA), pp. 358–361, May 2017. https://doi.org/10.23919/MVA.2017.7986875
3. Everingham, M., Van Gool, L., Williams, C.K.I., Winn, J., Zisserman, A.: The pascal visual object classes (VOC) challenge. Int. J. Comput. Vis. **88**(2), 303–338 (2010). https://doi.org/10.1007/s11263-009-0275-4
4. Goncu, C., Madugalla, A., Marinai, S., Marriott, K.: Accessible on-line floor plans. In: Proceedings of the 24th International Conference on World Wide Web, WWW 2015, Florence, Italy, 18–22 May 2015, pp. 388–398 (2015). http://doi.acm.org/10.1145/2736277.2741660
5. Gupta, G., Sharma, S.M., Vig, L.: Information extraction from hand-marked industrial inspection sheets. In: 14th IAPR International Conference on Document Analysis and Recognition, ICDAR 2017, Kyoto, Japan, 9–15 November 2017, pp. 33–38 (2017)
6. de las Heras, L.P., Ahmed, S., Liwicki, M., Valveny, E., Sánchez, G.: Statistical segmentation and structural recognition for floor plan interpretation. Int. J. Doc. Anal. Recognit. (IJDAR) **17**(3), 221–237 (2014). https://doi.org/10.1007/s10032-013-0215-2

7. Huang, J., et al.: Speed/accuracy trade-offs for modern convolutional object detectors. In: 2017 IEEE Conference on Computer Vision and Pattern Recognition (CVPR), pp. 3296–3297, July 2017. https://doi.org/10.1109/CVPR.2017.351
8. Huang, J., et al.: TensorFlow object detection API (2018). https://github.com/tensorflow/models/tree/master/research/object_detection
9. Julca-Aguilar, F.D., Hirata, N.S.T.: Symbol detection in online handwritten graphics using faster R-CNN. In: Proceedings of the 13th International Workshop on Document Analysis Systems, pp. 151–156 (2018)
10. Lin, T., et al.: Microsoft COCO: common objects. In: Proceedings of Computer Vision - ECCV 2014 - 13th European Conference Part V, Zurich, Switzerland, 6–12 September 2014, pp. 740–755 (2014)
11. Liu, W., et al.: SSD: single shot multibox detector (2016, to appear). http://arxiv.org/abs/1512.02325
12. Macé, S., Locteau, H., Valveny, E., Tabbone, S.: A system to detect rooms in architectural floor plan images. In: Proceedings of the 9th IAPR International Workshop on Document Analysis Systems, DAS 2010, pp. 167–174. ACM, New York (2010). http://doi.acm.org/10.1145/1815330.1815352
13. Madugalla, A., Marriott, K., Marinai, S.: Partitioning open plan areas in floor plans. In: 14th IAPR International Conference on Document Analysis and Recognition, ICDAR 2017, Kyoto, Japan, 9–15 November 2017, pp. 47–52 (2017). https://doi.org/10.1109/ICDAR.2017.17
14. Pacha, A., Eidenberger, H., Kwon-Young Choi, B.C., Ricquebourg, Y., Zanibbi, R.: Handwritten music object detection: open issues and baseline results. In: Proceedings of the 13th International Workshop on Document Analysis Systems, pp. 163–168 (2018)
15. Qin, X., Zhou, Y., He, Z., Wang, Y., Tang, Z.: A faster R-CNN based method for comic characters face detection. In: 14th IAPR International Conference on Document Analysis and Recognition, ICDAR 2017, Kyoto, Japan, 9–15 November 2017, pp. 1074–1080 (2017)
16. Redmon, J., Farhadi, A.: YOLO9000: better, faster, stronger. In: 2017 IEEE Conference on Computer Vision and Pattern Recognition, CVPR 2017, Honolulu, HI, USA, 21–26 July 2017, pp. 6517–6525 (2017). https://doi.org/10.1109/CVPR.2017.690
17. Ren, S., He, K., Girshick, R., Sun, J.: Faster R-CNN: towards real-time object detection with region proposal networks. IEEE Trans. Pattern Anal. Mach. Intell. **39**(6), 1137–1149 (2017). https://doi.org/10.1109/TPAMI.2016.2577031
18. Yi, X., Gao, L., Liao, Y., Zhang, X., Liu, R., Jiang, Z.: CNN based page object detection in document images. In: 14th IAPR International Conference on Document Analysis and Recognition, ICDAR 2017, Kyoto, Japan, 9–15 November 2017, pp. 230–235 (2017)

Historical Handwritten Document Segmentation by Using a Weighted Loss

Samuele Capobianco[(✉)], Leonardo Scommegna, and Simone Marinai

University of Florence, via di Santa Marta, 3, Firenze, Italy
{samuele.capobianco,simone.marinai}@unifi.it,
leonardo.scommegna@stud.unifi.it

Abstract. In this work we propose one deep architecture to identify text and not-text regions in historical handwritten documents. In particular we adopt the U-net architecture in combination with a suitable weighted loss function in order to put more emphasis on most critical areas. We define one weighted map to balance the pixel frequency among classes and to guide the training with local prior rules. In the experiments we evaluate the performance of the U-net architecture and of the weighted training on one benchmark dataset. We obtain good results using global metrics improving global and local classification scores.

Keywords: Convolutional Neural Networks · Page segmentation
Loss functions

1 Introduction

Understanding handwritten historical documents is a challenging task that includes several sub-problems. One of the first steps is to segment and extract text lines which could be recognized in subsequent phases to understand the document content. The layout analysis of handwritten documents can be very difficult, because of the variable layout structure, the presence of decorations, different writing styles and degradations due to the aging of the document. In the last years different techniques have been proposed to address this task [3,10,13]. In particular, to extract text lines from handwritten documents we can consider two types of related problems. Considering the page segmentation task the target is to split a document image into regions of interest [3]. On the other hand the text line extraction stage allows to localize and extract the text lines directly from the document image [1]. These approaches extract regions of interest which are considered as text lines and often provide similar results when considering handwritten documents. To clarify the goal of text extraction from historical document we show in Fig. 1 one example from one benchmark dataset together with the ground truth of the page.

Among several solutions proposed to solve this task some use assumptions to simplify the approach. In [8] the authors assume that for each text line there

© Springer Nature Switzerland AG 2018
L. Pancioni et al. (Eds.): ANNPR 2018, LNAI 11081, pp. 395–406, 2018.
https://doi.org/10.1007/978-3-319-99978-4_31

is one path from one side of the image to the other that crosses only one text line. Based on this assumption, they trace the text line after the blurred image transformation extracting directly the text lines.

Later, it has been proposed another solution [5] where the authors are able to extract text line from handwritten pages using Hough transform and the page structure as prior knowledge.

In the last years, many different CNN architectures have been presented to solve several computer vision tasks. One important task is the semantic segmentation of images whose goal is to classify pixels from different categories and subsequently to extract homogeneous regions. One interesting solution adopts Fully Convolutional Networks [12] composed only by convolution and pooling operations used to learn representations based on local spatial input to compute pixel-wise predictions. The FCNs with respect to CNNs architectures do not use fully connected layers and use upsampling layers as deconvolution operations.

In this paper we address the page segmentation using one Fully Convolutional Network with a weighting of the pixels used to compute the training loss designed to address our task. In this way, we aim at classifying with better results some areas of the image that are more critical to perform the text line extraction, without using dedicated post processing techniques. The main contributions of this paper are the use of the FCN to perform text segmentation and the design of the weighting schema.

The rest of the paper is organized as follows. In Sect. 2 a brief review of related works in semantic and page segmentation tasks. Then, in Sect. 3 we describe the architecture used to perform page segmentation. The proposed weighting is presented in Sect. 4. Experimental results are discussed in Sect. 5 and concluding remarks are in Sect. 6.

2   Related Works

In the field of document analysis, page segmentation task has gained a lot of attentions during the time. Several solutions use artificial neural networks as well as Convolutional Neural Networks which have been applied successfully to this task showing best results compared to handcrafted features solutions [3,9]. In the work [9] the authors use CNNs to extract text lines from historical documents classifying the central pixel from extracted image patch. After one suitable post-processing phase, using the watershed transform, it is possible to extract the text lines and also provide a page segmentation.

Instead, in [3] the authors propose to use a Convolution Network for the pixel labeling task. Using a superpixel algorithm to extract coherent patches, they are able to perform page segmentation using a trained CNN model to predict the semantic class for each extracted patch.

We recently proposed one solution [1] to detect text lines on the basis of the assumption that for each text line it is possible to define one separator line and one median line. The median line is the middle line between the top profile of the text and the bottom profile, while the separator line is the middle line

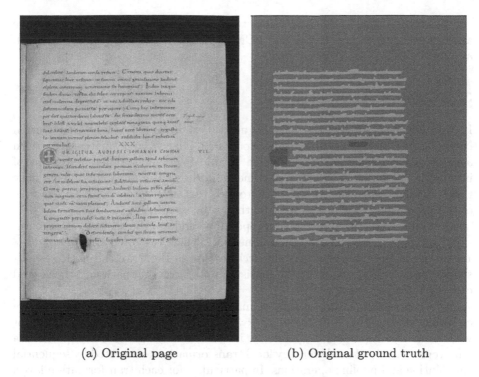

(a) Original page   (b) Original ground truth

Fig. 1. One document example from the dataset and its ground truth image. The background area is in red, decoration in blue and text in green. (Color figure online)

between two consecutive median lines in the text area. In [1] we used one suitable Convolutional Network to separate the text line areas from the document background.

Fully Convolutional Networks are largely adopted in semantic segmentation field. One adapted version of FCN named *U-net* [11] has been applied to biomedical image segmentation outperforming previous methods for cell tracking challenge by a large margin. Always in biomedical research area, it has been proposed a novel deep contour-aware network [2] to solve the gland segmentation task. This model is able to segment gland and separate the clustered objects into individual ones training a unified multi-task learning framework.

In scene parsing, using Pyramid Scene Parsing Network [15] is possible to merge together local and global representation defining a pyramid pooling module. In this way the authors prove as a global prior representation could improve the final segmentation result.

A Fully Convolution Network has been also used for page segmentation [14] where the neural network is used to provide a pixel-wise classification followed by post processing techniques to split a document image into regions of interest. The main focus is not the extraction of text lines, but the pixel classification and therefore the metrics used are related to semantic segmentation.

Recently, a very challenging dataset has been introduced by [7] where the authors evaluate various end-to-end FCN architectures to segment handwritten annotation in historical documents.

3   Network Architecture

In this work we address the page segmentation using one Fully Convolution Network that is trained directly on document image (large) patches to learn a pixel-wise classification model which is able to segment different regions of interest in the input image. The documents addressed in our experiments have three different semantic classes: background, text and decoration. An example of one input image with its ground truth can be seen in Fig. 1.

Several architectures have been proposed to address the semantic segmentation. In this work we propose a neural network which is strongly inspired by the *U-net* [11] model. By inspecting the architecture in Fig. 2 we can notice the U-shaped model where the first part consists in a contracting path and the second consists in an expansive path.

The contracting path consists of many encoding operations composed by convolution operators with kernel 3×3, stride 1, and max-pooling operator with kernel 2×2 stride 2, respectively. In this way the model is able to learn a data representation based on many local transformations computed by sequential convolution and pooling operations. In particular, for each transformation layer, we have two convolution operation followed by a pooling operation. The number of filters for each transformation layer is variable and we adapted these values to our problem. In particular, in the first layer we have 32, in the second 64, in the third 128, in the forth 256, and in the last 512 filters.

The expansive path consists of several decoding operations composed by upsampling and convolution operators. Having a look to Fig. 2, for each decoding step the features are concatenated with the computed feature maps from the contracting path (with the same shape). Still in the same decoding layer two convolution operations with kernel 3×3 and stride 1 are applied to the previously computed features. The expansive path proposes the same number of filters for each decoding layer, but in reverse order with respect to the contracting path. All the convolution operators use Rectified Linear Units (ReLUs) as activation function. In the final layer one single 1×1 convolution linear operator is used to map the last features into the number of desired output channels.

In order to map the features into a classification score we use the Softmax operator to predict the probability score related to the semantic segmentation. In particular, we compute pixel-wise classification scores to determine a class for each input pixel. In the basic approach we use the cross-entropy loss function to train the model from random weights initialized using the technique proposed by [6]. This loss function is then modified in order to take into account the peculiarities of the problem addressed in this paper.

To build the training set we randomly crop several patches with a fixed shape from each document image. To maximize the differences between training

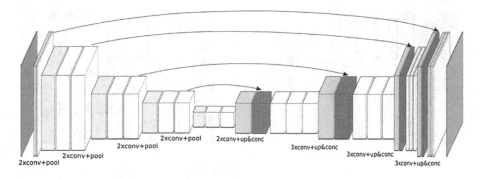

Fig. 2. The model architecture. Different transformations are depicted in different colors. The Input Layer is identified in gray, Convolutional Layers in white, Max Pooling in yellow, green for Upsampling Layers, blue for Combination Layers and red for the Softmax. (Color figure online)

patches the maximum overlap between patches is set to 25%. Like in [1] during the test phase, we systematically extract document patches from the input image with an overlap of 50%. For each pixel the final prediction is the average of the probability scores computed by the neural network for all the overlapping probability maps as illustrated in Fig. 3.

4 Weighting the Loss

In page segmentation there are several issues which make it difficult to obtain good performances. One significant problem is the unbalanced pixel class distribution. Having a look to Fig. 1 we can see that the pixel distribution is highly unbalanced for background pixels with respect to the foreground pixels (considering foreground as text and decoration parts). We can notice also that some background pixels are very important to segment text lines. Often the text lines are very close to each other and in this case some misclassification errors of pixels between two text lines could give rise to significant problems for properly segmenting contiguous text lines.

The model is trained using a categorical cross entropy. One possibility to give different cost values to the input during the training is to add one weighted map to the loss function

$$WCE = -\sum_{x \in \Phi} w(x) \log p_{q(x)}(x) \tag{1}$$

where $\Phi \subset \mathbb{Z}^2$ is the set of pixel positions, $q : \Phi \to 1, \ldots, K$ maps input pixels to the class label of the predicted distribution p (K is the number of classes), $w : \Phi \to \mathbb{R}^+$ is the weight function that maps each pixel x to a suitable weight.

Considering Eq. 1, we define a weighted map function $w(x)$ which assigns a cost to each pixel considering the class frequency and the contribution which could provide in the segmentation task. In particular, considering the set of

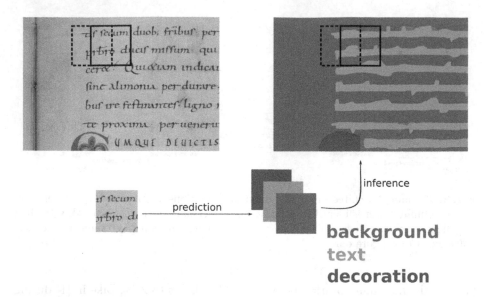

Fig. 3. Moving a sliding window over the input image, we compute a pixel-wise classification score for each patch. The results are combined by averaging the scores of overlapping patches.

pixels Φ in the training mini-batch that are used to compute the loss function, we define a weight map to balance the class frequency and also to put more attention in specific areas which are useful to segment different regions properly. The weight map therefore includes two aspects of the document, the background and foreground areas. Formally, the weighted map assigns to a pixel x one weight balancing the pixel class frequency with a factor α and managing the background pixels with a predefined weight mask $\beta(x)$ (described in Sect. 4.1), such that:

$$w(x) = \begin{cases} \alpha & x \in \Phi_f \\ \beta(x) & x \in \Phi_b \end{cases} \tag{2}$$

where the foreground pixels $\Phi_f \subset \Phi$ represent the text and the decoration areas, while $\Phi_b \subset \Phi$ represent the background pixels.

Considering Φ the set of pixels for each mini-batch, the foreground pixel frequency is a variable number (usually $|\Phi_f| < |\Phi_b|$). In order to balance the foreground areas we apply a factor α as $\frac{|\Phi_b|}{|\Phi_f|}$ computed for each mini-batch. Having a pixel weight related to the class frequency we can balance the loss function improving the training.

As previously mentioned, not all background pixels have the same importance with respect to the overall performance. In particular, misclassification errors between contiguous text lines could give rise to improper segmentation of the text lines. To address this problem, we define one training rule weighting more the background pixels between different regions (text lines or decorations). This

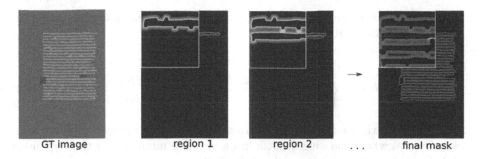

Fig. 4. Creation of the weighted mask of a ground truth page. In the third image, after merging the GT for region 1 and region 2 we can see pixels closer to both regions give a larger contribution to the weight mask. (Color figure online)

topological constraint is a rule which could be defined directly into the weighted map (Eq. 2) defining a weighed mask $\beta(x)$ for the background area as described in the following.

4.1 Weighting Background Pixels

The weighted mask $\beta(x)$ gives more emphasis on background pixels considering the distance between two contiguous lines. The background pixels have a classification cost inversely proportional to the distance between two contiguous text lines. To this purpose, the weight mask assigns to each background pixel one value considering the distance to the nearest line (a larger distance gives a smaller value and vice versa). For the others background pixels, the weight mask $\beta(x)$ returns a fixed (neutral) weight value.

To compute the weight mask $\beta(x)$ we first transform the ground truth image from three class to a two class representation by merging text and decorator as foreground and the rest are background pixels. Considering this representation, taking a text region per time, we compute the *distance transform* which designs level curves from the region borders to a defined maximum distance d. An example of these level curves is shown in Fig. 4 (region one) where the level curve value (in false colors) decreases when increasing the distance from the region border.

These level curves encode one information useful to consider the distance to the nearest regions. Iteratively, computing a level curve for each region and summing-up these values we can produce an overall weight mask. In this way, when the regions are close each other, the level curves are summed providing a larger value when the regions are closer. The largest value is obtained when the distance between two regions is only one pixel. We force the range of values for the level curves to be between 0 and 1. By using a factor λ to multiply α (Sect. 4) we obtain mask values larger than foreground weights.

Considering a binary representation I of the ground truth image, for each region r_i at time i, we compute the level curves on the basis of the distance transform $dist_d(r_i)$ limiting this representation until a max distance d. We can consider the area around all the region borders with a maximum distance d as a dilation operation with kernel d.

In this way, the mask for an image with N regions is:

$$\beta(x) = \begin{cases} 1 + \frac{\lambda\alpha}{2d} \cdot (\sum_i^N dist_d(r_i)) & x \in dilate_d(I) \\ 1 & otherwise \end{cases} \tag{3}$$

where $dilate_d$ is the morphological dilation operator with kernel d useful to consider the area where the weight mask has a variable number. For the remaining pixels in the page the weight mask maps pixels to a neutral value.

We illustrate in Fig. 4 the approach to compute the weight mask. Starting from a ground truth image we compute a binary representation with foreground regions and background. For each region, we compute the distance curve levels as $dist(r_i)$ which are sequentially summed with the next region representations. The final result is the computed mask for all the pixels $x \in dilate_d(I)$ which are the critical pixels where we want to put more emphasis during the training to learn background representation. To provide a better idea about the critical pixels, in Fig. 5 we highlighted in red the critical pixel areas.

5 Experiments

In this section we describe the experiments performed to test the proposed model to segment historical document images. The tests have been made on the *Saint Gall* dataset that consists of handwritten manuscript images that contain the hagiography *Vita sancti Galli* by Walafrid Strab. The manuscript has been most likely written by one single hand in Carolingian script with ink on parchment. Carolingian minuscules are predominant, but there are also some upper script letters that emphasize the structure of the text and some richly ornamented initials. Each page is written in a single column that contains 24 text lines. Altogether, the Saint Gall database includes 60 manuscript pages [4]. The database is freely downloadable and it is provided with layout descriptions in XML format. The document images in the original dataset have an average size of 3328×4992 pixels.

We evaluate the model performance using four metrics applied to semantic segmentation proposed by [12]. These measures are based on pixel accuracy and region intersection over union (IU). In particular, we evaluate the performance using: pixel accuracy, mean pixel accuracy, mean IU, and frequency weighted IU (f.w. IoU).

Let n_{ij} be the number of pixels of class i predicted to belong to class j (in total there are n_{cl} classes), and $t_i = \sum_j n_{ij}$ be the total number of pixels of class i. We can express the measures as:

- Pixel accuracy

$$pix.acc. = \frac{\sum_i n_{ii}}{\sum_i t_i} \tag{4}$$

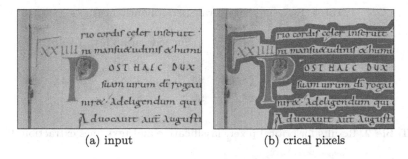

(a) input   (b) crical pixels

Fig. 5. Given a input page, we can define the critical pixel areas (red) around the semantic regions found into the ground truth representation. (Color figure online)

– Mean accuracy

$$mean.acc = \frac{1}{n_{cl}} \sum_i \frac{n_{ii}}{t_i} \qquad (5)$$

– Mean IoU (Intersection over Union)

$$mean.IoU = \frac{1}{n_{cl}} \sum_i \frac{n_{ii}}{t_i + \sum_j (n_{ji} - n_{ii})} \qquad (6)$$

– Frequency weighted IoU

$$f.w.IoU = \frac{1}{\sum_k t_k} \sum_i \frac{t_i n_{ii}}{t_i + \sum_j (n_{ji} - n_{ii})} \qquad (7)$$

The previous metrics are used to define a global evaluation for whole pages. To better evaluate the performance, we also define one local pixel accuracy considering only the area around the foreground regions. In Fig. 5 we depict in red the area around foreground regions where the local pixel accuracy is computed. This area is important to extract text lines because misclassification pixels in it could give rise to a wrong layout analysis.

In the experiments we trained the proposed model learning the parameters by Stochastic Gradient Descent algorithm and using the extracted patches from the original training pages as train set. The training dataset if composed by several patches of size 256×256 pixels randomly extracted from the input pages. Overall the training dataset contains $299,756$ patches. The different methods are compared evaluating the models on the test set pages.

In Table 1 we report the results for page segmentation on the *Saint gall* dataset and compare with previous results on the same dataset reported by Chen et al. [3]. The proposed model obtains good results with respect to [3] also by using the standard cross entropy loss. We improve only the mean IoU by using the proposed weighted loss. These metrics evaluate the page segmentation globally, but as we previously mentioned some misclassification errors have more importance in the final segmentation results.

Table 1. Results for overall performance measures.

Model	pix.acc.	mean.acc.	mean.IoU	f.w.IoU
Baseline (CE Loss)	98.07	95.12	90.80	96.27
Weighted (CE Loss)	98.03	94.75	90.88	96.18
Chen et al. [3]	98	90	87	96

Table 2. Results for critical pixel classification and text lines extraction [9].

Model	Critical pixel accuracy	DR	RA	FM
Baseline (CE Loss)	95.65	77.81	83.48	80.55
Weighted (CE Loss)	96.25	81.28	86.18	83.65

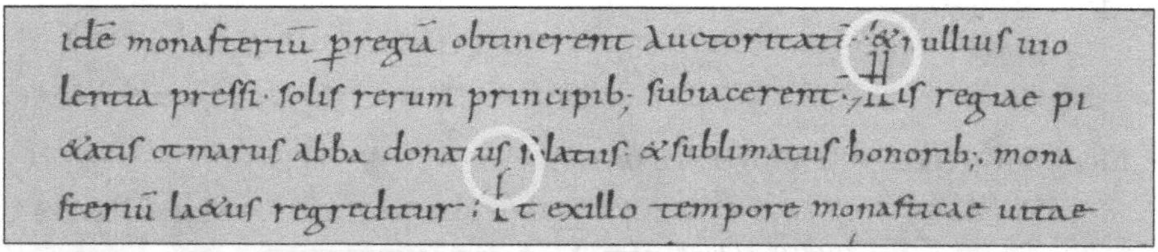

(a) Input

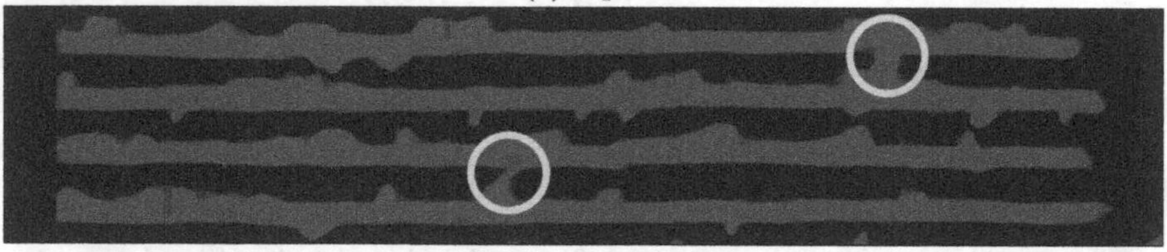

(b) CE

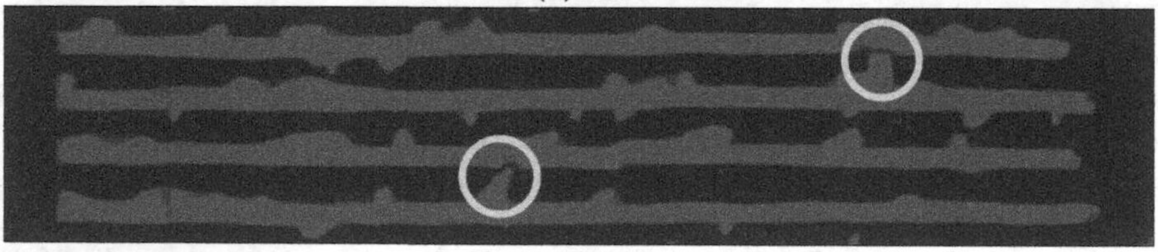

(c) Weighted CE

Fig. 6. Different results obtained with one model trained using cross entropy loss and one with weighted cross entropy loss.

The results reported in Table 2 detail the critical pixel accuracy. This measure is useful to evaluate the model behavior after the training done using different losses. Using the weighted loss we can obtain better results which could be useful to extract text line directly after the page segmentation.

For a qualitative evaluation of results we show in Fig. 6 one part of one page and two results, one from a model trained with cross entropy loss and the other from a model trained with the proposed weighted loss. We can notice that the model trained with the weighted map is able to better segment different text lines.

In order to evaluate the trained models with the measure proposed by [9] the model trained by weighted loss obtains Detection Rate (DR) and Recognition Accuracy (RA) respectively better than the model trained by cross entropy loss. Comparing these scores in Table 2 the model trained by the proposed approach is able to extract more accurate text lines than the baseline model.

6   Conclusions

In this work we addressed the segmentation of handwritten historical documents by means of deep architectures. We presented one approach to weight a cross entropy loss to improve the results in particular in critical regions. By weighting the pixels to obtain a balanced loss and putting more emphasis on the background pixel around text lines, we obtained better classification results more suitable to extract text line as a post-processing of the classification based on neural networks.

In the future research we will from one side address more challenging datasets and from the other side we will exploit the pixel classification produced by the model discussed in this paper. In particular, we will compare the performance using other FCNs architecture to explicitly extract the text-lines from document pages.

References

1. Capobianco, S., Marinai, S.: Text line extraction in handwritten historical documents. In: Digital Libraries and Archives - 13th Italian Research Conference on Digital Libraries, IRCDL 2017, Revised Selected Papers, Modena, Italy, 26–27 January 2017, pp. 68–79 (2017)
2. Chen, H., Qi, X., Yu, L., Dou, Q., Qin, J., Heng, P.: DCAN: deep contour-aware networks for object instance segmentation from histology images. Med. Image Anal. **36**, 135–146 (2017)
3. Chen, K., Seuret, M., Hennebert, J., Ingold, R.: Convolutional neural networks for page segmentation of historical document images. In: 14th IAPR International Conference on Document Analysis and Recognition, ICDAR 2017, Kyoto, Japan, 9–15 November 2017, pp. 965–970 (2017)
4. Fischer, A., Frinken, V., Fornés, A., Bunke, H.: Transcription alignment of Latin manuscripts using hidden Markov models. In: Proceedings of the 2011 Workshop on Historical Document Imaging and Processing, HIP 2011, pp. 29–36. ACM, New York (2011)
5. Gatos, B., Louloudis, G., Stamatopoulos, N.: Segmentation of historical handwritten documents into text zones and text lines. In: 14th International Conference on Frontiers in Handwriting Recognition, ICFHR 2014, Crete, Greece, 1–4 September 2014, pp. 464–469 (2014)

6. Glorot, X., Bengio, Y.: Understanding the difficulty of training deep feedforward neural networks. In: Proceedings of the International Conference on Artificial Intelligence and Statistics (AISTATS 2010). Society for Artificial Intelligence and Statistics (2010)
7. Kölsch, A., Mishra, A., Varshneya, S., Liwicki, M.: Recognizing challenging handwritten annotations with fully convolutional networks. CoRR abs/1804.00236 (2018)
8. Nicolaou, A., Gatos, B.: Handwritten text line segmentation by shredding text into its lines. In: 10th International Conference on Document Analysis and Recognition, ICDAR 2009, Barcelona, Spain, 26–29 July 2009, pp. 626–630 (2009)
9. Pastor-Pellicer, J., Afzal, M.Z., Liwicki, M., Castro-Bleda, M.J.: Complete system for text line extraction using convolutional neural networks and watershed transform. In: 2016 12th IAPR Workshop on Document Analysis Systems (DAS), pp. 30–35, April 2016
10. Renton, G., Chatelain, C., Adam, S., Kermorvant, C., Paquet, T.: Handwritten text line segmentation using fully convolutional network. In: First Workshop of Machine Learning, 14th IAPR International Conference on Document Analysis and Recognition, WML@ICDAR 2017, Kyoto, Japan, 9–15 November 2017, pp. 5–9 (2017)
11. Ronneberger, O., Fischer, P., Brox, T.: U-Net: convolutional networks for biomedical image segmentation. In: Proceedings of the 18th International Conference Medical Image Computing and Computer-Assisted Intervention, MICCAI 2015, Part III, Munich, Germany, 5–9 October 2015, pp. 234–241 (2015)
12. Shelhamer, E., Long, J., Darrell, T.: Fully convolutional networks for semantic segmentation. IEEE Trans. Pattern Anal. Mach. Intell. **39**(4), 640–651 (2017)
13. Vo, Q.N., Kim, S., Yang, H.J., Lee, G.: Text line segmentation using a fully convolutional network in handwritten document images. IET Image Process. **12**(3), 438–446 (2018)
14. Xu, Y., He, W., Yin, F., Liu, C.: Page segmentation for historical handwritten documents using fully convolutional networks. In: 14th IAPR International Conference on Document Analysis and Recognition, ICDAR 2017, Kyoto, Japan, 9–15 November 2017, pp. 541–546 (2017)
15. Zhao, H., Shi, J., Qi, X., Wang, X., Jia, J.: Pyramid scene parsing network. In: 2017 IEEE Conference on Computer Vision and Pattern Recognition, CVPR 2017, Honolulu, HI, USA, 21–26 July 2017, pp. 6230–6239 (2017)

Author Index

Printed in the United States
By Bookmasters